T0176052

CHAPMAN & HALL/CRC
Monographs and Surveys in
Pure and Applied Mathematics 123

# INVERSE BOUNDARY SPECTRAL PROBLEMS

CHAPMAN & HALL/CRC
Monographs and Surveys in
Pure and Applied Mathematics 123

# INVERSE BOUNDARY SPECTRAL PROBLEMS

ALEXANDER KATCHALOV
YAROSLAV KURYLEV
MATTI LASSAS

CRC Press
Taylor & Francis Group
Boca Raton London New York

CRC Press is an imprint of the
Taylor & Francis Group, an **informa** business
A CHAPMAN & HALL BOOK

CRC Press
Taylor & Francis Group
6000 Broken Sound Parkway NW, Suite 300
Boca Raton, FL 33487-2742
First issued in paperback 2019



ISBN-13: 978-1-58488-005-9 (hbk)
ISBN-13: 978-0-367-39705-0 (pbk)



**Library of Congress Cataloging-in-Publication Data**

**Kachalov, Alexander.**
**Inverse boundary spectral problems / Alexander Kachalov, Yaroslav Kurylev and Matti Lassas**
**p. cm.--(Chapman & Hall/CRC monographs and surveys in pure and applied mathematics)**
**Includes bibliographical references and index.**
**ISBN 1-58488-005-8 (alk. paper)**
**1. Inverse problems (Differential equations) 2. Boundary value problems. I. Kurylev, Yaroslav. II. Lassas, Matti. III. Title. IV. Series.**
**QA374 .K28 2001**
**515'.35—dc21** **2001032511**
**CIP**

**Catalog record is available from the Library of Congress**

**Library of Congress Card Number 20001032511**

**Visit the Taylor & Francis Web site at**
**http://www.taylorandfrancis.com**

**and the CRC Press Web site at**
**http://www.crcpress.com**

To our families

# Contents

# Introduction

Many physical processes are described by partial differential equations and systems of partial differential equations. The coefficients of these equations, which may depend on spatial and time variables, describe the properties of the medium where these processes take place. Solutions of these equations are called fields. The so-called direct problem consists of finding these fields when we know applied sources, initial and boundary conditions, and, of course, parameters of the medium. However, in physics and practical applications, the problem under consideration is often somewhat opposite. It is the unknown properties of the medium, described by the coefficients of the differential equations, that are to be determined. For example, in geophysical explorations, one wants to know the density and the Lame parameters that describe the properties of the Earth. These parameters may be used to find oil and other mineral fields. To find these parameters, we can utilize the fact that they appear in the coefficients of equations of elasticity. Henceforth, elastic waves, which propagate in the Earth, depend upon these parameters. Therefore, there is a natural problem of finding elastic parameters from the measurements of elastic fields. This problem is called the inverse problem of elasticity. Similar problems appear when one wants to find electric permittivity and conductivity and magnetic permeability from the measurements of electromagnetic fields, or acoustic velocity from the sound measurements, etc. These examples show the importance of inverse problems both for understanding physical phenomena and practical applications.

Inverse problems do not often have a unique solution. The data obtained from measurements may be the same for different physical models. For this reason, the goal of inverse problems is to find all equivalent models, i.e., those models that correspond to equal measurements. In spite of the non-uniqueness, the situation in real life is not as bad as it sounds. Measured data together with some additional *a priori* information about the model may be sufficient to find the unique model that describes the process. Indeed, in many cases, the non-uniqueness is due only to a natural freedom in describing the process. For instance, the domain filled with the medium may be described in different coordinate systems. Clearly, the resulting solution, which actually describes the same process, will be written in different forms. Therefore, in solving inverse problems, we need to take this into account and find their solutions in a coordinate invariant form. On the other hand, sometimes there are natural coordinates relevant to a particular inverse problem. These coordinates are often the travel time coordinates as, for example, in the described geophysical inverse problem. These travel time coordinates determine the natural distances between points. The distance function gives rise to a Riemannian structure, so it is natural to consider the domain as a Riemannian manifold. Therefore, the reconstruction of the Riemannian manifold and the corresponding metric is an essential part of solving inverse problems. This reconstruction of the manifold is a generalization of the reconstruction of the parameters of the medium in the travel time coordinates.

As seen from the above description, inverse problems have a wide range of practical applications and their mathematical theory is based upon an interaction of analysis and geometry with some additional algebraic ideas. Therefore, this book is written mainly for analysts, applied mathematicians and other applied scientists and engineers having an interest in the mathematical foundations of inverse problems. We believe that this book is also useful for geometricians who are interested in the interdisciplinary research.

It is clear that a large variety of physical phenomena described earlier give rise to numerous different types of mathematical inverse problems. To describe this variety and to understand the special nature of the problems that are described in this book, we will provide a rather sketchy classification of mathematical inverse problems. Our classification is based on the difference in various types of mathematical models that describe the corresponding physical processes, as well as various types of measured data used to reconstruct the unknown parameters of the medium. Before providing this classification, we note that the mathematical theory of inverse problems is still very far from its completion. As yet, there are no exact solutions to a number of practically important inverse problems. To solve such problems, mathematicians use various approximation methods. Moreover, in practice, given data is usually incomplete and errorprone. This also makes it necessary to utilize approximations and numerical techniques to study inverse problems. Corresponding methods are described in a number of books, e.g., [EnHaNu], [G], [BkGo]. The main aim of this book is to develop a rigorous theory to solve several types of inverse problems exactly, rather than to discuss applied and numerical aspects of these problems. However, we believe that this book will be useful for practitioners. This is not only because the book contains a new invariant approach to solve inverse problems, but also because of the algorithmic nature of the methods developed. This means that one of the main features of the book is to provide some procedures to reconstruct unknown parameters. Similarly, these procedures can be used to construct numerical solutions of inverse problems.

In the next few paragraphs we will give a classification of inverse problems and discuss briefly those problems that will be considered later in the book. First, we distinguish between the one-dimensional, i.e., with one spatial variable, and the multidimensional inverse problems. The one-dimensional case has many specific properties that do not exist in the multidimensional one. Historically, the study of mathematical inverse problems has been started with the one-dimensional case. Works of a number of outstanding mathematicians ([Bo], [Lv], [GeLe], [Ma1]-[Ma4], [Kr1]-[Kr4]) in the 1940s and '50s resulted in thorough understanding of these problems and provided several powerful methods to solve them. These methods were further developed later (see, for example, such classical books as [Ma4], [ChSa], [PoTr], [BeBl2], etc).

This book deals mainly with the multidimensional inverse problems, so that our further classification and references to literature will deal only with the multidimensional case. However, the methods developed in this book are perfectly applicable to one-dimensional inverse problems. Since these methods are easier to understand in the one-dimensional case, we provide the one-dimensional variant of the method in Chapter 1. We believe that this is a good introduction to study inverse problems in the multidimensional case.

Physical processes in a medium are described by various fields, e.g., electromagnetic fields, elastic fields, etc. These fields are often of vectorial nature. This means that the corresponding equations are vector equations, or, in other words, systems of equations. However, sometimes, the physical nature of the process makes it possible to reduce the vector equations to scalar ones. This occurs, for example, in gravitation or acoustics. Clearly, the vector case is significantly more difficult to analyze and the mathematical theory of the vector inverse problems is substantially less developed (see, however, textbooks [CoKs], [Ki], [Ro1] and historical remarks to Chapter 4. In this book we will consider only the scalar inverse problems, although some of the developed methods can be extended to the vector case.

As mentioned, our classification of inverse problems is based on differentiation between various types of mathematical models that give rise to the corresponding inverse problems, e.g., one-dimensional versus multidimensional, scalar versus vector, and also differentiation among various types of inverse data.

In general, data used in inverse problems contains some given or measured information about the corresponding fields. Usually, this information describes the behaviour of these fields on the boundary of the domain, occupied by the medium, or at infinity. In the first case, we speak about inverse boundary problems, and in the second, about inverse scattering problems. Inverse scattering problems are of great importance due to their role in quantum mechanics and gravitational theory, and, historically, are the first inverse problems where mathematically rigorous results were obtained (see, for example, [CoKs], [Nw], [Fa3], [LxPh], [Me]). Inverse boundary problems, which have a wide range of important applications to geosciences, medical imaging, non-destructive testing, process monitoring, etc., are quite different from inverse scattering problems and require rather different methods and techniques. This book is devoted to inverse boundary problems.

According to the type of time dependence, physical processes are divided into stationary and non-stationary ones. Therefore, the corresponding inverse problems are also divided into stationary and non-stationary ones. Non-stationary inverse problems are also called dynamical or evolutionary problems. Depending on the hyperbolic or parabolic nature of the fields, the dynamical inverse problems are subdivided into hyperbolic or parabolic ones (see, e.g., [Ki], [Ro1], [Is1], [Sh1]). Stationary inverse boundary problems are subdivided into inverse boundary spectral problems and fixed frequency inverse problems. Fixed frequency inverse problems use boundary measurements at a finite number of frequencies. Moreover, the vast majority of results for these problems are obtained when the frequency is equal to 0, i.e., for the static case. These problems go back to the famous paper by Calderon [Cl] with fundamental mathematical theory developed in [SyU], [NvKh] with the crucial role of Faddeev's Green function [Fa1]. During the last two decades, this theory has been significantly extended and developed, see e.g., [Na1], [Na2], [NkU1], [OPaS], and Notes in Chapter 4. A good exposition of the fixed frequency inverse problems is given in [Is1].

The main goal of this book is to study the inverse boundary spectral problems. However, the approach to inverse problems, which will be developed, is a dynamical approach. It is based on the consideration of the corresponding wave equations and involves various techniques to study an initial-boundary value problem for the wave equation. Henceforth, many results that are obtained for the inverse boundary spectral problems are valid for the dynamical inverse boundary problems as described in Chapter 4.

Next we will describe, without going into details, the principal features and ingredients of the approach developed in the book.

1. Control theory for hyperbolic partial differential equations.

2. Geometry of geodesics and metric properties of Riemannian manifolds.

3. Asymptotic solutions of hyperbolic partial differential equations and, in particular, Gaussian beams.

4. Coordinate and gauge invariance of inverse problems and corresponding groups of transformations.

In our approach, we combine these ideas whose history can be briefly described as follows:

1. The importance of control theory for inverse problems was first understood by Belishev [Be1]. He used control theory to develop the first variant of the boundary control (BC) method, which is the analytical backbone of the book (see [Be2] for recent developments).

2. Later, the ideas based on control theory were combined with the geometrical ones. The importance of geometry for inverse problems follows the fact that any elliptic second-order differential operator gives rise to a Riemannian metric in the corresponding domain. The role of this metric becomes clearer if we consider the solutions of the corresponding wave equation. Indeed, these waves propagate with the unit speed along geodesics of this Riemannian metric. These geometric ideas were introduced to the boundary control method in [BeKu3] and [Ku4].

3. The close relation between the wave equation and the corresponding Riemannian metric becomes particularly clear when we consider some special classes of asymptotic solutions of this equation, like Gaussian beams. This asymptotic solution of the wave equation has the form of a wave packet. It propagates like a particle along a geodesic of the Riemannian metric, which prompts the name "quasiphoton" for this solution. This property makes Gaussian beams a very convenient technique to solve dynamical inverse problems. This observation was made in [BeKa1] and [KaKu1].

4. In the study of inverse boundary problems for general elliptic differential operators, it is necessary to take into account their non-uniqueness. In fact, there is a group of transformations of the operator, which preserve the boundary spectral data. This group of transformations was first analyzed in [Ku1], [Ku2].

In this book, we study the inverse boundary spectral problem for a general self-adjoint second-order elliptic differential operator on a compact manifold with boundary. In particular, our considerations cover the case of elliptic operators in bounded domains in Euclidean spaces. Moreover, we show that the developed approach is applicable to the study of the dynamical inverse boundary problems for the corresponding wave and heat equations.

The book consists of two unequal parts. The first part, Chapter 1, is devoted to the one-dimensional inverse boundary spectral problem on a finite interval. In this part, we introduce the principal ideas and techniques of the approach. The one-dimensional inverse problems are very interesting by themselves and have numerous important applications. In connection with this, we give a self-contained exposition of the approach for this case. Those readers who are interested only in the one-dimensional inverse problems can read Chapter 1 only. In the one-dimensional case, these principal ideas become much simpler and clearer than in the multidimensional case. Therefore, the study of the one-dimensional case is very instructive for further development of the theory in the multidimensional case.

The main part of the book, Chapters 2–4, are devoted to the multidimensional inverse boundary problem. In Chapter 3, we will give a solution for the Gel'fand inverse boundary spectral problem for a general second order self-adjoint elliptic differential operator on a compact manifold with boundary. To this end, we will develop an approach to the multidimensional inverse problems based on the boundary control method. We will describe the group of gauge transformations of an operator that do not change the boundary spectral data. We will show that any orbit of this group contains a unique (Riemannian) Schrödinger operator. We will describe an algorithm to construct this Schrödinger operator and the underlying differential manifold from the boundary spectral data of any operator in the orbit.

Chapter 4 is devoted to various generalizations and extensions to other inverse boundary problems of the approach developed in Chapter 3. There are four main subjects discussed in this chapter. First, we will analyze the relations between the inverse boundary spectral problems and the dynamical inverse boundary problems for the wave and heat equations. This will make it possible to reduce the dynamical inverse problems to the inverse boundary spectral ones. Second, we will show that it is possible to solve the dynamical inverse problem for the wave equation directly, without reducing it to the inverse boundary spectral problem. The corresponding method is analogous to the method developed in Chapter 3. However, there is a significant difference. Namely, if we possess boundary data only on a finite interval of time, then we can reconstruct the operator only in a collar neighborhood of the boundary. The width of this collar neighborhood is determined by the metric generated by the operator. Third, we will consider the inverse boundary spectral problem with data given only on a part of the boundary. This kind of problem is often encountered in practice. Fourth, we will consider the inverse boundary spectral problems in bounded domains of Euclidean spaces. Using the results in Chapter 3, we will describe the groups of transformations in this case. Furthermore, we will show that additional *a priori* information about the structure of operators makes the inverse problem uniquely solvable. This takes place, for example, for some isotropic operators. We also provide a reconstruction algorithm. Considerations in Chapter 4 are given in less detail than those in Chapter 3.

Chapter 2 is of an auxiliary nature. Here we provide an analytical and geometrical background of the approach developed in Chapters 3 and 4.

This book is mainly aimed at readers with a background in analysis. This determines the choice of material and the level of exposition in Chapter 2. We expect our readers to know the basics of distribution theory, Sobolev spaces and partial differential equations. When presented material goes beyond the standard textbooks, we will prove the corresponding results. Otherwise, we will provide only the rigorous formulations of the statements with references to the literature in Notes at the end of the chapter. Chapter 2 contains a rather wide range of ideas and techniques from modern analysis and geometry. The authors believe that this diversity reflects the interdisciplinary nature of the mathematical theory of inverse problems and corresponds to the current state of their development.

Each chapter consists of several titled sections subdivided into subsections, which we still call sections. Numeration of formulae and statements, i.e., theorems, lemmas and corollaries, is unified throughout a chapter. In references to a formula, the first number refers to the chapter and the second to the numeration of this formula in the chapter. There are exercises in Chapters 1–3. They are not very difficult. Usually they refer to those parts of exposition, that can be proven by considerations similar to those that are given earlier in the text. There are also sections in Chapters 2 and 3 that are marked by ★. These sections may be skipped in the first reading. At the beginning of each chapter, there is a short description of its structure and contents. At the end of each chapter, there is also a section called Notes. It provides references to the results used in the chapter and also gives references to the relevant literature on the subjects covered in the chapter. This sections also contains some historical commentaries about the results discussed in the chapter.

# Acknowledgements

The authors are grateful to various people and organizations that helped us to write this monograph. We express our special gratitude to Professors Gunther Uhlmann, Lassi Päivärinta and Erkki Somersalo for numerous discussions on inverse problems that have been very useful in writing this book.

We are thankful to our students, Pekka Tietäväinen and Robert Marlow, whose comments have been very valuable, and to Nina Katchalova for the enormous job of preparing the pictures. We are grateful to Dr. Niculae Mandache, who suggested some important modifications to the proofs and, in particular, to Kenrick Bingham, who read the manuscript carefully and pointed out various errors and misprints, and suggested many improvements to the text. Also, we are thankful to Sharon Taylor for her editorial guidance.

We thank Tet-A-Tet program at Euler International Mathematical Institute St. Petersburg, Russia, and the Research in Pairs program at Mathematisches Forschugsinstitute, Oberwolfach, Germany, and also Rolf Nevanlinna institute (Finland), University of Loughborough (UK), St. Petersburg Division of Steklov Mathematical Institute (Russia), University of Oulu (Finland) and Helsinki University of Technology (Finland), all of which made it possible for us to write this monograph in a creative, friendly atmosphere. We are also grateful to the Finnish Academy, Finland; the Russian Academy of Sciences, Russia (grant RFFI 99-01-00107); EPSRC, UK (grant GR/M36595); the Royal Society, UK; London Mathematical Society, UK; TEKES, Finland; and Volkswagen Stiftung, Germany for their financial support.

Alexander Katchalov
Yaroslav Kurylev
Matti Lassas

# Chapter 1

# One-dimensional inverse problem

In this chapter, we will study the one-dimensional inverse boundary spectral problem. In sections 1.1.1–1.1.3 we will give a precise formulation of this problem. In the rest of section 1.1 we will describe the admissible transformations that preserve the boundary spectral data and reduce the inverse problem for a general operator to that for a Schrödinger operator. In section 1.2, we will consider the initial-boundary value problem for the corresponding wave equation. In particular, we will obtain a formula for the Fourier coefficients of the waves in terms of the boundary spectral data. In section 1.3, we will describe the necessary controllability results for the wave equation and give a solution to the inverse problem for the Schrödinger operator. To do that, we will introduce the so-called slicing procedure. In section 1.4, we will describe the one-dimensional Gaussian beams. We will later apply them to obtain an alternative solution of the inverse problem.

## 1.1. Inverse problem and main result

**1.1.1.** In many applied sciences, there appear elliptic ordinary differential operators with real smooth coefficients,

$$\mathcal{A}y = -a(x)y''(x) + b(x)y'(x) + c(x)y(x), \quad a > 0, \tag{1.1}$$

for a function $y(x)$ with $x$ varying in the interval $[0, l]$. For example, in modeling harmonic oscillations of an inhomogeneous string, we deal with the equation

$$\mathcal{A}y = -ay'' + by' + cy = \omega^2 y.$$

Here $a$, $b$ and $c$ are related to the density $\rho$, Hooke constant $\mu$ and stiffness $f$ by the formulae

$$a = \mu\rho^{-1}, \quad b = -\mu'\rho^{-1}, \quad c = f,$$

and $\omega$ is the frequency of oscillations.

In the direct problems, we know the material parameters and our goal is to find $y(x)$. However, for many practical purposes, the goal is just the opposite. We have information about solutions $y(x)$ and would like to use this information to find the material parameters. Such problems are called inverse problems. In practice, information about the solutions $y(x)$ comes from the measurements, and different types of measurements give rise to different types of inverse problems. In this chapter, we will study the one-dimensional inverse boundary spectral problems related to operator (1.1).

**1.1.2.** To define an operator $\mathcal{A}$ of form (1.1) rigorously, it is necessary to make some assumptions on its coefficients and also functions $y(x)$. In this chapter, we assume that $a$, $b$, and $c$ are real-valued, smooth, i.e., infinitely differentiable, functions on the closed interval $[0, l]$. The operator $\mathcal{A}$ is then defined on the functions $y \in H^2([0, l])$, which means that $y$, $y'$ are continuous and $y'' \in L^2([0, l])$, which satisfies the Dirichlet boundary conditions

$$y|_{x=0} = y|_{x=l} = 0. \tag{1.2}$$

The functions satisfying all these conditions form the domain of the operator $\mathcal{A}$, which is denoted by $\mathcal{D}(\mathcal{A})$,

$$\mathcal{D}(\mathcal{A}) = \{y \in H^2([0, l]) : \; y(0) = y(l) = 0\}. \tag{1.3}$$

Any operator of form (1.1), which is often called a Sturm-Liouville operator, can be rewritten as

$$\mathcal{A}y = -m^{-1}g^{-1/2}(mg^{-1/2}y')' + cy \tag{1.4}$$

with some positive smooth functions $m(x)$ and $g(x)$.

**Exercise 1.1** *Find $m(x)$ and $g(x)$ in terms of $a$ and $b$.*

The function $m(x)$ determines a weighted space $L^2([0,l], dV)$,

$$dV = m dV_g = m g^{1/2} dx$$

with the inner product of the form

$$\langle y, z\rangle = \int_0^l y(x)\overline{z(x)} m(x) g^{1/2} dx. \tag{1.5}$$

**Remark** Representation (1.4) for the operator $\mathcal{A}$ and the volume elements $dV$ and $dV_g$ are taken in the form suitable for the multi-dimensional case. In particular, the metric, or length element, on the interval $[0,l]$ that corresponds to the operator $\mathcal{A}$ has the form

$$ds^2 = g(x) dx^2.$$

Integrating by parts, we can show that

$$\langle \mathcal{A}y, z\rangle = \langle y, \mathcal{A}z\rangle, \tag{1.6}$$

for any $y, z \in \mathcal{D}(\mathcal{A})$. This means that $\mathcal{A}$ is symmetric in $L^2([0,l], dV)$. Moreover, it is known that any operator $\mathcal{A}$ of form (1.4), (1.3) is self-adjoint in $L^2([0,l], dV)$. Its spectrum consists of the isolated eigenvalues $\lambda_1 < \lambda_2 < \cdots < \lambda_k < \cdots$, $\lambda_k \to +\infty$. The corresponding eigenfunctions $\varphi_k(x)$, $\varphi_k \in \mathcal{D}(\mathcal{A})$ satisfy

$$\mathcal{A}\varphi_k = \lambda_k \varphi_k,$$

or,

$$-a(x)\varphi_k''(x) + b(x)\varphi_k'(x) + c(x)\varphi_k(x) = \lambda_k \varphi_k(x), \quad x \in [0,l],$$

$$\varphi_k(0) = 0, \quad \varphi_k(l) = 0.$$

can be chosen to form an orthonormal basis in $L^2([0,l], dV)$, i.e., $\langle \varphi_k, \varphi_l\rangle = \delta_{kl}$.

**1.1.3.** We are now in the position to formulate the inverse boundary spectral problem for the operator $\mathcal{A}$.

**Problem 1** *Let all eigenvalues, $\lambda_1, \lambda_2, \ldots$, be known as well as the values at the boundary points $x = 0$ and $x = l$ of the scaled derivatives of the normalized eigenfunctions $g^{-1/2}(0)\varphi_1'(0)$, $g^{-1/2}(0)\varphi_2'(0), \ldots$ and $g^{-1/2}(l)\varphi_1'(l)$, $g^{-1/2}(l)\varphi_2'(l), \ldots$. Is it possible to find the coefficients $m$, $g$ and $c$ from these data?*

**Problem 2** *Let all eigenvalues $\lambda_1, \lambda_2, \ldots$, be known as well as the values at one boundary point, say $x = 0$, of the scaled derivatives of the normalized eigenfunctions $g^{-1/2}(0)\varphi_1'(0)$, $g^{-1/2}(0)\varphi_2'(0), \ldots$. Is it possible to find the coefficients $m$, $g$ and $c$ from these data?*

The data used in the formulations of Problems 1 and 2 are called the boundary spectral data.

**Definition 1.2** *The collection*

$$\{\lambda_j,\ g^{-1/2}(0)\varphi_j'(0),\ g^{-1/2}(l)\varphi_j'(l) :\ j = 1, 2, \ldots\}$$

*is called the boundary spectral data of operator $\mathcal{A}$ corresponding to Problem 1. The collection*

$$\{\lambda_j,\ g^{-1/2}(0)\varphi_j'(0) :\ j = 1, 2, \ldots\}$$

*is called the boundary spectral data of operator $\mathcal{A}$ corresponding to Problem 2.*

**Remark** The factor $g^{-1/2}$ in the formulation of the boundary spectral data appears due to the equation

$$\frac{dy}{ds} = g^{-1/2}\frac{dy}{dx}.$$

This makes the definition of the boundary spectral data suitable for the multidimensional case also.

In the following, we will concentrate on the more complicated Problem 2. However, in the multidimensional case, we will be mostly preoccupied with the multidimensional analog of Problem 1. [1]

[1] In the one-dimensional case, problems 1, 2 are practically equivalent. In fact, if we know the boundary data $\lambda_n$, $\alpha_n = g^{-1/2}(0)\varphi_n'(0)$, $n = 1, 2, \ldots$, of Problem 1 and $\lambda_n \neq 0$ for all $n$, then $\beta_n = g^{-1/2}(l)\varphi_n'(l)$ can be found using formulae $\beta_n = C\alpha_n^{-1}\lambda_n \prod_{k\neq n}(1 - \lambda_n/\lambda_k)^{-1}$. Due to the non-uniqueness of the inverse boundary spectral problem, constant $C$ can be arbitrary positive (or negative) constant. The sign of the constant depends on the solution of an auxiliary problem.

**1.1.4.** As we are going to show, Problems 1 and 2 do not have a unique solution, $m$, $g$, and $c$. Indeed, we can change $m$, $g$ and $c$, so that the boundary spectral data remains unchanged. To see this, let us consider two types of transformations: changes of coordinates and gauge transformations.

i) *Changes of coordinates.* Let $\widetilde{X}$ be a diffeomorphism, $\widetilde{X} : [0, l] \to [0, \widetilde{l}]$, i.e., $\widetilde{X} \in C^\infty([0, l])$, $\widetilde{X}' > 0$, $\widetilde{X}(0) = 0$, $\widetilde{X}(l) = \widetilde{l}$ with its inverse $\widetilde{X}^{-1}$ denoted by $X$. This diffeomorphism corresponds to the change of coordinates from $x$ to $\widetilde{x} = \widetilde{X}(x)$. In the new coordinates $\widetilde{x}$, the function $y$ becomes the function $\widetilde{y} = y \circ X$, i.e. $\widetilde{y}(\widetilde{x}) = y(X(\widetilde{x}))$. To preserve the inner product, we introduce new functions $\widetilde{m}$, $\widetilde{g}$,

$$\widetilde{m}(\widetilde{x}) = m(X(\widetilde{x})), \quad \widetilde{g}(\widetilde{x}) = g(X(\widetilde{x}))[X'(\widetilde{x})]^2. \tag{1.7}$$

Then

$$\langle y, z \rangle = \int_0^l y(x)\overline{z(x)}dV = \int_0^{\widetilde{l}} \widetilde{y}(\widetilde{x})\overline{\widetilde{z}(\widetilde{x})}d\widetilde{V} = \langle \widetilde{y}, \widetilde{z} \rangle, \tag{1.8}$$

where $d\widetilde{V} = \widetilde{m}\widetilde{g}^{1/2}d\widetilde{x}$. In these coordinates, the operator $\mathcal{A}$ becomes the operator $\widetilde{\mathcal{A}}$,

$$\widetilde{\mathcal{A}}\widetilde{y}(\widetilde{x}) = -\widetilde{m}^{-1}\widetilde{g}^{-1/2}(\widetilde{m}\widetilde{g}^{-1/2}\widetilde{y}')' + \widetilde{c}\widetilde{y}, \tag{1.9}$$

where $\widetilde{c} = c \circ X$. More precisely, this means that

$$(\mathcal{A}y)(X(\widetilde{x})) = \widetilde{\mathcal{A}}\widetilde{y}(\widetilde{x}).$$

In particular,

$$\widetilde{\mathcal{A}}\widetilde{\varphi}_k = \lambda_k \widetilde{\varphi}_k.$$

Clearly, $\widetilde{\varphi}_k$ satisfy the Dirichlet boundary conditions at $\widetilde{x} = 0$ and $\widetilde{x} = \widetilde{l}$ and, henceforth, $\widetilde{\varphi}_k$ are the eigenfunctions of operator $\widetilde{\mathcal{A}}$. Due to identity (1.8), $\widetilde{\varphi}_k$, $k = 1, 2, \dots$, remain orthonormalized, i.e. $\langle \widetilde{\varphi}_k, \widetilde{\varphi}_l \rangle = \delta_{kl}$. Moreover,

$$\widetilde{g}^{-1/2}(0)\widetilde{\varphi}_k'(0) = g^{-1/2}(0)\varphi_k'(0), \quad \widetilde{g}^{-1/2}(\widetilde{l})\widetilde{\varphi}_k'(\widetilde{l}) = g^{-1/2}(l)\varphi_k'(l),$$

for all $k = 1, 2, \dots$. This means that the boundary spectral data of $\mathcal{A}$ and $\widetilde{\mathcal{A}}$ are the same.

**1.1.5.** *ii) Gauge transformations.* Let $\kappa$ be a smooth positive function, $\kappa \in C^\infty([0,l])$, $\kappa > 0$. This function determines a transformation $\mathcal{A}_\kappa$ of the operator $\mathcal{A}$, which is called a gauge transformation, of $\mathcal{A}$

$$\mathcal{A}_\kappa u = \kappa \mathcal{A}(\kappa^{-1} u). \tag{1.10}$$

Operator $\mathcal{A}_\kappa$ has form (1.4) with $m$, $g$ and $c$ replaced by $m_\kappa$, $g_\kappa$, and $c_\kappa$,

$$m_\kappa = \kappa^{-2} m, \quad g_\kappa = g, \quad c_\kappa = \mathcal{A}(\mathbf{1}) \tag{1.11}$$

where $\mathbf{1}$ is the constant function $\mathbf{1}(x) = 1$.

**Exercise 1.3** *Prove equations (1.11).*

Then the functions $\psi_k$,

$$\psi_k = \kappa \varphi_k, \quad k = 1, 2, \dots, \tag{1.12}$$

where $\varphi_k$ are the eigenfunctions of $\mathcal{A}$, satisfy the equations

$$\mathcal{A}_\kappa \psi_k = \lambda_k \psi_k, \quad k = 1, 2, \dots,$$

as well as the Dirichlet boundary conditions. This means that $\lambda_k$ and $\psi_k$ are the eigenvalues and eigenfunctions of the operator $\mathcal{A}_\kappa$, which is self-adjoint in $L^2([0,l], dV_\kappa)$, where $dV_\kappa = m_\kappa g_\kappa^{1/2} dx$. Moreover, the functions $\psi_k$ form an orthonormal basis in the space $L^2([0,l], dV_\kappa)$. Henceforth, the boundary spectral data of $\mathcal{A}_\kappa$ are given by $\lambda_k$ and

$$g_\kappa^{-1/2}(0)\psi_k'(0) = \kappa(0)[g^{-1/2}(0)\varphi_k'(0)], \tag{1.13}$$

$$g_\kappa^{-1/2}(l)\psi_k'(l) = \kappa(l)[g^{-1/2}(l)\varphi_k'(l)],$$

where $k = 1, 2, \dots$ If $\kappa(0) = \kappa(l) = 1$, then the boundary spectral data of $\mathcal{A}$ and $\mathcal{A}_\kappa$, used in Problem 1, are the same and, if $\kappa(0) = 1$, then the boundary spectral data of $\mathcal{A}$ and $\mathcal{A}_\kappa$ used in Problem 2 are the same.

Summarizing, we see that any function $\kappa \in C^\infty([0,l])$, $\kappa > 0$ determines a transformation of an operator $\mathcal{A}$ that given by formula (1.10). This transformation is called the gauge transformation corresponding to $\kappa$.

When $\kappa$ satisfies additional boundary conditions $\kappa(0) = \kappa(l) = 1$ or $\kappa(0) = 1$, we call the corresponding gauge transformation the normalized gauge transformation related to Problems 1 or 2.

**1.1.6.** In the previous section, we described two types of transformations of an operator $\mathcal{A}$. Next, we use them to make $\mathcal{A}$ as simple as possible. Our goal is to obtain a Schrödinger operator,

$$\mathcal{A}_0 \equiv -\frac{d^2}{dx^2} + q(x). \tag{1.14}$$

**Lemma 1.4** *For any operator $\mathcal{A}$ of form (1.4) there exists a gauge transformation, followed by a change of coordinates that transform $\mathcal{A}$ into a Schrödinger operator*

$$\mathcal{A}_0 = -\frac{d^2}{d\widetilde{x}^2} + q(\widetilde{x})$$

*on the interval $[0, \widetilde{l}]$. For this end, we should take*

$$\kappa(x) = m^{1/2}(x), \tag{1.15}$$

$$\widetilde{x} = \widetilde{X}(x) = \int_0^x g^{1/2}(x')dx', \tag{1.16}$$

*so that*

$$\widetilde{l} = \widetilde{X}(l).$$

**Proof.** According to formula (1.11), gauge transformation (1.15) makes $m_\kappa = 1$, $g_\kappa = g$. According to formulae (1.7) and (1.9), the change of coordinates (1.16) keeps $\widetilde{m} = 1$ and makes $\widetilde{g} = 1$. □

**Exercise 1.5** *Find the potential $q$ in terms of $m$, $g$ and $c$.*

We note that we can change the order of the transformations, making first the change of coordinates (1.16) followed by the gauge transformation which corresponds to $\widetilde{\kappa}(\widetilde{x}) = \kappa(X(\widetilde{x}))$.

**1.1.7.** Later, we will describe a method of the reconstruction of the potential $q$ of a Schrödinger operator from the boundary spectral data. Lemma 1.4 shows that any operator of form (1.4), (1.3) can be transformed into a Schrödinger operator. Unfortunately, gauge transformations (1.15) are not, in general, normalized. This means that the boundary spectral data of the Schrödinger operator $\mathcal{A}_0$, which is used for the reconstruction of $q$, differs from the given boundary spectral data of the original operator $\mathcal{A}$ of form (1.4).

However, there is a method to find the boundary values of $m$ from the boundary spectral data of $\mathcal{A}$. To this end, we use the asymptotics of the eigenvalues $\lambda_k$ and eigenfunctions $\widetilde{\varphi}_k$ of the Schrödinger operator $\mathcal{A}_0$ which corresponds to the operator $\mathcal{A}$, (see Notes at the end of the chapter)

$$\lambda_k = \left(\frac{\pi k}{\widetilde{l}}\right)^2 + O(1),$$

$$\widetilde{\varphi}_k(\widetilde{x}) = \sqrt{\frac{2}{\widetilde{l}}}\sin\left(\frac{k\pi\widetilde{x}}{\widetilde{l}}\right) + O(k^{-1}), \tag{1.17}$$

$$\widetilde{\varphi}_k'(\widetilde{x}) = \sqrt{\frac{2}{\widetilde{l}}}\frac{k\pi}{\widetilde{l}}\cos\left(\frac{k\pi\widetilde{x}}{\widetilde{l}}\right) + O(k^{-1}).$$

Using the transformations described in Lemma 1.4, we see that

$$\widetilde{\varphi}_k(\widetilde{X}(x)) = m^{1/2}(x)\varphi_k(x).$$

Therefore, the boundary spectral data for the operator $\mathcal{A}$ has the form

$$\widetilde{\varphi}_k'(0) = m^{1/2}(0)[g^{-1/2}(0)\varphi_k'(0)], \tag{1.18}$$

$$\widetilde{\varphi}_k'(\widetilde{l}) = m^{1/2}(l)[g^{-1/2}(l)\varphi_k'(l)].$$

These formulae together with formulae (1.17) show that

$$m(0) = \lim_{k\to\infty}\frac{2\lambda_k^{3/2}}{\pi k[g^{-1/2}(0)\varphi_k'(0)]^2},$$

$$m(l) = \lim_{k\to\infty}\frac{2\lambda_k^{3/2}}{\pi k[g^{-1/2}(l)\varphi_k'(l)]^2}.$$

Combining these formulae with the equations (1.18), we obtain the boundary spectral data for the Schrödinger operator $\mathcal{A}_0$, which corresponds to an arbitrary general operator $\mathcal{A}$.

Summarizing considerations of sections 1.1.6, 1.1.7, we see that we can reduce the inverse boundary spectral problem for a general operator to the inverse boundary spectral problem for the corresponding Schrödinger operator $\mathcal{A}_0$.

**1.1.8.** Now we are in the position to formulate our main result for the one-dimensional inverse boundary spectral Problems 1 and 2. The key result is the following theorem.

**Theorem 1.6** *Assume that* $\{\lambda_1, \lambda_2, \dots, \varphi_1'(0), \varphi_2'(0), \dots\}$ *are the boundary spectral data at* $x = 0$ *of a Dirichlet Schrödinger operator,* $\mathcal{A}_0 = -\frac{d^2}{dx^2} + q$ *on an interval* $[0, l]$. *Then these data determine* $l$ *and* $q(x)$ *uniquely.*

Theorem 1.6 gives an affirmative answer to Problem 2.

This theorem is proven in sections 1.2 and 1.3 below. In fact, we will describe a procedure how to find $l$ and reconstruct $q$. Obviously, Theorem 1.6 also gives an answer to Problem 1 in the case of a Schrödinger operator.

In the case of a general operator of form (1.4), we will first construct the corresponding Schrödinger operator. The operator $\mathcal{A}$ lies in the class of operators that can be obtained from this Schrödinger operator $\mathcal{A}_0$ by a gauge transformation with $\kappa(0) = m^{1/2}(0)$, $\kappa(l) = m^{1/2}(l)$, followed by a change of coordinates. All operators of this class have the same boundary spectral data and, henceforth, are indistinguishable from the boundary data. Moreover, Theorem 1.6 implies the following theorem for the general operators.

**1.1.9.**

**Theorem 1.7** *Two operators* $\mathcal{A}$ *and* $\mathcal{B}$ *of form (1.4), (1.3) have the same boundary spectral data, if and only if they can be obtained from one another by a change of coordinates and a normalized gauge transformation.*

**Proof.** i) The part "if" of the theorem is already proven in sections 1.1.4, 1.1.5.

ii) To prove the part "only if", we first note that, due to (1.11), $m_{\mathcal{A}}(0) = m_{\mathcal{B}}(0)$, where we have denoted by $m_{\mathcal{A}}$, $l_{\mathcal{A}}$, etc. the corresponding quantities for the operator $\mathcal{A}$ and by $m_{\mathcal{B}}$, $l_{\mathcal{B}}$, etc. the corresponding quantities for the operator $\mathcal{B}$. Hence, the boundary spectral data of the corresponding Schrödinger operators $\mathcal{A}_0$ and $\mathcal{B}_0$ are the same and, by Theorem 1.6, $\mathcal{A}_0 = \mathcal{B}_0$.

In view of the remark at the end of section 1.1.6, the operator $\mathcal{A}_0$ is obtained from $\mathcal{A}$ by the following consecutive transformations.

First we make a change of coordinates and obtain an operator $\widehat{\mathcal{A}}$. Then we make a gauge transformation corresponding to $\kappa_{\mathcal{A}}$ with $\kappa_{\mathcal{A}}(0) = m_{\mathcal{A}}^{1/2}(0)$ and obtain $\mathcal{A}_0$. Analogously, to obtain $\mathcal{B}_0$ from $\mathcal{B}$, we first make a change of coordinates and obtain an operator $\widehat{\mathcal{B}}$. Then we make a gauge transformation corresponding to $\kappa_{\mathcal{B}}$ with $\kappa_{\mathcal{B}}(0) = m_{\mathcal{B}}^{1/2}(0)$.

Since $\widehat{\mathcal{A}}$ and $\widehat{\mathcal{B}}$ are gauge equivalent to the same Schrödinger operator $\mathcal{A}_0 = \mathcal{B}_0$, they are still defined on the same interval $[0, l]$. We have

$$\mathcal{A}_0 = \kappa_{\mathcal{A}} \widehat{\mathcal{A}} \kappa_{\mathcal{A}}^{-1} = \kappa_{\mathcal{B}} \widehat{\mathcal{B}} \kappa_{\mathcal{B}}^{-1},$$

which yields that

$$\widehat{\mathcal{B}} = (\kappa_{\mathcal{B}}^{-1}\kappa_{\mathcal{A}})\widehat{\mathcal{A}}(\kappa_{\mathcal{B}}^{-1}\kappa_{\mathcal{A}})^{-1} = \kappa \widehat{\mathcal{A}} \kappa^{-1}.$$

Since $m_{\mathcal{A}}(0) = m_{\mathcal{B}}(0)$, this is a normalized gauge transformation. Therefore, the operator $\mathcal{B}$ is obtained from $\mathcal{A}$ by change of coordinates, a normalized gauge transformation and one more change of coordinates. Finally, we observe that the resulting transformation can be obtained as a combination of one normalized gauge transformation and one change of coordinates. □

**1.1.10.** The previous discussion shows that there is no chance to reconstruct all coefficients of a general operator (1.4) from its boundary spectral data. However, in many cases that are important for applications, we have some additional information about the form of the operator $\mathcal{A}$. Sometimes, this information is sufficient for the unique reconstruction of $\mathcal{A}$, when the class of gauge equivalent operators contain only one operator having the specified form. An obvious example is the Schrödinger operator. Other examples are given by the operators

$$\mathcal{A} = -c^2(x)\frac{d^2}{dx^2}, \quad \mathcal{B} = -\frac{d}{dx}\mu(x)\frac{d}{dx},$$

which appear, e.g., in the study of the inverse problem for an inhomogeneous string.

**Exercise 1.8** *Show that the boundary spectral data of the operators $\mathcal{A}$ and $\mathcal{B}$ determine $c$ and $\mu$ uniquely.*

## 1.2. Wave equation

**1.2.1.** This book is devoted to the inverse boundary spectral problems. However, our approach to solving this problem is based on the properties of the corresponding wave equation. For an operator $\mathcal{A}$ of form (1.1) or (1.4), the corresponding wave equation is

$$\partial_t^2 w + \mathcal{A}w \equiv \partial_t^2 w - a(x)\partial_x^2 w + b(x)\partial_x w + c(x)w =$$

$$= \partial_t^2 w - m^{-1} g^{-1/2} \partial_x (m g^{-1/2} \partial_x w) + cw = 0, \quad x \in [0,l] \quad (1.19)$$

where $w = w(x,t)$. A well-known and important property of the wave equation is the finite velocity of the wave propagation. For equation (1.19), this velocity, $v(x)$ is given by formula,

$$v(x) = a^{1/2}(x) = g^{-1/2}(x). \quad (1.20)$$

The velocity $v$ determines the travel time $\tau$ between any two points $x_1$ and $x_2$, $0 \le x_1 \le x_2 \le l$,

$$\tau(x_1, x_2) = \int_{x_1}^{x_2} \frac{dx}{v(x)} = \int_{x_1}^{x_2} g^{1/2}(x)dx. \quad (1.21)$$

From the point of view of physics, $\tau(x_1, x_2)$ is the time necessary for a perturbation at the point $x_1$ to reach the point $x_2$. The travel time can be interpreted as the physically meaningful distance between points. We can use the travel time, $\tau(x) = \tau(0, x)$ from the boundary point 0 to a variable point $x$ as a new coordinate. The coordinate transformation from $x$ to $\tau(x)$ is exactly the change of coordinates, which transforms a general operator to a Schrödinger one. Using also gauge transformation (1.15), which has the physical meaning of re-scaling of the dependent variable, we transform wave equation (1.19) into the wave equation

$$\partial_t^2 u + \mathcal{A}_0 u \equiv \partial_t^2 u - \partial_\tau^2 u + qu = 0, \quad (1.22)$$

for the function $u(x,t) = \kappa(x)w(x,t)$.

**1.2.2.** The remaining part of Chapter 1 is mainly devoted to the consideration of Problem 2 for the Dirichlet Schrödinger operator

$\mathcal{A}_0$ of form (1.14), (1.3). Let us consider the initial-boundary value problem for equation (1.22),

$$\begin{cases} \partial_t^2 u - \partial_x^2 u + q(x)u = F(x,t), \\ u\big|_{x=0} = 0, \quad u\big|_{x=l} = 0, \\ u\big|_{t=0} = \partial_t u\big|_{t=0} = 0, \end{cases} \tag{1.23}$$

with $F \in L^2([0,l] \times [0,T])$ for any $T > 0$.

We start our considerations with the case $F \in C_0^\infty([0,l] \times [0,T])$. It is known that, in this case, initial-boundary value problem (1.23) has a unique solution $u \in C^\infty([0,l] \times [0,T])$ and $u(x,t) = 0$ for $x > t$, when $F(x,t) = 0$ for $x > t$. The last property follows from the finite velocity of the wave propagation, which is equal to 1 for wave equation (1.22) (see Notes in the end of the chapter). In the following, we need some less regular solutions. For this end, we prove estimates for the solution.

**Lemma 1.9** *Let $u(x,t)$ be the solution of the initial-boundary value problem (1.23) for $F \in C_0^\infty([0,l] \times [0,T])$. Then*

$$H_u(t) \le C(T)\|F\|^2_{L^2([0,l]\times[0,T])}, \quad 0 \le t \le T, \tag{1.24}$$

$$|u(x,t)| \le C(T)\|F\|_{L^2([0,l]\times[0,T])}, \quad 0 \le x \le l, \quad 0 \le t \le T, \tag{1.25}$$

*where*

$$H_u(t) = \frac{1}{2}\int_0^l (|\partial_t u(x,t)|^2 + |\partial_x u(x,t)|^2 + |u(x,t)|^2)dx. \tag{1.26}$$

**Proof.** Using the wave equation in (1.23) and integration by parts, we obtain

$$\frac{dH_u}{dt} = \Re \int_0^l \Big(\partial_t u(x,t)\,\overline{\partial_t^2 u(x,t)} + \partial_t\partial_x u(x,t)\,\overline{\partial_x u(x,t)} +$$

$$+\partial_t u(x,t)\,\overline{u(x,t)}\Big)\, dx =$$

$$= \Re \int_0^l \partial_t u(x,t)(\overline{F(x,t)} + (1 - q(x))\overline{u(x,t)})dx +$$

$$+\Re(\partial_t u(l,t)\overline{\partial_x u(l,t)} - \partial_t u(0,t)\overline{\partial_x u(0,t)}).$$

The last term in this equation is equal to zero due to the boundary conditions in (1.23). Then,

$$\frac{dH_u}{dt} \leq \frac{1}{2}\|F(\cdot,t)\|^2 + (C_q+2)H_u, \tag{1.27}$$

where

$$C_q = \max_{x\in[0,l]} |q(x)|. \tag{1.28}$$

This inequality can be rewritten as

$$\frac{d}{dt}\left(e^{-(C_q+2)t}H_u\right) \leq \frac{1}{2}e^{-(C_q+2)t}\|F(\cdot,t)\|^2. \tag{1.29}$$

In view of the initial conditions in (1.23), $H_u(0) = 0$, so that inequality (1.29) implies that

$$H_u(t) \leq \frac{1}{2}\int_0^t e^{(C_q+2)(t-t')}\|F(\cdot,t')\|^2 dt' \tag{1.30}$$

$$\leq C(T)\|F\|^2_{L^2([0,l]\times[0,T])}.$$

In particular,

$$\int_0^l |\partial_x u(x,t)|^2 dx \leq C(T)\|F\|^2_{L^2([0,l]\times[0,T])}, \tag{1.31}$$

so that, by the Cauchy inequality,

$$|u(x,t)| \leq \int_0^x |\partial_x u(x,t)|dx \leq C(T)\|F\|_{L^2([0,l]\times[0,T])}. \tag{1.32}$$

□

Estimate (1.25) makes it possible to define the solution of initial-boundary value problem (1.23) when $F \in L^2([0,l]\times[0,T])$. Indeed, let $F_n \to F$ in $L^2([0,l]\times[0,T])$ and $F_n \in C_0^\infty([0,l]\times[0,T])$. Denote by $u_n(x,t)$ the solution of initial-boundary value problem (1.23) with $F$ replaced by $F_n$. Then, by inequality (1.25), there is a function $u(x,t) \in C([0,l]\times[0,T])$ such that

$$u_n(x,t) \to u(x,t), \quad 0 \leq x \leq l, \quad 0 \leq t \leq T.$$

This function $u(x,t)$ is called the solution of initial-boundary value problem (1.23). The previous considerations imply the following result

**Corollary 1.10** *For $F \in L^2([0,l] \times [0,T])$ there is a unique solution $u \in C(([0,l] \times [0,T])$ for which estimates (1.24), (1.25) are valid. Moreover, if $F(x,t) = 0$ for $x > t$ then $u(x,t) = 0$ for $x > t$.*

**Remark.** It is known that, when $F \in L^2([0,l] \times [0,T])$, the function $u(x,t)$ coincides with the unique weak solution of initial-boundary value problem (1.23). Moreover, estimate (1.24) means that this weak solution belongs to the energy class. The properties and uniqueness of the weak solutions can be found in details from references given in Notes.

**1.2.3.** The initial-boundary value problem for the equation (1.22), which is related to the inverse boundary spectral problem, is the problem

$$\begin{cases} \partial_t^2 u - \partial_x^2 u + q(x)u = 0, \\ u|_{x=0} = f(t), \quad u|_{x=l} = 0, \\ u|_{t=0} = \partial_t u|_{t=0} = 0, \end{cases} \tag{1.33}$$

for $f \in L^2([0,T])$. We denote $u(x,t) = u^f(x,t)$ to indicate the Dirichlet boundary value of $u$. Function $f$ is also called the boundary source.

When $q = 0$, the solution $u_0^f(x,t)$ of this problem, for $0 < t < T$, is given by the formula

$$u_0^f(x,t) = \sum_{j=0}^{\infty} [f(t - x - 2lj) - f(t + x - 2l(j+1))], \tag{1.34}$$

when we continue $f(t)$ to be 0 for $t < 0$. We point out that, for $t < T$, the sum in the right-hand side is finite. Indeed, all terms for $j > T/l$ vanish.

To construct the solution for the initial-boundary value problem (1.33), we consider the solution $u_1^f(x,t)$ of the following initial-boundary value problem,

$$\begin{cases} (\partial_t^2 - \partial_x^2 + q(x))u_1^f = -qu_0^f, \\ u_1^f|_{x=0} = 0, \quad u_1^f|_{x=l} = 0, \\ u_1^f|_{t=0} = \partial_t u_1^f|_{t=0} = 0. \end{cases} \tag{1.35}$$

By Corollary 1.10, $u_1^f \in C([0,l] \times [0,T])$. Then the unique solution of initial-boundary value problem (1.33) is given by the formula,

$$u^f(x,t) = u_0^f(x,t) + u_1^f(x,t). \tag{1.36}$$

Denote by $C([0,T], L^2([0,l]))$ the space of functions of $t$ with values in $L^2([0,l])$, which depend continuously on $t$. We denote the norm of the space $L^2([0,l])$ by $\|\cdot\|$.

**Lemma 1.11** *For any $f \in L^2([0,T])$, initial-boundary value problem (1.33) has a unique solution $u^f(x,t) \in C([0,T], L^2([0,l]))$, such that*

$$\|u^f(\cdot,t)\| \le C(T)\|f\|, \quad t < T.$$

**Proof.** Each term in the right-hand side of formula (1.34) is in the space $C([0,T], L^2([0,l]))$. Since the sum in (1.34) is finite, $u_0^f \in C([0,T], L^2([0,l]))$ and

$$\|u_0^f(\cdot,t)\| \le C(T)\|f\|.$$

Then the right-hand side in (1.35), i.e., the function $-qu_0^f \in L^2([0,l] \times [0,T])$. By Corollary 1.10, $u_1^f \in C([0,l] \times [0,T])$ and satisfies estimate (1.32) with $F = -qu_0^f$. Therefore,

$$|u_1^f(x,t)| \le C(T)\|qu_0^f\|_{L^2([0,l]\times[0,T])} \le C_1(T)\|f\|.$$

□

**1.2.4.** In this section, we will construct Green's function $G(x,t,t')$, $t' > 0$, for problem (1.33). $G(x,t,t')$ is the weak solution of (1.33) with the boundary source $f(t) = \delta(t-t')$, where $\delta(t)$ is the Dirac delta-function,

$$\begin{cases} \partial_t^2 G - \partial_x^2 G + qG = 0, & 0 < x < l, \quad t > 0, \\ G|_{x=0} = \delta(t-t'), & G|_{x=l} = 0, \\ G|_{t=0} = \partial_t G|_{t=0} = 0. \end{cases} \tag{1.37}$$

**Lemma 1.12** *Let $G(x,t,t') = G(x,t-t')$ be*

$$G(x,t,t') = \delta(t-t'-x) -$$

$$\frac{1}{2}\int_0^x q(x')dx' \cdot H(t-t'-x) + G_0(x,t-t'). \tag{1.38}$$

*Here $H(t)$ is the Heaviside function and $G_0(x,t)$ a continuous function satisfying the initial-boundary value problem,*

$$\begin{cases} \partial_t^2 G_0 - \partial_x^2 G_0 + qG_0 = (-\frac{1}{2}q'(x) + \frac{q(x)}{2}\int_0^x q(x')dx')H(t-x) \\ G_0|_{x=0} = G_0|_{x=l} = 0 \\ G_0|_{t=0} = \partial_t G_0|_{t=0} = 0. \end{cases} \tag{1.39}$$

*Then $G(x,t,t')$ is the unique solution of initial-boundary value problem (1.37) for $0 \le t \le l$.*

**Proof.** *Step 1.* By Lemma 1.11, initial-boundary value problem (1.39) has a unique solution $G_0(x,t) \in C([0,l] \times [0,l])$. Moreover, $G_0(x,t-t')$ is the unique solution of problem (1.39) with $H(t-x)$ replaced by $H(t-t'-x)$. It is clear that $G(x,t,t')$ of form (1.38) satisfies the initial and boundary conditions required in (1.37). The fact that $G$ satisfies the wave equation (1.37) can be verified by direct substitution.

*Step 2.* Let $G(x,t,t')$ be the solution of problem (1.37). Then,

$$\widetilde{G}_0(x,t,t') = G(x,t,t') - \delta(t-t'-x) + \frac{1}{2}\int_0^x q(x')dx' \cdot H(t-t'-x),$$

satisfies system (1.39) with $H(t-x)$ replaced by $H(t-t'-x)$. Therefore, by Lemma 1.9,

$$\widetilde{G}_0(x,t,t') = G_0(x,t-t'),$$

which proves the uniqueness. □

In view of Corollary 1.10, $G_0(x,t) = 0$ for $x > t$. Therefore, representation (1.38) implies that $G(x,t,t') = 0$ for $x > t - t'$. We use Green's function only for $0 \le t \le l$, when there is yet no reflection from the right end $x = l$.

Green's function $G(x,t)$ can be used to represent the solution of initial-boundary value problem (1.33) with an arbitrary $f \in L^2([0,T])$,

$$u^f(x,t) = \int_0^t G(x,t,t')f(t')dt' = \tag{1.40}$$

$$= f(t-x) + \int_0^t G_1(x,t')f(t-t')dt',$$

which is valid for $t < l$. Here

$$G_1(x,t) = -\frac{1}{2}\int_0^x q(x')dx' \cdot H(t-x) + G_0(x,t). \tag{1.41}$$

The function $G_1$ is continuous with respect to $(x,t)$ in the domain $x < t$ and $G_1(x,t) = 0$ for $x > t$.

When $f \in C_0^\infty([0,T])$, we can differentiate representation (1.40) with respect to $t$. Then,

$$\partial_t^n u^f(x,t) = f^{(n)}(t-x) + \int_0^t G_1(x,t')f^{(n)}(t-t')dt', \tag{1.42}$$

so that $u^f(x,t) \in C^\infty([0,T], L^2([0,l]))$.

**1.2.5.** In this section, we will obtain a spectral representation of the solution $u^f(x,t)$ of initial-boundary value problem (1.33).

The orthonormalized eigenfunctions $\varphi_k(x)$, $k = 1,2,\ldots$, of the Schrödinger operator $\mathcal{A}_0$ form an orthonormal basis in $L^2([0,l])$. Since the solution $u^f(x,t)$, $f \in L^2([0,T])$, lies in $L^2([0,l])$ for any $t$, we can represent it in the form

$$u^f(x,t) = \sum_{k=1}^{\infty} u_k^f(t)\varphi_k(x). \tag{1.43}$$

Here

$$u_k^f(t) = \int_0^l u^f(x,t)\varphi_k(x)dx \tag{1.44}$$

are the Fourier coefficients of $u^f(\cdot,t)$. A simple but very important result is a representation of $u_k^f(t)$ in terms of the boundary spectral data of $\mathcal{A}_0$.

**Lemma 1.13** *Let $u^f(x,t)$ be the solution of initial-boundary value problem (1.33) with $f \in L^2([0,T])$. Then, for $k = 1,2,\ldots$,*

$$u_k^f(t) = \int_0^t f(t')s_k(t-t')dt'\varphi_k'(0), \tag{1.45}$$

*where*

$$s_k(t) = \begin{cases} \frac{\sin\sqrt{\lambda_k}t}{\sqrt{\lambda_k}}, & \lambda_k > 0, \\ t, & \lambda_k = 0, \\ \frac{\sinh\sqrt{|\lambda_k|}t}{\sqrt{|\lambda_k|}}, & \lambda_k < 0. \end{cases} \tag{1.46}$$

**Proof.** We will give the proof only for $f \in C_0^\infty([0,T])$ and leave the generalization to $f \in L^2([0,T])$ as an exercise. Using (1.44) and integration by parts, we see that

$$\frac{d^2}{dt^2}u_k^f(t) = \int_0^l \partial_t^2 u^f(x,t)\varphi_k(x)dx =$$

$$= \int_0^l \{\partial_x^2 u^f(x,t) - q(x)u^f(x,t)\}\varphi_k(x)dx =$$

$$= [\partial_x u^f(l,t)\varphi_k(l) - u^f(l,t)\partial_x\varphi_k(l)] -$$

$$-[\partial_x u^f(0,t)\varphi_k(0) - u^f(0,t)\partial_x\varphi_k(0)] +$$

$$+ \int_0^l u^f(x,t)\{\partial_x^2\varphi_k(x) - q(x)\varphi_k(x)\}dx, \qquad (1.47)$$

where the second equality follows from wave equation (1.22). The boundary conditions for $u^f(x,t)$ and $\varphi_k(x)$ and the fact that $\varphi_k$ is the eigenfunction of $\mathcal{A}_0$ that corresponds to the eigenvalue $\lambda_k$, mean that equation (1.47) takes the form

$$\frac{d^2}{dt^2}u_k^f(t) = -\lambda_k u_k^f(t) + \varphi_k'(0)f(t). \qquad (1.48)$$

The initial conditions in (1.33) yield that

$$u_k^f(0) = \frac{d}{dt}u_k^f(0) = 0. \qquad (1.49)$$

Solving equation (1.48) with initial conditions (1.49), we obtain formula (1.45) for the Fourier coefficients $u_k^f(t)$. □

**Exercise 1.14** *By Lemma 1.11, the map $f \mapsto u^f(\cdot,t)$ is a continuous map from $L^2([0,T])$ into $L^2([0,l])$ for any $0 < t < T$. Use this fact to show the validity of (1.45) for $f \in L^2([0,T])$.*

**1.2.6.** Lemma 1.13 makes it possible to calculate the inner products of any two waves from the boundary, i.e., solutions of initial-boundary value problem (1.33). Indeed, let $u^f(x,t)$ and $u^h(x,t)$ be the solutions of (1.33) that correspond to the boundary sources $f$ and $h$. Then, for any $t, s \geq 0$,

$$\langle u^f(\cdot,t), u^h(\cdot,s)\rangle = \sum_{k=1}^{\infty} u_k^f(t)\overline{u_k^h(s)}, \tag{1.50}$$

where the Fourier coefficients in the right-hand side of (1.50) are given by formula (1.45).

## 1.3. Controllability and projectors

**1.3.1.** Representation (1.43), (1.45) makes it possible to find the wave $u^f(\cdot,T)$ for arbitrary $f$. Let us change our point of view and try to answer the following question.

Given a function $a \in L^2([0,l])$, we want to find a source $f \in L^2([0,T])$ such that $u^f(\cdot,T) = a$. Problems of this type are often called problems of controllability. They play a crucial role in the considerations of this book.

First, we note that, due to the finite velocity of the wave propagation, which is equal to 1 in our case,

$$\text{supp } u^f(\cdot,T) \subset [0,T]. \tag{1.51}$$

Hence, for the positive answer to the controllability problem, it is necessary that supp $a \subset [0,T]$.

When $q = 0$, representation (1.40), where $G_1(x,t) = 0$ for $t < l$, shows that $u^f(x,T) = f(T-x)$, $T < l$. This gives an immediate answer to the controllability problem for an arbitrary $a \in L^2([0,T])$, i.e., for $a \in L^2([0,l])$, supp $a \subset [0,T]$, the solution is given by

$$f(t) = a(T-t).$$

This result can be extended to the general case.

**Lemma 1.15** *For any $0 \leq T \leq l$ and any $a \in L^2([0,T])$ there exists a unique boundary source $f \in L^2([0,T])$ such that*

$$u^f(x,T) = a(x). \tag{1.52}$$

**Proof.** Using representation (1.40), we rewrite equation (1.52) in the form

$$a(x) = f(T - x) + \int_0^T G_1(x, t')f(T - t')dt', \quad 0 \le x \le T. \quad (1.53)$$

Changing variables to $y = T - x$, equation (1.53) takes the form

$$a(T - y) = f(y) + \int_0^T G_1(T - y, t')f(T - t')dt'$$

Denote $G_1(T-y, T-t')$ by $G_T(y, t')$. Then, due to the note after the proof of Lemma 1.12, $G_T(y, t') = 0$ for $t' > y$. Moreover, by Lemma 1.12, $G_T(y, t')$ is continuous as a function of $(y, t')$ for $y > t'$. Now we can rewrite equation (1.53) as a Volterra equation of the second kind,

$$f(y) + \int_0^y G_T(y, t')f(t')dt' = a(T - y).$$

This equation has a unique solution $f(y)$, $0 \le y \le T$. It can be obtained by iterations,

$$f_n(y) = a(T - y) - \int_0^y G_T(y, t')f_{n-1}(t')dt',$$

where $f_0(y) = 0$. □

For $T > l$, the solution of controllability problem (1.52) is no longer unique. However, we can still prove the existence.

**Corollary 1.16** *For any $T > l$ and any $a \in L^2([0, l])$, there exists a boundary source $f \in L^2([0, T])$, such that*

$$u^f(x, T) = a(x). \quad (1.54)$$

**Proof.** By Lemma 1.15, for any $a \in L^2([0, l])$ there exists a unique $f \in L^2([0, l])$, such that

$$u^f(x, l) = a(x).$$

Introduce the function

$$\widetilde{f}(t) = \begin{cases} 0, & \text{for} \quad 0 < t < T - l, \\ f(t - (T - l)), & \text{for} \quad T - l < t < T, \end{cases}$$

$\widetilde{f} \in L^2([0,T])$. Since $q$ is time-independent,

$$u^{\widetilde{f}}(x,t) = u^f(x, t-(T-l)),$$

for any $t > 0$. In particular,

$$u^{\widetilde{f}}(x,T) = u^f(x,l) = a(x).$$

□

**1.3.2.** In this section, we will introduce one of the main tools that we use throughout this book to solve the inverse boundary spectral problem. This is the so-called slicing procedure.

To define a slice, we first fix a point $x_0$ on the boundary which, in the one-dimensional case, may be either $x = 0$ or $x = l$. Then we take two arbitrary positive number $\tau_2 > \tau_1 \geq 0$. The corresponding slice, $\mathcal{M}(x_0; \tau_1, \tau_2)$, is the set of all points in the domain, for which the travel time to $x_0$ lies between $\tau_1$ and $\tau_2$.

Since the coordinate $x$ in wave equation (1.22) is the travel time coordinate,

$$\mathcal{M}(0; \tau_1, \tau_2) = [\tau_1, \tau_2] \cap [0, l], \tag{1.55}$$

$$\mathcal{M}(l; \tau_1, \tau_2) = [l-\tau_2, l-\tau_1] \cap [0, l].$$

For general wave equation (1.19),

$$\mathcal{M}(0; \tau_1, \tau_2) = \{x : \tau_1 \leq \int_0^x g^{1/2}(x')dx' \leq \tau_2\} \cap [0, l],$$

$$\mathcal{M}(l; \tau_1, \tau_2) = \{x : l-\tau_2 \leq \int_x^l g^{1/2}(x')dx' \leq l-\tau_1\} \cap [0, l]. \tag{1.56}$$

Since we deal with Problem 2 only, we use notation $\mathcal{M}(\tau_1, \tau_2) = \mathcal{M}(0, \tau_1, \tau_2)$.

Any slice $\mathcal{M}(\tau_1, \tau_2)$ corresponds to a subspace $L^2(\mathcal{M}(\tau_1, \tau_2)) \subset L^2([0, l])$. It consists of all functions with support in $\mathcal{M}(\tau_1, \tau_2)$. The orthoprojector, $P_{\tau_1,\tau_2} = P_{\mathcal{M}(\tau_1,\tau_2)}$ in $L^2([0, l])$ onto $L^2(\mathcal{M}(\tau_1, \tau_2))$ has the form

$$(P_{\tau_1,\tau_2}a)(x) = \chi_{\tau_1,\tau_2}(x)a(x), \tag{1.57}$$

where $\chi_{\tau_1,\tau_2} = \chi_{\mathcal{M}(\tau_1,\tau_2)}(x)$ is the characteristic function of the slice $\mathcal{M}(\tau_1, \tau_2)$,

$$\chi_{\tau_1,\tau_2}(x) = \begin{cases} 1, & \text{for } x \in \mathcal{M}(\tau_1, \tau_2), \\ 0, & \text{for } x \notin \mathcal{M}(\tau_1, \tau_2). \end{cases}$$

Then the orthoprojector $P_{\tau_1,\tau_2}$ determines the Gram-Schmidt matrix $M_{\tau_1,\tau_2}$,

$$(M_{\tau_1,\tau_2})_{jk} = \langle P_{\tau_1,\tau_2}\varphi_j, \varphi_k\rangle, \quad j, k = 1, 2, \dots. \tag{1.58}$$

We will show shortly that, for any slice $\mathcal{M}(\tau_1, \tau_2)$, the Gram-Schmidt matrix (1.58) can be constructed from the boundary spectral data. The procedure of constructing this matrix from the boundary spectral data is called the slicing procedure.

**1.3.3.** We precede the construction of the Gram-Schmidt matrices $M_{\tau_1,\tau_2}$ with the demonstration of their importance for the solution of the inverse boundary spectral problem. In fact, in the one-dimensional case, the knowledge of $\langle P_{\tau_1,\tau_2}\varphi_j, \varphi_j\rangle$ as a function of $(\tau_1, \tau_2)$ for any $j$ is sufficient for the reconstruction of the potential $q$.

**Lemma 1.17** *Assume that for some $j$ and any $\tau > 0$, we know $\langle P_{0,\tau}\varphi_j, \varphi_j\rangle$, where $\varphi_j$ is a normalized eigenfunction of the Schrödinger operator $\mathcal{A}_0$. Then these data uniquely determine $l$ and $q(x)$, $0 < x < l$.*

**Proof.** By the definition of $P_{0,\tau}$,

$$\langle P_{0,\tau}\varphi_j, \varphi_j\rangle = \int_0^{\min(\tau,l)} \varphi_j^2(x)dx.$$

By this formula, $\langle P_{0,\tau}\varphi_j, \varphi_j\rangle < 1$ for $\tau < l$ and $\langle P_{0,\tau}\varphi_j, \varphi_j\rangle = 1$ for $\tau \geq l$. This observation makes it possible to find $l$. For $\tau < l$ we use the formula

$$\frac{d}{d\tau}\langle P_{0,\tau}\varphi_j, \varphi_j\rangle = \varphi_j^2(\tau). \tag{1.59}$$

As $\varphi_j(x)$ has no multiple zeros for $0 < x < l$, equation (1.59) determines function $\varphi_j(x)$, $0 \leq x \leq l$ to within a multiplication by $\pm 1$. Then, outside a finite number of points,

$$q(x) = \frac{\partial_x^2\varphi_j(x) + \lambda_j\varphi_j(x)}{\varphi_j(x)}.$$

Since $q(x)$ is smooth, this determines $q$ on the whole interval $[0, l]$.
□

**1.3.4.** In this section, we will construct the Gram-Schmidt matrix $M_{\tau_1,\tau_2}$ from the boundary spectral data.

**Theorem 1.18** *Let $\{\lambda_j, \varphi_j'(0) : j = 1, 2, \dots\}$ be the boundary spectral data of a Schrödinger operator $\mathcal{A}_0$. Then these data determine the Gram-Schmidt matrix $M_{\tau_1,\tau_2}$ for any $0 < \tau_1 < \tau_2 < l$.*

**Proof.** The proof will be divided into several steps.

*Step 1.* By the definition of $M_{\tau_1,\tau_2}$,

$$M_{\tau_1,\tau_2} = M_{0,\tau_2} - M_{0,\tau_1}.$$

Therefore, it is sufficient to construct $M_{0,\tau}$ for arbitrary $\tau > 0$.

*Step 2.* Let us take any orthonormal basis $\{\alpha_k(t); k = 1, 2, \dots\}$ in $L^2([0, \tau])$. Denote by $u^{\alpha_k}(x, t)$ the solutions of initial-boundary value problem (1.33) with $f(t) = \alpha_k(t)$, $k = 1, 2, \dots$.

In view of Lemma 1.15, the finite linear combinations of functions $u^{\alpha_k}(\cdot, \tau)$, $k = 1, 2, \dots$, form a dense set in $L^2([0, \min(\tau, l)])$. Moreover, by means of the construction in section 1.2.6, we can find the inner products $\langle u^{\alpha_k}(\cdot, \tau), u^{\alpha_j}(\cdot, \tau)\rangle$, $j, k = 1, 2, \dots$. In general,

$$U_{kj} = \langle u^{\alpha_k}(\cdot, \tau), u^{\alpha_j}(\cdot, \tau)\rangle \neq \delta_{kj}, \tag{1.60}$$

so that $u^{\alpha_k}(\cdot, \tau)$ are not orthonormal.

*Step 3.* Using the Gram-Schmidt orthogonalization procedure for the matrix $\{U_{kj}\}_{k,j=1}^{\infty}$, we can find finite linear combinations of $u^{\alpha_k}(x, \tau)$,

$$v_l(x) = \sum_{k=1}^{p(l)} d_{kl} u^{\alpha_k}(x, \tau),$$

which form an orthonormal basis in $L^2([0, \min(\tau, l)])$. Since initial-boundary value problem (1.33) is linear,

$$v_l(x) = u^{\beta_l}(x, \tau), \tag{1.61}$$

where

$$\beta_l(t) = \sum_{k=1}^{p(l)} d_{kl} \alpha_k(t).$$

*Step 4.* Since $\{v_l(x), l = 1, 2, \dots\}$ form an orthonormal basis in $L^2([0, \min(\tau, l)])$, then, for any $a \in L^2([0, l])$,

$$P_{0,\tau}a = \sum_{l=1}^{\infty} \langle a, v_l \rangle v_l(x). \tag{1.62}$$

In particular, this formula implies that

$$\langle P_{0,\tau}\varphi_j, \varphi_k \rangle = \sum_{l=1}^{\infty} \langle \varphi_j, v_l \rangle \overline{\langle \varphi_k, v_l \rangle}. \tag{1.63}$$

Furthermore, the inner products $\langle v_l, \varphi_j \rangle$ are the Fourier coefficients of $v_l(x)$ with respect to the basis $\{\varphi_j(x), j = 1, 2, \dots\}$. By Lemma 1.13,

$$\langle v_l, \varphi_j \rangle = \int_0^{\tau} \beta_l(t') s_j(\tau - t') dt' \varphi_j'(0),$$

where $s_j(t)$ are given by formulae (1.46). □

Hence Theorem 1.6 is proven.

**1.3.5.** To conclude section 1.3, we will show that the Gram-Schmidt matrix $M_{\tau_1,\tau_2}$ determines the inner products $\langle P_{\tau_1,\tau_2} u^f(\cdot, t), u^g(\cdot, s) \rangle$ for any sources $f$, $g$ and $t, s > 0$. This result is used later to provide another solution of the one-dimensional inverse boundary spectral problem.

**Lemma 1.19** *For any sources $f, g \in L^2([0, \infty))$,*

$$\langle P_{\tau_1,\tau_2} u^f(\cdot, t), u^g(\cdot, s) \rangle = \sum_{j,k=1}^{\infty} (M_{\tau_1,\tau_2})_{jk} u_j^f(t) \overline{u_k^g(s)}, \tag{1.64}$$

*where $u_j^k(t)$ and $u_k^g(s)$ are given by formulae (1.45).*

**Exercise 1.20** *Prove formula (1.64).*

## 1.4. Gaussian beams

**1.4.1.** In this section, we will introduce the second main tool, which is also used later to solve the multidimensional inverse boundary spectral problem. As it was shown in section 1.3, the slicing procedure is sufficient to reconstruct the potential $q$. To illustrate the difference between the one-dimensional and the multidimensional cases, we mention two things. Both of them are of geometric character.

i) In the one-dimensional case, the slice $\mathcal{M}(\tau_1, \tau_2)$ is known a priori for each $\tau_1$ and $\tau_2$. This follows from the fact that, in the one-dimensional case, any operator can be transformed to a Schrödinger operator. Then, in the travel-time coordinates, the corresponding slice is just the interval $\tau_1 \le x \le \tau_2$. In the multidimensional case, it is, in principle, impossible to transform a general second-order operator to the form

$$-\left((\frac{\partial}{\partial x^1})^2 + \cdots + (\frac{\partial}{\partial x^m})^2\right) + q,$$

so that the boundary spectral data remain intact. Henceforth, the geometry of slices are *a priori* unknown.

ii) In the one-dimensional case, the slices $\mathcal{M}(\tau_1, \tau_2)$ shrink to $x_0$, if $\tau_1 < x_0 < \tau_2$ and $\tau_1, \tau_2 \to x_0$. This makes it possible to isolate the point $x_0$. We have used this localization to find $q(x)$. In the multidimensional case, when $\tau_1, \tau_2 \to \tau_0$, the corresponding slices $\mathcal{M}(\tau_1, \tau_2)$ do not, in general, tend to one point. There is a way to develop an analogous slicing procedure to get localization in the multidimensional case. However, we prefer to overcome this difficulty by using some special solutions of the wave equation that have the localization property. In this section, we will construct some special solutions of the wave equation which, at any time $t$, are localized near a point $x = x(t)$. These solutions are called Gaussian beams or quasiphotons. The point of localization, $x = x(t)$, moves along a characteristic of the wave equation.

**1.4.2.** Consider initial-boundary value problem (1.33), where the source $f$ is of the form

$$f(t) = f_\epsilon(t; a) = (\pi\epsilon)^{-1/4} \exp\{-(i\epsilon)^{-1}\theta(t-a)\}\chi_a(t). \qquad (1.65)$$

Here

$$\theta(t) = -t + \frac{i}{2}t^2, \qquad (1.66)$$

is called the phase function and $a$ and $\epsilon$ are positive constants, and the function $\chi_a(t)$ is given by the formula,

$$\chi_a(t) = \chi\left(\frac{t-a}{a}\right),$$

where $\chi(t)$ is a usual smooth cut-off function,

$$\chi(t) = \begin{cases} 1, & \text{for } |t| < 1/2, \\ 0, & \text{for } |t| > 1 \end{cases}, \quad 0 \le \chi(x) \le 1, \quad \chi \in C^\infty(\mathbf{R}).$$

**Definition 1.21** *The Gaussian beam $u_\epsilon(x,t;a)$ is the solution of initial-boundary value problem (1.33) with the source $f_\epsilon$ of the form (1.65), (1.66).*

Usually, Gaussian beams are defined as a special class of asymptotic solutions of the wave equation. We use only Gaussian beams of a special type. Their asymptotic properties are described in the following theorem.

**Theorem 1.22** *Let $u_\epsilon(x,t;a)$ be a Gaussian beam. Then for $t < l+a$*

$$u_\epsilon(x,t;a) = \tag{1.67}$$

$$(\pi\epsilon)^{-1/4}\chi_a(t-x)\exp\{-(i\epsilon)^{-1}\theta(t-x-a)\}(1+i\epsilon u_1(x,t-a))+$$

$$+R_\epsilon(x,t;a),$$

*where*

$$u_1(x,t) = \frac{1}{2\theta'(t-x)}\int_0^x q(x')dx'. \tag{1.68}$$

*The remainder $R_\epsilon(x,t;a)$ satisfy the estimate*

$$\|R_\epsilon(\cdot,t;a)\| \le C\epsilon^2, \tag{1.69}$$

*where the constant $C$ may depend on $a$ and $t < l+a$.*

**Proof.** We construct the Gaussian beam by using the anzatz,

$$U_\epsilon^N(x,t) = (\pi\epsilon)^{-1/4} \exp\{-(i\epsilon)^{-1}\theta(t-x)\} \sum_{n=0}^{N} u_n(x,t)(i\epsilon)^n. \quad (1.70)$$

Following the ideology of asymptotic methods, we consider $\epsilon$ as an independent small parameter.

*Step 1.* When we substitute $U_\epsilon^N(x,t)$ into wave equation (1.22) and obtain a polynomial of $\epsilon$ with coefficients depending on $(x,t)$,

$$\partial_t^2 U_\epsilon^N - \partial_x^2 U_\epsilon^N + q(x) U_\epsilon^N =$$

$$= \pi^{-1/4}\epsilon^{-5/4} \exp\{-(i\epsilon)^{-1}\theta(t-x)\} \sum_{n=0}^{N+1} v_n(x,t)(i\epsilon)^n, \qquad (1.71)$$

where

$$v_n = i\{2\theta'(t-x)(\partial_x + \partial_t)u_n(x,t) - (\partial_t^2 - \partial_x^2 + q(x))u_{n-1}(x,t)\},$$

for $n = 0, 1, \ldots, N+1$, and $u_{-1} \equiv 0$, $u_{N+1} \equiv 0$.

Next we will find the $N+1$ unknown functions $u_0, \ldots, u_N$, using the $N+1$ equations $v_0 = 0, \ldots, v_N = 0$. Note that we do not pose any conditions for $v_{N+1}$. This will minimize the right-hand side of equation (1.71) when $\epsilon \to 0$. The requirement that $v_n(x,t) = 0$ for $n = 0, 1, \ldots, N$, yields a recurrent system of equations for $u_n(x,t)$,

$$\frac{\partial u_n}{\partial t}(y,t) = \frac{-1}{2\theta'(-y)} \left\{ 2\frac{\partial^2 u_{n-1}}{\partial y \partial t} - \frac{\partial^2 u_{n-1}}{\partial t^2} - q(y+t)u_{n-1} \right\} \quad (1.72)$$

where we use the coordinates $y = x - t$ and $t$.

To satisfy boundary condition (1.65), we require that

$$u_0(y,-y) = 1, \quad u_n(y,-y) = 0, \quad \text{for } n = 1, \ldots, N. \qquad (1.73)$$

which in coordinates $(x,t)$ correspond to the conditions

$$u_0(0,t) = 1, \quad u_n(0,t) = 0, \quad \text{for} \quad n = 1, \ldots, N.$$

Equations (1.72) are ordinary differential equations in $t$ for the functions $u_n(y,t)$, where $y$ is considered as a parameter. Therefore, system (1.72), (1.73) constitutes a Cauchy problem for $u_n$. Since

$u_{-1} \equiv 0$, we can solve these equations recurrently, starting from $n = 0$. In particular, we obtain that $u_0 \equiv 1$ and $u_1$ have form (1.68).

Next we use $U_\epsilon^N$ to construct the Gaussian beam $u_\epsilon(x, t; a)$ and prove estimate (1.69).

*Step 2.* Consider the function $\chi_a(t - x)U_\epsilon^2(x, t - a)$ that corresponds to $N = 2$. This function is the solution of the initial-boundary value problem,

$$(\partial_t^2 - \partial_x^2 + q)\,[\chi_a(t - x)U_\epsilon^2(x, t - a)] = \mathcal{F}_{\epsilon,a}(x, t), \qquad (1.74)$$

$$\chi_a(t - x)U_\epsilon^2(x, t - a)\big|_{x=0} = f_\epsilon(t, a), \quad \chi_a(t - x)U_\epsilon^2(x, t - a)\big|_{x=l} = 0,$$

$$\chi_a(t - x)U_\epsilon^2(x, t - a)\big|_{t=0} = \partial_t[\chi_a(t - x)U_\epsilon^2(x, t - a)]\big|_{t=0} = 0,$$

for any $t \le l$. Here $\mathcal{F}_{\epsilon,a}(x, t)$ is a function that satisfies the inequality,

$$\|\mathcal{F}_{\epsilon,a}\|_{L^2([0,l]\times[0,l])} \le C_a\epsilon^2. \qquad (1.75)$$

Estimate (1.75) follows from two observations.

First, for any $M > 0$,

$$|\exp\{-(i\epsilon)^{-1}\theta(y)\}(1 - \chi(y))| \le C\epsilon^M$$

and the same is true for all derivatives of this function. Therefore, multiplication of $U_\epsilon^2(x, t)$ by $\chi_a(t - x)$ gives an error of order $\epsilon^M$ for any $M > 0$ to (1.33).

Second, due to the constructions of step 1, $U_\epsilon^2$ satisfies the equation

$$(\partial_t^2 - \partial_x^2 + q)\,U_\epsilon^2 = \pi^{-\frac{1}{4}}\epsilon^{-\frac{5}{4}}\exp\{-(i\epsilon)^{-1}\theta(t - x - a)\}v^3(x, t)(i\epsilon)^3.$$

*Step 3.* Comparing initial-boundary value problem (1.33) with $f = f_\epsilon(t, a)$ of form (1.65) and initial-boundary value problem (1.74), it follows from Lemma 1.11 that,

$$\|u_\epsilon(\cdot, t; a) - \chi_a(t - \cdot)U_\epsilon^2(\cdot, t - a)\|_{L^2([0,l])} \le C_a\epsilon^2,$$

for any $t \le l$.

At last, we observe that in the sum (1.70) with $N = 2$, the last term satisfies

$$\|\epsilon^{-\frac{1}{4}}\chi_a(t - x)\exp\{-(i\epsilon)^{-1}\theta(x - t + a)\}(i\epsilon)^2 u_2(x, t)\|_{L^2([0,l])} \le C_a\epsilon^2$$

for any $t \leq l$.

Summarizing, we obtain representation (1.67), (1.68) with the remainder estimate (1.69). □

Using the representation of the Gaussian beam given by Theorem 1.22, we see that

$$\|u_\epsilon(\cdot, t; a)\| = 1 + O(\epsilon^2), \quad \text{for } a < t < l. \tag{1.76}$$

**Exercise 1.23** *Prove estimate (1.76).*

**1.4.3.** In this and the next sections we will combine the slicing procedure and the technique of Gaussian beams to find $q(x)$.

**Lemma 1.24** *Let $u_\epsilon(x, t; a)$ be a Gaussian beam. Then, for any $t$, $\tau$, $a$, $\epsilon$, such that*

$$0 < a < t < l, \quad \tau < t - a, \quad 0 < \epsilon < 1,$$

*we have*

$$\langle P_{0,\tau} u_\epsilon(\cdot, t; a), u_\epsilon(\cdot, t; a)\rangle = O(\epsilon^2), \quad \tau < t - a,$$

$$\langle P_{0,\tau} u_\epsilon(\cdot, t; a), u_\epsilon(\cdot, t; a)\rangle = 1 + O(\epsilon^2), \quad \tau > t - a, \tag{1.77}$$

$$\langle P_{0,\tau} u_\epsilon(\cdot, t; a), u_\epsilon(\cdot, t; a)\rangle = \frac{1}{2} + \frac{\epsilon^{3/2}}{2\sqrt{\pi}} \int_0^\tau q(x)dx + O(\epsilon^2), \quad \tau = t - a.$$

Note that the left-hand side is continuous with respect to $t$, $\tau$, $a$, $\epsilon$. On the other hand, the right-hand side is discontinuous with respect to $t$, $\tau$, $a$. This means that asymptotic formula (1.77) is not uniform with respect to $t$, $\tau$, $a$. It actually describes the behaviour of the inner product as a function of $\epsilon$, $\epsilon \to 0$, for fixed parameters $t$, $\tau$, $a$.

**Proof.** The first two inequalities of formula (1.77) for $\tau < t - a$ and $\tau > t - a$, can be obtained by the following considerations.

First, in view of the exponential factor $\exp\{-(i\epsilon)^{-1}\theta(t - x - a)\}$, the Gaussian beam $u_\epsilon(x, t; a)$ decays exponentially when $\epsilon \to 0$ and $x \neq t - a$. Second, by formula (1.76), the $L^2$-norm of the Gaussian beam is $1 + O(\epsilon^2)$.

To prove formula (1.77) for $\tau = t - a$, we use the asymptotic expansion given by Theorem 1.22. Due to estimate (1.69) of $R_\epsilon(x,t;a)$ and the exponential decay for the Gaussian beam,

$$\langle P_{0,t-a}u_\epsilon(\cdot,t;a), u_\epsilon(\cdot,t;a)\rangle =$$

$$= \int_{-\infty}^{t-a} |U_\epsilon^1(x,t-a)|^2 dx + O(\epsilon^2). \tag{1.78}$$

Here $U_\epsilon^1$ is given by formula (1.70) with $N = 1$. By means of representation (1.67), (1.68), the integral in the right-hand side of formula (1.78) can be written in the form

$$(\pi\epsilon)^{-1/2} \int_{-\infty}^{0} e^{-y^2/\epsilon}(1 - 2\epsilon\Im u_1(y+t-a, t-a))dy + O(\epsilon^2) =$$

$$=(\pi\epsilon)^{-\frac{1}{2}} \int_{-\infty}^{0} e^{-\frac{y^2}{\epsilon}}(1 - \epsilon\Im\left(\frac{1}{\theta'(-y)}\int_0^{y+t-a} q(x)dx\right))dy + O(\epsilon^2), \tag{1.79}$$

where $y = x - t + a$. Using representation (1.66) for the phase function $\theta$, we see that

$$\Im\left(\frac{1}{\theta'(-y)}\int_0^{y+t-a} q(x)dx\right) = y\int_0^{t-a} q(x)dx + O(y^2). \tag{1.80}$$

Combining estimates (1.79), (1.80), we obtain that

$$(\pi\epsilon)^{-1/2} \int_{-\infty}^{0} e^{-y^2/\epsilon}\left[1 - \epsilon\Im\left(\frac{1}{\theta'(-y)}\int_0^{y+t-a} q(x)dx\right)\right] dy =$$

$$= \frac{1}{2} + \frac{\epsilon^{3/2}}{2\sqrt{\pi}}\int_0^{t-a} q(x)dx + O(\epsilon^2).$$

Summarizing the previous calculations, we prove formula (1.77) for $\tau = t - a$.

□

**1.4.4.** We are now in the position to describe a procedure to solve the inverse boundary spectral problem.

i) By means of the slicing procedure described in section 1.3.4, we find the Gram-Schmidt matrix $M_{0,\tau}$ for any $\tau > 0$.

ii) By means of the procedure described in section 1.3.5, we calculate the inner products

$$\langle P_{0,\tau} u_\epsilon(\cdot, \tau + a; a), u_\epsilon(\cdot, \tau + a; a)\rangle.$$

iii) Using these inner products, we obtain from Lemma 1.24 that

$$\int_0^t q(x)dx = \lim_{\epsilon \to 0} \frac{2\sqrt{\pi}}{\epsilon^{3/2}} \left\{ \langle P_{0,t} u_\epsilon(\cdot, t + a; a), u_\epsilon(\cdot, t + a; a)\rangle - \frac{1}{2} \right\} \tag{1.81}$$

for $0 < t < l - a$.

iv) Differentiating integral (1.81), we find $q(x)$ for $0 \leq x \leq l - a$. Since $a > 0$ is arbitrary, $q(x)$ is found for $0 \leq x \leq l$.

**Notes.** There are numerous textbooks on the spectral theory of the Sturm-Liouville operators. The properties of the eigenvalues and eigenfunctions of these operators, and, in particular, their asymptotic properties, which are used in sections 1.1.2 and 1.1.7, can be found, e.g., in [Ki], section 4.3, [LeSa], Chapter 1, [ReRg], Section 10.2, [ChCoPaRn], Chapter 3. The precise formulation of the initial-boundary value problem for the wave equation and its main properties, such as existence, uniqueness, regularity and finite velocity of the wave propagation, can be found, e.g., in [Ld], Chapter 4, [Ev], Section 7.2. These textbooks deal with the multidimensional case and we use them also as our standard references for the multidimensional wave equation. The theory of the one-dimensional Gaussian beams, which is described in section 1.4, is self-contained. For those readers who are interested in further developments of this theory, we can recommend [BaBuMo], Chapter. 3. The basic theory of Volterra equations used in Chapter 1 can be found, e.g., in [ChCoPaRn].

The one-dimensional inverse boundary spectral problem goes back to the classical works [Am], [Bo], [Lv], [GeLe], [Kr1]–[Kr4], [Ma1]–[Ma4]. There are now numerous efficient methods to solve this problem and also other types of the one-dimensional inverse problems. We refer only to some major classical and modern textbooks on the subject, such as [PoTr], [Ma1], [Ma2], [LeSa], [Nz], [Le], [ChSa], [BeBl1], [Ki]. Our approach, which is developed in this chapter, has its roots in the method of Krein-Blagovestchenskii, [Kr1]–[Kr4] and [Bl1]–[Bl3].

# Chapter 2

# Basic geometrical and analytical methods for inverse problems

This chapter is of an auxiliary nature. Here, we describe fundamental geometrical and analytical results that are the mathematical background of the approach to the inverse problems developed in Chapter 3 and 4. We believe that the geometrical and analytical ideas described in this chapter are by themselves important for the mathematical theory of inverse problems.

Section 2.1 is devoted to the basic concepts of differential and Riemannian geometry. These concepts are rather standard for textbooks in geometry. We, however, pay special attention to the case of a manifold with boundary, which often lies beyond the framework of the standard textbooks in geometry.

Section 2.2 deals with the elliptic self-adjoint second-order differential operators on manifolds. We pay special attention to the gauge transformations of such operators. These transformations are important for the further development of the invariant theory of inverse problems.

In section 2.3, we consider the initial-boundary value problem for the wave equations on manifolds with boundary. Our exposition differs somewhat from many standard textbooks on the subject due to the invariant formulation of the results.

In section 2.4, we develop the multidimensional theory of Gaussian beams. This theory is rather new and not presented in the textbooks. For these reasons, our exposition is self-contained.

Section 2.5 is devoted to the Carleman estimates and unique continuation for the wave equation. As this material is rather advanced and is not contained in the existing textbooks, we provide complete proofs of the results. Our exposition differs from other works on the subject in the way we use only energy-types estimates, rather then microlocal analysis.

## 2.1. Basic tools of Riemannian geometry for inverse problems

In this section, we will briefly describe the foundations of differential and Riemannian geometry for a reader unfamiliar with the subject. Our choice of material is dictated by further applications to inverse problems. Readers who are interested in a more systematic exposition of Riemannian geometry are referred to the literature cited in Notes to the chapter. We provide proofs only if they cannot be found in the standard textbooks.

**2.1.1.** In this book we consider compact manifolds $\mathcal{N}$ of dimension $m \geq 1$. In fact, any compact $m$-dimensional manifold can be seen as a compact $m$-dimensional surface in $\mathbf{R}^k$ for sufficiently large $k$. Any compact manifold $\mathcal{N}$ can be covered by a finite collection of open sets $U_i$ with one-to-one homeomorphic mappings $X_i$ on $U_i$, $X_i : U_i \to U_i' \subset \mathbf{R}^m$. The sets $U_i$ and the maps $X_i$ are called the coordinate charts and the coordinate mappings on $\mathcal{N}$ (see Fig. 2.1). In this book we deal with smooth manifolds, i.e., $C^\infty$-manifolds, which means that $X_i \circ X_j^{-1} : X_j(U_i \cap U_j) \to X_i(U_i \cap U_j)$ are smooth mappings. In the other words, a compact manifold consists of a finite number of pieces of an Euclidean space that are glued together. The functions $X_i(\mathbf{x}) = (x_i^1(\mathbf{x}), \ldots, x_i^m(\mathbf{x})) \in \mathbf{R}^m$ are called local coordinates on $U_i$. Sometimes, when there is no danger of misunderstanding, we also write $\mathbf{x} = (x^1, \ldots, x^m)$ identifying a point $\mathbf{x} \in \mathcal{N}$ with its representation in some local coordinates. All manifolds in this book are assumed to be compact and connected.

Figure 2.1: Coordinate charts and coordinate mappings

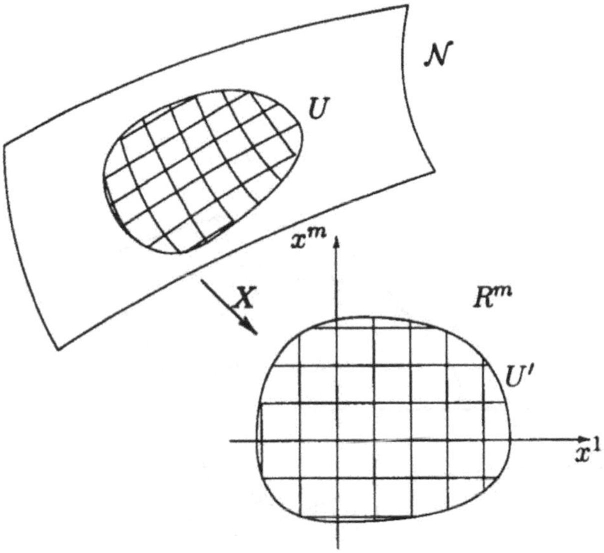


**2.1.2.** Our next goal is to introduce a tangent space at a point $\mathbf{x} \in \mathcal{N}$. When a manifold $\mathcal{N}$ is interpreted as a surface in the higher dimensional space $\mathbf{R}^k$, the tangent space is just the "plane", which is tangent to the surface. To define the tangent space in an abstract way, consider a path $\mu$ through $\mathbf{x}$, $\mu : (-\delta, \delta) \to \mathcal{N}$, $\mu(0) = \mathbf{x}$. In local coordinates $(U, X)$, the path $\mu$ has the form $\mu(t) = (x^1(t), \dots, x^m(t))$. The velocity vector of this path (at the point $\mathbf{x}$) is

$$\mathbf{v} = (v^1, \dots, v^m) = \left.\frac{d\mu}{dt}\right|_{t=0} = \left(\frac{dx^1}{dt}(0), \dots, \frac{dx^m}{dt}(0)\right).$$

If we use some other local coordinates $(\widetilde{U}, \widetilde{X})$ near $\mathbf{x}$, then the components of the vectors $\mathbf{v}$ are changed according to the rule,

$$\widetilde{v}^j = \frac{\partial \widetilde{x}^j}{\partial x^k} v^k. \tag{2.1}$$

Here, we use Einstein's summation convention by omitting sum-signs for indexes that appear both up and down in formulae. For instance, we denote $a^k b_k = \sum_k a^k b_k$.

Figure 2.2: Tangent space

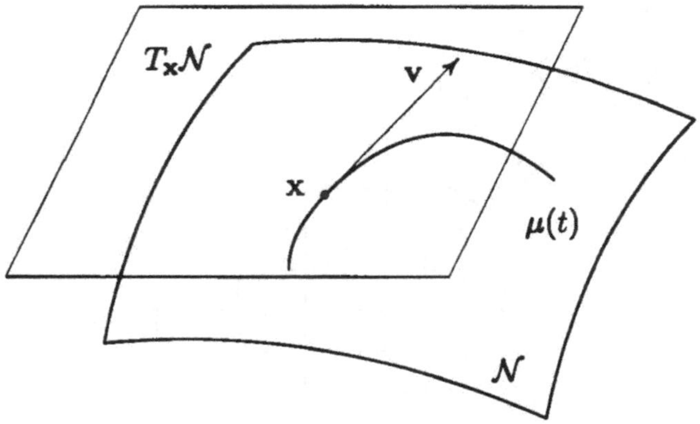


The tangent space $T_{\mathbf{x}}\mathcal{N}$ at $\mathbf{x}$ is the vector space of all these velocity vectors, which are also called tangent vectors (Fig. 2.2). Each tangent vector can also be considered as differentiation on functions $f \in C^{\infty}(\mathcal{N})$, i.e., the functions $f$, which are smooth in local coordinates. Precisely, we define

$$\mathbf{v}(f) = \frac{d}{dt}(f(\mu(t)))\bigg|_{t=0} = \partial_i f \frac{dx^i}{dt} = v^i \partial_i f,$$

where

$$\partial_i f(\mathbf{x}) = \frac{\partial f}{\partial x^i}(\mathbf{x}).$$

Clearly, $T_{\mathbf{x}}\mathcal{N}$ is a $m$-dimensional vector space. The basis of the tangent space $T_{\mathbf{x}}\mathcal{N}$, which corresponds to the local coordinates $(U, X)$ is

$$\mathbf{e}_1 = (1, 0, \ldots, 0), \ldots, \ \mathbf{e}_m = (0, 0, \ldots, 1).$$

Vectors $\mathbf{e}_k$ are also denoted by

$$\mathbf{e}_k = \frac{\partial}{\partial x^k}$$

so that

$$\mathbf{v} = v^k \frac{\partial}{\partial x^k}.$$

The cotangent space $T^*_{\mathbf{x}}\mathcal{N}$ is the space of the linear functionals on $T_{\mathbf{x}}\mathcal{N}$. Its elements are called covectors or 1-forms and denoted by $\mathbf{p}$.

The basis of the cotangent space $T^*_{\mathbf{x}}\mathcal{N}$, which corresponds to local coordinates $(U, X)$, is

$$dx^1, \dots, dx^m,$$

with

$$dx^k\left(\frac{\partial}{\partial x^j}\right) = \delta^k_j.$$

Then $\mathbf{p} = p_j dx^j$. The duality between $T^*_{\mathbf{x}}\mathcal{N}$ and $T_{\mathbf{x}}\mathcal{N}$ is denoted by

$$(\mathbf{p}, \mathbf{v}) = p_j v^j.$$

If we use other coordinates $(\widetilde{U}, \widetilde{X})$ near $\mathbf{x}$, then the components $p_j$ are changed according to the rule

$$\widetilde{p}_k = \frac{\partial x^j}{\partial \widetilde{x}^k} p_j.$$

**2.1.3.** The disjoint union of the tangent spaces,

$$T\mathcal{N} = \bigcup_{\mathbf{x}\in\mathcal{N}} T_{\mathbf{x}}\mathcal{N}$$

is called the tangent bundle of $\mathcal{N}$. Respectively, the cotangent bundle $T^*\mathcal{N}$ is the union of the spaces $T^*_{\mathbf{x}}\mathcal{N}$, $\mathbf{x} \in \mathcal{N}$.

These bundles can be considered as smooth manifolds of dimension $2m$. Local coordinates on $T\mathcal{N}$ are constructed from the local coordinates on $\mathcal{N}$. Precisely, local coordinates $(U, X)$ on $\mathcal{N}$ induce the corresponding local coordinates $(TU, \mathcal{X})$ on $T\mathcal{N}$, where

$$TU = \bigcup_{\mathbf{x}\in U} T_{\mathbf{x}}\mathcal{N}, \quad \mathcal{X} : (\mathbf{x}, \mathbf{v}) \mapsto (x^1, \dots, x^m, v^1, \dots, v^m) \in \mathbf{R}^{2m}$$

where $(x^1, \dots, x^m)$ are local coordinates of $\mathbf{x}$ and $(v^1, \dots, v^m)$ is the representation of $\mathbf{v}$ in basis $\frac{\partial}{\partial x^j}$. Coordinates of $T^*\mathcal{N}$ are defined analogously.

A smooth mapping $V : \mathcal{N} \to T\mathcal{N}$ such that $\mathbf{v} = V(\mathbf{x}) \in T_{\mathbf{x}}\mathcal{N}$ is called a smooth vector field on $\mathcal{N}$, i.e., a vector field consists of vectors attached to all points of the manifold. In local coordinates, smooth vector fields are given by $V(\mathbf{x}) = (v^1(\mathbf{x}), \dots, v^m(\mathbf{x}))$ where $v^j(\mathbf{x}) \in C^\infty(U)$.

Analogously to the definition of a vector field, smooth mappings $\mathbf{p} : \mathcal{N} \to T^*\mathcal{N}$ such that $\mathbf{p}(\mathbf{x}) \in T^*_{\mathbf{x}}\mathcal{N}$ are called differential 1-forms on $\mathcal{N}$. The space of vector fields is denoted by $\mathcal{F}\mathcal{N}$ and the space of differential 1-forms by $\Lambda^1(\mathcal{N})$.

In addition to the tangent and cotangent spaces, we need their tensor products. A multilinear mapping

$$T : \underbrace{T^*_{\mathbf{x}}\mathcal{N} \times \cdots \times T^*_{\mathbf{x}}\mathcal{N}}_{l} \times \underbrace{T_{\mathbf{x}}\mathcal{N} \times \cdots \times T_{\mathbf{x}}\mathcal{N}}_{k} \to \mathbf{R} \tag{2.2}$$

is called $(l,k)$-tensor, or $l$ covariant and $k$ contravariant tensor. The space of $(l,k)$-tensors is denoted by $\mathcal{T}^{l,k}_{\mathbf{x}}\mathcal{N}$. Its basis is given by

$$\frac{\partial}{\partial x^{i_1}} \otimes \cdots \otimes \frac{\partial}{\partial x^{i_l}} \otimes dx^{j_1} \otimes \cdots \otimes dx^{j_k}, \tag{2.3}$$

where $i_1, \dots, i_l,\ j_1, \dots, j_l \in \{1, \dots, m\}$. For example, an operator $S : T_{\mathbf{x}}\mathcal{N} \to T_{\mathbf{x}}\mathcal{N}$ can be identified with a $(1,1)$-tensor. Analogously, we can operate with complex tensors that are $\mathbf{C}$-valued mappings.

Smooth tensor fields on $\mathcal{N}$ are defined similarly to smooth vector fields. The space of smooth $(l,k)$-tensor fields is denoted by $\mathcal{F}^{l,k}\mathcal{N}$. For instance, vector fields may be considered as 1-contravariant and 1-forms are 1-covariant tensor fields.

Let $f : \mathcal{N} \to \widehat{\mathcal{N}}$ be a smooth map between two manifolds $\mathcal{N}$ and $\widehat{\mathcal{N}}$, $\dim \mathcal{N} = m$, $\dim \widehat{\mathcal{N}} = \widehat{m}$. At any $\mathbf{x} \in \mathcal{N}$, the map $f$ defines a linear mapping

$$df|_{\mathbf{x}} : T_{\mathbf{x}}\mathcal{N} \to T_{f(\mathbf{x})}\widehat{\mathcal{N}}.$$

If $(f^1(x^1, \dots, x^m), \dots, f^{\widehat{m}}(x^1, \dots, x^m))$ is the representation of $f$ in local coordinates $(U, X)$ near $\mathbf{x}$ and $(\widehat{U}, \widehat{X})$ near $f(\mathbf{x})$, then

$$df|_{\mathbf{x}}(\mathbf{v}) = (\partial_k f^1 v^k, \dots, \partial_k f^{\widehat{m}} v^k), \quad k = 1, \dots, m,$$

Hence $f$ defines the mapping $df : T\mathcal{N} \to T\widehat{\mathcal{N}}$, which is called the differential of $f$. Note that when $\widehat{\mathcal{N}} = \mathbf{R}$, the map $f$ is actually a function, $f \in C^\infty(\mathcal{N})$. As for any $s \in \mathbf{R}$ the tangent space $T_s(\mathbf{R})$ can be identified with $\mathbf{R}$, $T_s(\mathbf{R}) = \mathbf{R}$, then $df|_{\mathbf{x}} \in T^*_{\mathbf{x}}(\mathcal{N})$. Thus, the differential $df$ of a function is a differential 1-form, i.e. $df \in \Lambda^1(\mathcal{N})$, and

$$(df|_{\mathbf{x}}, \mathbf{v}) = \mathbf{v}(f), \quad \mathbf{v} \in T_{\mathbf{x}}(\mathcal{N}).$$

The mapping $f$ also determines the pull-back $f^* : C^\infty(\widehat{\mathcal{N}}) \to C^\infty(\mathcal{N})$,

$$f^* g(\mathbf{x}) = g(f(\mathbf{x})). \tag{2.4}$$

Let $\mathcal{N}$ and $\widehat{\mathcal{N}}$ be two manifolds and $f : \mathcal{N} \to \widehat{\mathcal{N}}$ be a smooth bijective mapping such that $f^{-1} : \widehat{\mathcal{N}} \to \mathcal{N}$ is also smooth. Then $f$ is a diffeomorphism from $\mathcal{N}$ to $\widehat{\mathcal{N}}$ and $\mathcal{N}$ and $\widehat{\mathcal{N}}$ are diffeomorphic manifolds. Diffeomorphic manifolds are identical from the point of view of differential geometry.

**2.1.4.** Our interest is focused on Riemannian manifolds, which are manifolds equipped with metric structure. Precisely, a Riemannian manifold $(\mathcal{N}, g)$ is a manifold $\mathcal{N}$ with a positive-definite 2-covariant tensor field $g$ called the metric tensor. In local coordinates, $g$ is given by a smooth, positive-definite, symmetric matrix function $g(\mathbf{x}) = [g_{ij}(\mathbf{x})]_{i,j=1}^m$. For any $\mathbf{x} \in \mathcal{N}$, the metric tensor defines an inner product on $T_{\mathbf{x}}\mathcal{N}$,

$$(\mathbf{v}, \mathbf{w})_g = g_{ij}(\mathbf{x}) v^i w^j, \quad \mathbf{v}, \mathbf{w} \in T_{\mathbf{x}}\mathcal{N},$$

and on $T^*_{\mathbf{x}}\mathcal{N}$

$$(\mathbf{p}, \mathbf{q})_g = g^{ij}(\mathbf{x}) p_i q_j, \quad \mathbf{p}, \mathbf{q} \in T^*_{\mathbf{x}}\mathcal{N}.$$

Here $[g^{ij}]$ is the inverse matrix of $[g_{ij}]$. The length $|\mathbf{v}|_g$ of a vector $\mathbf{v}$ is given by $|\mathbf{v}|_g^2 = (\mathbf{v}, \mathbf{v})_g$. A Riemannian metric induces a canonical transformation

$$I_g : T^*_{\mathbf{x}}\mathcal{N} \to T_{\mathbf{x}}\mathcal{N}, \tag{2.5}$$

which maps a covector $\mathbf{p} \in T^*_{\mathbf{x}}\mathcal{N}$ to a vector $\mathbf{v} \in T_{\mathbf{x}}\mathcal{N}$, so that

$$(\mathbf{p}, \mathbf{w}) = (\mathbf{v}, \mathbf{w})_g, \quad \text{for any} \quad \mathbf{w} \in T_{\mathbf{x}}\mathcal{N}.$$

In local coordinates,

$$(I_g\mathbf{p})^j = v^j = g^{ij}p_i.$$

This transformation defines a norm in $T^*_{\mathbf{x}}\mathcal{N}$, $|\mathbf{p}|_g = |I_g\mathbf{p}|_g$.

For any function $f$ on $\mathcal{N}$, the canonical transformation $I_g$ makes it possible to define the gradient vector field, Grad $f$,

$$\text{Grad}\, f|_{\mathbf{x}} = I_g\, df|_{\mathbf{x}},$$

so that

$$(\text{ Grad}\, f|_{\mathbf{x}}, \mathbf{v})_g = df|_{\mathbf{x}}(\mathbf{v}), \quad \mathbf{v} \in T_{\mathbf{x}}\mathcal{N}.$$

In local coordinates,

$$\text{Grad}\, f = g^{jk}\partial_k f \frac{\partial}{\partial x^j}.$$

**2.1.5.** The metric $g$ defines the length $|\mu[a,b]|$ of any path $\mu : [a,b] \to \mathcal{N}$,

$$|\mu([a,b])| = \int_a^b \left|\frac{d\mu(t)}{dt}\right|_g dt.$$

The lengths of paths define also the distance $d(\mathbf{x},\mathbf{y}) = d_{\mathcal{N}}(\mathbf{x},\mathbf{y})$ between any points $\mathbf{x},\mathbf{y} \in \mathcal{N}$, which is the infimum of lengths of paths connecting $\mathbf{x}$ and $\mathbf{y}$. In this chapter we assume that all manifolds are complete metric spaces with respect to this metric.

A path $\mu : [a,b] \to \mathcal{N}$ with endpoints $\mu(a) \neq \mu(b)$ is called the shortest path between its endpoints if

$$|\mu([a,b])| = d(\mu(a),\mu(b)).$$

By the Hopf-Rinow theorem for any points $\mathbf{x},\mathbf{y}$ on the complete manifold $\mathcal{N}$ can be connected by a shortest path. The metric tensor $g$ induces the Riemannian volume $dV_g$,

$$dV_g = g^{1/2}(\mathbf{x})dx^1 \ldots dx^m,$$

where $g = \det\,(g_{ij})$.

**2.1.6.** Our next goal is to define an invariant, i.e., independent of the choice of local coordinates, differentiation of the vector fields.

Let $\mathbf{x} \in \mathcal{N}$, $\mathbf{v} \in T_{\mathbf{x}}\mathcal{N}$, and $W \in \mathcal{FN}$. The covariant derivative of $W$ in the direction $\mathbf{v}$ at the point $\mathbf{x}$ is a vector in $T_{\mathbf{x}}\mathcal{N}$ which, in local coordinates $(x^1, \dots, x^m)$ near $\mathbf{x}$, has the form

$$\nabla_{\mathbf{v}} W = v^p \partial_p W^l(\mathbf{x}) \frac{\partial}{\partial x^l} + \Gamma^l_{pq}(\mathbf{x}) v^p W^q(\mathbf{x}) \frac{\partial}{\partial x^l}. \tag{2.6}$$

Here $\Gamma^k_{ij}$ are the Christoffel symbols,

$$\Gamma^k_{ij}(\mathbf{x}) = \frac{1}{2} g^{kp} \left( \frac{\partial g_{jp}}{\partial x^i} + \frac{\partial g_{ip}}{\partial x^j} - \frac{\partial g_{ij}}{\partial x^p} \right).$$

**Exercise 2.1** *Show that the transformation rule of the Christoffel symbols $\Gamma^k_{ij}$, induced by the change of coordinates $(x^1, \dots, x^m) \to (\tilde{x}^1, \dots, \tilde{x}^m)$, is given by the formula.*

$$\tilde{\Gamma}^k_{ij} = \frac{\partial \tilde{x}^k}{\partial x^q} \frac{\partial x^r}{\partial \tilde{x}^i} \frac{\partial x^s}{\partial \tilde{x}^j} \Gamma^q_{rs} + \frac{\partial \tilde{x}^k}{\partial x^s} \frac{\partial^2 x^s}{\partial \tilde{x}^i \partial \tilde{x}^j}. \tag{2.7}$$

*In particular, formula (2.7) implies that the Christoffel symbols $\Gamma^k_{ij}$ do not form a tensor.*

**Exercise 2.2** *Using formula (2.7), show that $\nabla_{\mathbf{v}} W$ given by (2.6) is a vector.*

In particular, if $V, W$ are vector fields then $\nabla_V W$ is a vector field.

**Exercise 2.3** *Show that the Christoffel symbols are compatible with the Riemannian metric tensor $g$, i.e.,*

$$X((Y, Z)_g) = (\nabla_X Y, Z)_g + (Y, \nabla_X Z)_g. \tag{2.8}$$

Covariant differentiation defines the divergence , Div $W$, of a vector field $W$ on a Riemannian manifold. Div $W$ is a scalar function,

$$\text{Div } W = \sum_{j=1}^{m} \nabla_{h_j} (W, h_j)_g,$$

where $h_j$ are vector fields which, at any point $\mathbf{x}$, form an orthonormal basis of $T_{\mathbf{x}}\mathcal{N}$. In local coordinates near $\mathbf{x}$,

$$\text{Div } W = \partial_p W^p(\mathbf{x}) + \Gamma^p_{pq}(\mathbf{x}) W^q(\mathbf{x}) = g^{-1/2}(\mathbf{x})\partial_p(g^{1/2}(\mathbf{x})W^p(\mathbf{x})).$$

The definition of the covariant derivatives can be easily generalized to the case of a path $\mu$ and a vector field $W$ along $\mu$, i.e. a mapping $W : [a, b] \to T\mathcal{N}$ with $W(t) \in T_{\mu(t)}\mathcal{N}$. The covariant derivative $\frac{DW}{dt}$ of $W$ along $\mu$ is given by

$$\frac{DW}{dt} = \nabla_{\mathbf{v}} W = \left(\frac{dW^k}{dt} + \Gamma^k_{pq} v^p W^q\right) \frac{\partial}{\partial x^k},$$

where $\mathbf{v}(t)$ is the velocity vector along the path $\mu$.

Similarly, we can define the covariant derivative $\nabla_V T$ of a tensor field $T$. In particular, the covariant derivative of the $(1,1)$-tensor field $S$, $S \in \mathcal{F}^{1,1}\mathcal{N}$ in the direction $\mathbf{v}$ at the point $\mathbf{x}$ is given by

$$\nabla_{\mathbf{v}} S = (v^k \partial_k S^i_j + \Gamma^i_{pq} v^p S^q_j - \Gamma^q_{jp} v^p S^i_q) \frac{\partial}{\partial x^i} \oplus dx^j \in \mathcal{F}^{1,1}_{\mathbf{x}}\mathcal{N}.$$

Clearly, this formula is written in local coordinates near $\mathbf{x}$.

Let $(\mathcal{N}, g)$ and $(\widehat{\mathcal{N}}, \widehat{g})$ be two Riemannian manifolds and let $f : \mathcal{N} \to \widehat{\mathcal{N}}$ be a diffeomorphism. If, in addition, for any $\mathbf{x}, \mathbf{y} \in \mathcal{N}$, we have $d_{\mathcal{N}}(\mathbf{x}, \mathbf{y}) = d_{\widehat{\mathcal{N}}}(f(\mathbf{x}), f(\mathbf{y}))$, then $f$ is said to be an isometry from $\mathcal{N}$ to $\widehat{\mathcal{N}}$ and $\mathcal{N}$ and $\widehat{\mathcal{N}}$ are isometric manifolds. Isometric manifolds are identical from the point of view of Riemannian geometry.

**2.1.7.** When the vector fields $V$ and $W$ are considered as differentiations, their commutator defines the vector field

$$[V, W] = VW - WV \in \mathcal{F}\mathcal{N},$$

which is also called the Poisson bracket of $V$ and $W$.

The covariant differentiation defines another important geometrical object called the curvature operator. Let $V, W, Z \in \mathcal{F}\mathcal{N}$. Then the expression,

$$R(V, W)Z = (\nabla_V \nabla_W - \nabla_W \nabla_V)Z - \nabla_{[V,W]}Z, \tag{2.9}$$

defines a vector field, $R(V, W)Z \in T\mathcal{N}$. Although the expression on the right-hand side of (2.9) looks like a second-order differential operator on $Z$, it does not contain any differentiation at all. In local coordinates,

$$R(V,W)Z = R^i_{jkl}V^jW^kZ^l\frac{\partial}{\partial x^i}, \tag{2.10}$$

where $R^i_{jkl}$ form a tensor called the curvature tensor. Its components can be expressed in terms of the Christoffel symbols, namely

$$R^i_{jkl} = \partial_k\Gamma^i_{lj} - \partial_l\Gamma^i_{kj} + \Gamma^r_{lj}\Gamma^i_{kr} - \Gamma^r_{kj}\Gamma^i_{lr}. \tag{2.11}$$

Formula (2.10) shows that $R(V,W)Z|_{\mathbf{x}}$ depends only on the values $V$, $W$, and $Z$ at the point $\mathbf{x}$ and, henceforth, that the curvature operator $R(\mathbf{v},\mathbf{w})\mathbf{z}$ is defined for $\mathbf{v},\mathbf{w},\mathbf{z} \in T_{\mathbf{x}}\mathcal{N}$, $\mathbf{x} \in \mathcal{N}$.

**2.1.8.** The notion of a shortest path between points gives rise to the concept of a geodesic on the Riemannian manifold. A path $\mu([a,b]) \to \mathcal{N}$ is called a geodesic if, for any $a_1, b_1 \in [a,b]$ with sufficiently small $|b_1 - a_1|$, the path $\mu([a_1,b_1])$ is a shortest path between its endpoints, i.e.

$$|\mu([a_1,b_1])| = d(\mu(a_1),\mu(b_1)).$$

In the future, we will denote a geodesic path $\mu$ by $\gamma$ and parameterize $\gamma$ with its arclength $s$ from a point $\mathbf{x} = \mu(a)$, so that $|d\gamma/ds|_g = 1$. Let $\mathbf{x}(s)$,

$$\mathbf{x}(s) = (x^1(s),\ldots,x^m(s))$$

be the representation of the geodesic $\gamma$ in local coordinates $(U,X)$. Then $\mathbf{x}(s)$ satisfies the second-order differential equations

$$\frac{d^2x^k(s)}{ds^2} = -\Gamma^k_{ij}(\mathbf{x}(s))\frac{dx^i(s)}{ds}\frac{dx^j(s)}{ds}, \tag{2.12}$$

where $\Gamma^k_{ij}$ are the Christoffel symbols.

**Exercise 2.4** *Let* $\mathbf{v}(s)$ *be the velocity vector field of the geodesic* $\gamma$. *Show that*

$$\nabla_{\mathbf{v}}\mathbf{v} = 0.$$

Equations (2.12) supplemented with initial conditions

$$\gamma(0) = \mathbf{y} \in \mathcal{N}, \ \frac{d\gamma(0)}{ds} = \mathbf{w} \in T_{\mathbf{y}}\mathcal{N}, \ |\mathbf{w}|_g = 1,$$

determine the unique geodesic $\gamma_{\mathbf{y},\mathbf{w}}$ that starts at the point $\mathbf{y}$ in the direction $\mathbf{w}$. The subsets $S\mathcal{N} \subset T\mathcal{N}$ and $S^*\mathcal{N} \subset T^*\mathcal{N}$,

$$S\mathcal{N} = \{(\mathbf{y}, \mathbf{w}) \in T\mathcal{N} : \ |\mathbf{w}|_g = 1\},$$

$$S^*\mathcal{N} = \{(\mathbf{y}, \mathbf{p}) \in T^*\mathcal{N} : \ |\mathbf{p}|_g = 1\}.$$

are called the sphere bundle and co-sphere bundle of $\mathcal{N}$.

The following results are well known.

**Lemma 2.5** *The shortest paths on a Riemannian manifold $\mathcal{N}$ have the following properties,*

*i) Any shortest path between points $\mathbf{x}, \mathbf{y} \in \mathcal{N}$ is a geodesic.*

*ii) Let $\gamma([0, b])$ be a shortest geodesic between its end points. Then*

$$\gamma([a_1, b_1]) \subset \gamma([0, b]), \quad 0 \le a_1 < b_1 < b,$$

*is the unique shortest geodesic between $\gamma(a_1)$ and $\gamma(b_1)$.*

**Exercise 2.6** *Prove the result ii) in the above lemma. Vice versa, prove that if $\gamma([0, b_1])$ is not a shortest geodesic, then $\gamma([0, b])$, $b > b_1$ is not a shortest geodesic between its endpoints.*

**2.1.9.** The set of geodesics, starting at a point $\mathbf{y}$, determines a mapping $\exp_{\mathbf{y}} : T_{\mathbf{y}}\mathcal{N} \to \mathcal{N}$,

$$\exp_{\mathbf{y}}(\mathbf{v}) = \gamma_{\mathbf{y},\mathbf{w}}(|\mathbf{v}|_g), \quad \mathbf{w} = \frac{\mathbf{v}}{|\mathbf{v}|_g},$$

which is called the exponential mapping (see Fig. 2.3).

Since the tangent space of the tangent space $T_{\mathbf{y}}\mathcal{N}$ at the point $(\mathbf{y}, \mathbf{w})$ is itself an $m$-dimensional vector space, it can be considered as identical to the space $T_{\mathbf{y}}\mathcal{N}$. More rigorously, we identify the velocity vector of the path $\mu(t) = \mathbf{w} + t\mathbf{v}$ in $T_{\mathbf{y}}\mathcal{N}$ with the vector $\mathbf{v} \in T_{\mathbf{y}}\mathcal{N}$. Thus, we consider the differential of the exponential mapping as a mapping $d\exp_{\mathbf{y}}|_{\mathbf{v}} : T_{\mathbf{y}}\mathcal{N} \to T_{\mathbf{x}}\mathcal{N}$, where $\mathbf{x} = \exp_{\mathbf{y}} \mathbf{v}$.

Varying $\mathbf{y} \in \mathcal{N}$, we obtain the exponential mapping $\exp : T\mathcal{N} \to \mathcal{N}$,

$$\exp(\mathbf{y}, \mathbf{v}) = \exp_{\mathbf{y}}(\mathbf{v}).$$

Figure 2.3: Exponential mapping

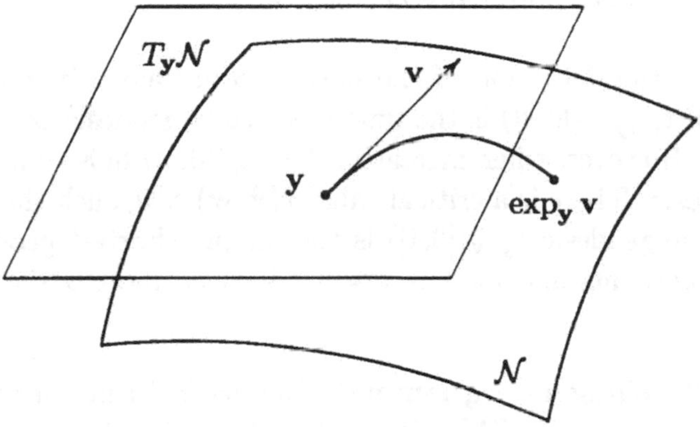


**2.1.10.** The exponential mapping is used to define a natural system of coordinates near any given point $\mathbf{y} \in \mathcal{N}$. In these coordinates, the Riemannian manifold $\mathcal{N}$ near $\mathbf{y}$ looks almost like a Euclidean space.

To introduce these coordinates, we start with an orthonormal basis $\mathbf{e}_j$, $j = 1, \ldots, m$,

$$(\mathbf{e}_j, \mathbf{e}_k)_g = \delta_{jk}$$

in $T_{\mathbf{y}}\mathcal{N}$. Then, for any vector $\mathbf{v} \in T_{\mathbf{y}}\mathcal{N}$,

$$\mathbf{v} = v^j \mathbf{e}_j, \; (v^1, \ldots, v^m) \in \mathbf{R}^m,$$

the length is given by $|\mathbf{v}|_g^2 = \sum |v^j|^2$.

Denote by $\mathcal{B}_\rho$ an open ball of radius $\rho$ in $T_{\mathbf{y}}\mathcal{N}$. Then,

$$\exp_{\mathbf{y}}(\mathcal{B}_\rho) = B_\rho(\mathbf{y}),$$

where

$$\{\mathbf{x} \in \mathcal{N} : \; d(\mathbf{x}, \mathbf{y}) < \rho\} = B_\rho(\mathbf{y})$$

is the metric ball in $\mathcal{N}$. For $\rho$ sufficiently small, $(B_\rho(\mathbf{y}), \exp_{\mathbf{y}}^{-1})$ is a local coordinate chart near $\mathbf{y}$ and

$$\exp_{\mathbf{y}}^{-1}(\mathbf{x}) = (v^1, \ldots, v^m) \tag{2.13}$$

are coordinate functions on $B_\rho(\mathbf{y})$. Such coordinates are called the (Riemannian) normal coordinates centered at $\mathbf{y}$.

**2.1.11.** From the definition of the normal coordinates it is clear that, for small $t$, $\gamma_{\mathbf{y},\mathbf{w}}([0,t])$ is the unique shortest geodesic between its endpoints. However, when increasing $t$, $\gamma_{\mathbf{y},\mathbf{w}}([0,t])$ fails to be the shortest geodesic. There is a critical value $\tau(\mathbf{y},\mathbf{w}) > 0$, such that for $t < \tau(\mathbf{y},\mathbf{w})$ the geodesic $\gamma_{\mathbf{y},\mathbf{w}}([0,t])$ is the unique shortest geodesic between its endpoints and for $t > \tau(\mathbf{y},\mathbf{w})$ it is no more a shortest geodesic.

**Theorem 2.7 (Klingenberg lemma)** *The critical value function $\tau(\mathbf{y},\mathbf{w})$ is continuous on $S\mathcal{N}$. When $t$ is the critical value, $t = \tau(\mathbf{y},\mathbf{w})$ then, either*

*i) there is another shortest geodesic $\gamma_{\mathbf{y},\mathbf{w}'}([0,t])$, $\mathbf{w} \neq \mathbf{w}'$ such that $\gamma_{\mathbf{y},\mathbf{w}}(t) = \gamma_{\mathbf{y},\mathbf{w}'}(t)$, $t = \tau(\mathbf{y},\mathbf{w})$;*

*or*

*ii) the differential of the exponential mapping $d\exp_{\mathbf{y}}|_{t\mathbf{w}}$ has non-trivial kernel, i.e. is degenerate, when $t = \tau(\mathbf{y},\mathbf{w})$.*

*Furthermore, when $t < \tau(\mathbf{y},\mathbf{w})$, $d\exp_{\mathbf{y}}|_{t\mathbf{w}}$ is not degenerate.*

**2.1.12.** A point $\mathbf{x} = \gamma_{\mathbf{y},\mathbf{w}}(\tau(\mathbf{y},\mathbf{w}))$, is called a cut point corresponding to $\mathbf{y}$. When $d\exp_{\mathbf{y}}|_{t\mathbf{w}}$ is degenerate, the point $\mathbf{x} = \gamma_{\mathbf{y},\mathbf{w}}(t)$ is called a point conjugate to $\mathbf{y}$ along the geodesic $\gamma = \gamma_{\mathbf{y},\mathbf{w}}$. The Klingenberg lemma claims that a cut point $\mathbf{x}$ is either the point that is connected to $\mathbf{y}$ by more then one shortest geodesic or is the first point conjugate to $\mathbf{y}$ along the shortest geodesic $\gamma_{\mathbf{y},\mathbf{w}}$.

The symmetry of the conjugate points means that if $\mathbf{x} = \gamma_{\mathbf{y},\mathbf{w}}(t)$ is a point conjugate to $\mathbf{y}$ along $\gamma$, then also $\mathbf{y}$ is a point conjugate to $\mathbf{x}$ along $\gamma_{\mathbf{x},\mathbf{v}}$, where $\mathbf{v} = -\frac{d}{dt}\gamma_{\mathbf{y},\mathbf{w}}(t)$. Actually, $\gamma_{\mathbf{x},\mathbf{v}}$ is just the original geodesic $\gamma$ run in the opposite direction. Exercise 2.6 implies the symmetry of the cut points, i.e., if $\mathbf{x}$ is a cut point for $\mathbf{y}$, then $\mathbf{y}$ is a cut point for $\mathbf{x}$.

The union of all cut points of $\mathbf{y}$ is called the cut locus $\omega(\mathbf{y})$ of $\mathbf{y}$,

$$\omega(\mathbf{y}) = \{\gamma_{\mathbf{y},\mathbf{w}}(\tau(\mathbf{y},\mathbf{w})) : \ \mathbf{w} \in S_{\mathbf{y}}\mathcal{N}\}.$$

Since $\tau(\mathbf{y},\mathbf{w})$ is a continuous function, $\omega(\mathbf{y})$ has measure zero. The Klingenberg lemma implies that the Riemannian normal coordinates centered in $\mathbf{y}$ are well defined on $\mathcal{N} \setminus \omega$,

$$\exp_{\mathbf{y}}^{-1} : \mathcal{N} \setminus \omega(\mathbf{y}) \to \{t\mathbf{w} \in T_{\mathbf{y}}\mathcal{N} : \quad t < \tau(\mathbf{y},\mathbf{w}), \quad \mathbf{w} \in S_{\mathbf{y}}\mathcal{N}\}.$$

Thus, $\mathcal{N}$ can be represented as a union of one coordinate chart $\mathcal{N} \setminus \omega(\mathbf{y})$ and a "thin" set $\omega(\mathbf{y})$ that has volume zero.

Let

$$\Omega = \{(\mathbf{x},\mathbf{y}) \in \mathcal{N} \times \mathcal{N} : \mathbf{x} \in \omega_{\mathbf{y}}\} = \{(\mathbf{x},\mathbf{y}) \in \mathcal{N} \times \mathcal{N} : \mathbf{y} \in \omega_{\mathbf{x}}\}.$$

The following result is a corollary of the Klingenberg lemma.

**Corollary 2.8** *The set $\Omega$ is closed in $\mathcal{N} \times \mathcal{N}$. On $\mathcal{N} \times \mathcal{N} \setminus \Omega$ the distance function has the following properties,*

*i) $d^2(\mathbf{x},\mathbf{y}) \in C^\infty(\mathcal{N} \times \mathcal{N} \setminus \Omega)$.*

*ii) $|\,\mathit{Grad}_{\mathbf{x}} d^2(\mathbf{x},\mathbf{y})| = 2d(\mathbf{x},\mathbf{y})$, and $|\,\mathit{Grad}_{\mathbf{y}} d^2(\mathbf{x},\mathbf{y})| = 2d(\mathbf{x},\mathbf{y})$.*

**Proof.** Corollary follows from the Klingenberg lemma and the fact that $d^2(\mathbf{x},\mathbf{y}) = |\mathbf{v}|_g^2$, where $\mathbf{v} = \exp_{\mathbf{y}}^{-1}(\mathbf{x})$. □

**2.1.13.** In the normal coordinates (2.13), the metric tensor $g_{ij}$ satisfies the following properties,

$$g_{ij}(\mathbf{y}) = \delta_{ij}, \quad \Gamma^i_{jl}(\mathbf{y}) = 0. \tag{2.14}$$

Using equations (2.14), one can prove the "cosine" theorem in a small neighborhood of $\mathbf{y}$. Let $\mathbf{v}_1, \mathbf{v}_2 \in T_{\mathbf{y}}\mathcal{N}$ and $\mathbf{x}_1 = \exp_{\mathbf{y}}(\mathbf{v}_1)$, $\mathbf{x}_2 = \exp_{\mathbf{y}}(\mathbf{v}_2)$. Then

$$d^2(\mathbf{x}_1,\mathbf{x}_2) = |\mathbf{v}_1|^2 + |\mathbf{v}_2|^2 - 2(\mathbf{v}_1,\mathbf{v}_2)_g + \mathcal{O}(|\mathbf{v}_1|^4 + |\mathbf{v}_1|^4). \tag{2.15}$$

**Exercise 2.9** *Using normal coordinates, prove formula (2.15).*

Figure 2.4: Coordinate charts and coordinate mappings near a boundary point

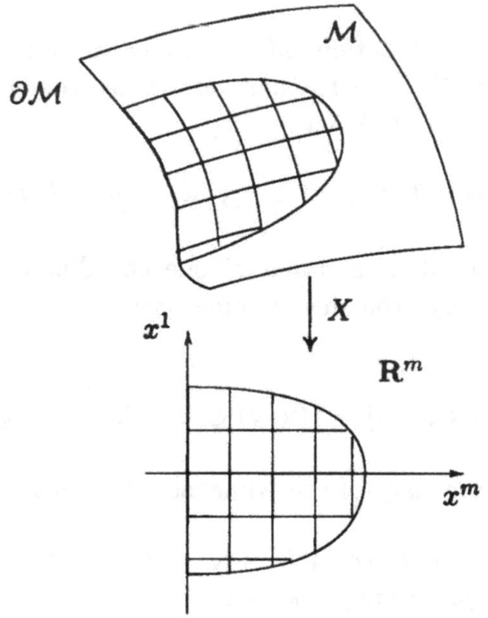


**2.1.14.** In this book, we mainly deal with manifolds $\mathcal{M}$ with boundary. A manifold $\mathcal{M}$ with boundary is a closed subset $\mathcal{M} \subset \mathcal{N}$, which has a $C^\infty$-boundary $\partial\mathcal{M}$. This means that, for $\mathbf{y} \in \partial\mathcal{M}$, there is a chart $(U, X)$ on $\mathcal{N}$, $\mathbf{y} \in U$, $X(U) = U' \subset \mathbf{R}^m$, such that

$$X(U \cap \mathcal{M}) \subset \mathbf{R}^m_+, \quad X(U \cap \partial\mathcal{M}) \subset \{\mathbf{x} \in \mathbf{R}^m : \ x^m = 0\},$$

where

$$\mathbf{R}^m_+ = \{\mathbf{x} \in \mathbf{R}^m : \ x^m \geq 0\}$$

(see Fig.2.4).

The definitions of the tangent and cotangent spaces, metric tensors, Christoffel symbols and curvature tensor remain valid for the manifolds with boundary. The boundary $\partial\mathcal{M}$ is itself an $(m-1)$-dimensional manifold without boundary. When $\mathbf{y} \in \partial\mathcal{M}$, we have two tangent spaces at $\mathbf{y}$, namely, $T_\mathbf{y}\mathcal{M}$ and $T_\mathbf{y}(\partial\mathcal{M})$ (see Fig. 2.5). Clearly, $T_\mathbf{y}(\partial\mathcal{M}) \subset T_\mathbf{y}\mathcal{M}$.

Figure 2.5: Tangent spaces of manifold and its boundary

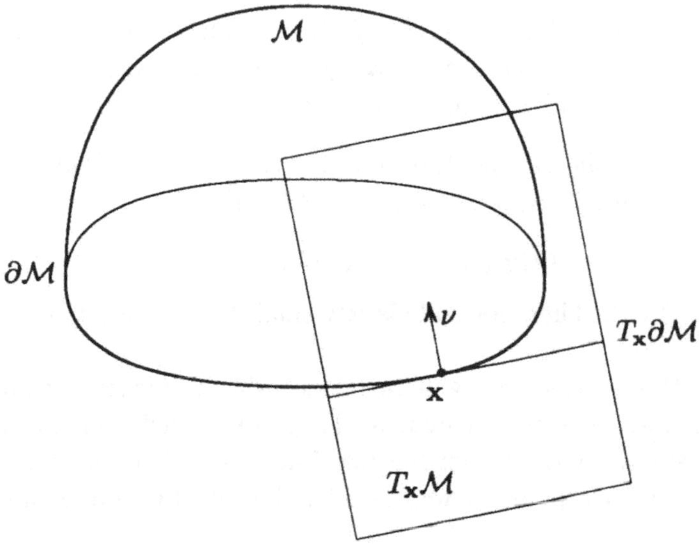


Hence, we can define the metric tensor on $\partial\mathcal{M}$ as a restriction of the metric tensor $g$ on $T(\partial\mathcal{M})$. Since $\dim(T_{\mathbf{y}}\mathcal{M}) = m$ and $\dim(T_{\mathbf{y}}(\partial\mathcal{M})) = m - 1$, there is a linear subspace $N_{\mathbf{y}} \subset T_{\mathbf{y}}\mathcal{N}$ of dimension 1, which is orthogonal to $T_{\mathbf{y}}(\partial\mathcal{M})$. We denote by $\nu = \nu_{\mathbf{y}} \in N_{\mathbf{y}}$ the unit normal vector, $|\nu|_g = 1$ which points inside $\mathcal{M}$.

Any geodesic on $\mathcal{N}$ that starts at $\mathbf{y} \in \mathcal{M}$ is simultaneously a geodesic of $\mathcal{M}$ as long as it remains in $\mathcal{M}$. A geodesic with initial point $\mathbf{z} \in \partial\mathcal{M}$ and initial velocity $\nu$, $\gamma = \gamma_{\mathbf{z},\nu}$ is called a normal geodesic.

**2.1.15.** Normal geodesics are closely related to the distance function to the boundary,

$$d(\mathbf{y}, \partial\mathcal{M}) = \min_{\mathbf{z}\in\partial\mathcal{M}} d(\mathbf{y}, \mathbf{z}).$$

**Lemma 2.10** *For any* $\mathbf{y} \in \mathcal{M}$ *there is* $\mathbf{z} \in \partial\mathcal{M}$ *such that*

$$\mathbf{y} = \gamma_{\mathbf{z},\nu}(s),$$

*where* $s = d(\mathbf{y}, \mathbf{z}) = d(\mathbf{y}, \partial\mathcal{M})$.

In the other words, this means that, for any point $\mathbf{y} \in \mathcal{M}$, there is a shortest geodesic to the boundary that is normal to $\partial\mathcal{M}$.

**Exercise 2.11** *According to the Hopf-Rinow theorem (see section 2.1.6), any* $\mathbf{x}, \mathbf{y} \in \mathcal{N}$ *can be connected by a shortest geodesic. Using this theorem and formula (2.15), prove Lemma 2.10.*

Analogously to the exponential mapping $\exp_\mathbf{y}$, we can define the boundary exponential mapping $\exp_{\partial\mathcal{M}} : \partial\mathcal{M} \times \mathbf{R}_+ \to \mathcal{N}$,

$$\exp_{\partial\mathcal{M}}(\mathbf{z}, t) = \gamma_{\mathbf{z},\nu}(t),$$

where $\mathbf{R}_+ = [0, \infty)$. Then, for sufficiently small $t$, $\exp_{\partial\mathcal{M}}(\mathbf{z}, t) \in \mathcal{M}$.

**2.1.16.** In this section, we will introduce the boundary normal (also called semi-geodesic) coordinates. They are defined analogously to the Riemannian normal coordinates, but, instead of the set of geodesics started at a point $\mathbf{y}$, we consider the set of geodesics normal to $\partial\mathcal{M}$.

As for the Riemannian normal coordinates, we start with the introduction of the boundary normal coordinates in a small neighborhood of $\partial\mathcal{M}$. Denote by

$$\mathcal{C}_\rho = \partial\mathcal{M} \times [0, \rho)$$

a collar neighborhood of $\partial\mathcal{M} \times \{0\}$ in the boundary cylinder $\partial\mathcal{M} \times \mathbf{R}_+$ (see Fig. 2.6). Then

$$\exp_{\partial\mathcal{M}}(\mathcal{C}_\rho) = \{\mathbf{x} \in \mathcal{M} : \ d(\mathbf{x}, \partial\mathcal{M}) < \rho\} = \mathcal{M}^\rho$$

is a collar neighborhood of $\partial\mathcal{M}$ in $\mathcal{M}$. When $\rho$ is sufficiently small, then $(\mathcal{M}^\rho, \exp_{\partial\mathcal{M}}^{-1})$ are coordinates in $\mathcal{M}$, with

$$\exp_{\partial\mathcal{M}}^{-1}(\mathbf{x}) = (\mathbf{z}, s). \tag{2.16}$$

Here $s = s(\mathbf{x}) = d(\mathbf{x}, \partial\mathcal{M})$, and $\mathbf{z} = \mathbf{z}(\mathbf{x}) \in \partial\mathcal{M}$ is the unique boundary point such that $d(\mathbf{x}, \mathbf{z}) = d(\mathbf{x}, \partial\mathcal{M})$. In local coordinates $(z^1, \dots, z^{m-1})$ on $\partial\mathcal{M}$ near $\mathbf{z}$, then we understand $\exp_{\partial\mathcal{M}}^{-1}(\mathbf{x})$ as $(z^1(\mathbf{z}), \dots, z^{m-1}(\mathbf{z}), s)$. Although $\gamma_{\mathbf{z},\nu}([0, s])$ is the unique shortest geodesic to $\partial\mathcal{M}$, when $s < \rho$, $\gamma_{\mathbf{z},\nu}([0, s])$ fails to be the shortest geodesic to $\partial\mathcal{M}$, when $s$ becomes large.

There is a critical value $\tau(\mathbf{z}) = \tau_{\partial\mathcal{M}}(\mathbf{z})$, such that, for $t < \tau_{\partial\mathcal{M}}(\mathbf{z})$, the geodesic $\gamma_{\mathbf{z},\nu}([0, t])$ is the unique shortest geodesic between from $\gamma_{\mathbf{z},\nu}(t)$ to $\partial\mathcal{M}$ and, for $t > \tau_{\partial\mathcal{M}}(\mathbf{z})$, it is no more the shortest one.

Figure 2.6: Boundary exponential mapping and collar neighborhoods

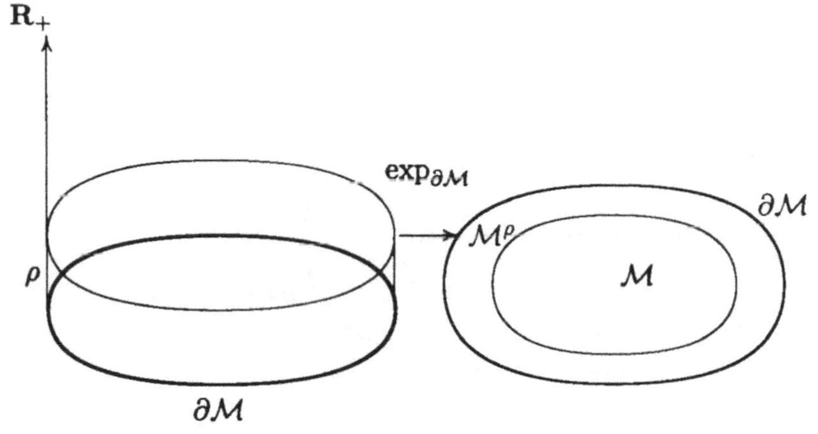


**2.1.17.** Next we formulate the Klingenberg lemma for the boundary normal coordinates.

**Theorem 2.12** *The function $\tau_{\partial\mathcal{M}}(\mathrm{z})$ is continuous on $\partial\mathcal{M}$. When $t$ is the critical value, $t = \tau_{\partial\mathcal{M}}(\mathbf{z})$, then either*

*i) there is another shortest geodesic $\gamma_{\mathbf{z}',\nu}([0,t])$, $\mathbf{z}' \neq \mathbf{z}$ such that $\gamma_{\mathbf{z},\nu}(t) = \gamma_{\mathbf{z}',\nu}(t)$, $t = \tau_{\partial\mathcal{M}}(\mathbf{z}) = \tau_{\partial\mathcal{M}}(\mathbf{z}')$,*

*or*

*ii) the differential of the exponential mapping $\exp_{\partial\mathcal{M}}$,*

$$d\exp_{\partial\mathcal{M}}|_{(\mathbf{z},t)} : T_{\mathbf{z}}(\partial\mathcal{M}) \times \mathbf{R} \to T_{\mathbf{y}}\mathcal{N}, \quad \mathbf{y} = \gamma_{\mathbf{z},\nu}(t),$$

*has a non-trivial kernel, i.e., is degenerate at $t = \tau_{\partial\mathcal{M}}(\mathbf{z})$.*

*Moreover, when $t < \tau_{\partial\mathcal{M}}(\mathbf{z})$, $d\exp_{\partial\mathcal{M}}|_{(\mathbf{z},t)}$ is not degenerate.*

A point $\mathbf{x} = \gamma_{\mathbf{z},\nu}(\tau_{\partial\mathcal{M}}(\mathbf{z}))$ is called a cut point corresponding to $\partial\mathcal{M}$. When $d\exp_{\partial\mathcal{M}}|_{(\mathbf{z},t)}$ is degenerate, the point $\mathbf{x} = \gamma_{\mathbf{z},\nu}(t)$ is called a focal point. The Klingenberg lemma claims that a cut point $\mathbf{x}$ is either point, which is connected to $\partial\mathcal{M}$ by more than one shortest

Figure 2.7: Boundary cut locus

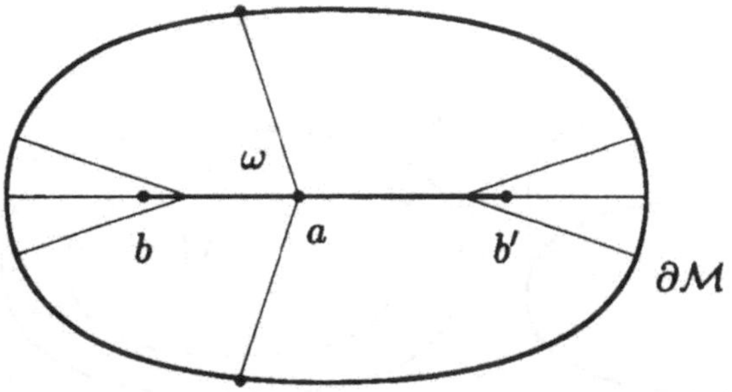


geodesic, or is the first focal point on the shortest geodesic $\gamma_{\mathbf{z},\nu}$ (see Fig. 2.7, where the point $a$ corresponds to the case $i$) and $b$ and $b'$ to the case $ii$)).

The union of all cut points is called the boundary cut locus,

$$\omega = \omega_{\partial\mathcal{M}} = \{\gamma_{\mathbf{z},\nu}(\tau_{\partial\mathcal{M}}(\mathbf{z})) : \ \mathbf{z} \in \partial\mathcal{M}\}.$$

Since $\tau_{\partial\mathcal{M}}$ is a continuous function, we see that $\omega$ has measure zero. As for the Riemannian normal coordinates, we can extend the mapping

$$\exp_{\partial\mathcal{M}}^{-1} : \mathcal{M}\backslash\omega \to \{(\mathbf{z}, t) \in \partial\mathcal{M} \times \mathbf{R}_+ : \ t < \tau(\mathbf{z}), \ \mathbf{z} \in \partial\mathcal{M}\},$$

from the set $\mathcal{M}^\rho$ to the set $\mathcal{M}\backslash\omega$ (see Fig. 2.8). This formula defines boundary normal coordinates on $\mathcal{M}\backslash\omega$. The above considerations also imply that

$$\mathcal{M} \subset \exp_{\partial\mathcal{M}}(\partial\mathcal{M} \times [0, T^*]),$$

where

$$T^* = \max_{\mathbf{z}\in\partial\mathcal{M}} \tau(\mathbf{z}). \tag{2.17}$$

Figure 2.8: Boundary normal coordinates

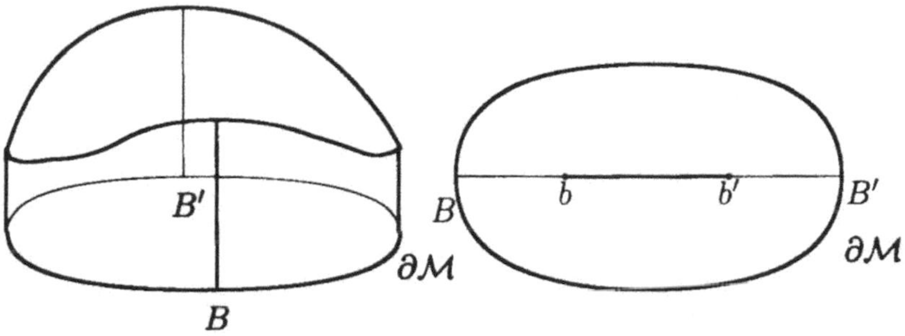


**2.1.18.** Let $(z^1, \ldots, z^{m-1})$ be some local coordinates near $\mathbf{z} \in \partial\mathcal{M}$. We denote the corresponding boundary normal coordinates by $(z^1, \ldots, z^m) = (z^1, \ldots, z^{m-1}, t)$. Then the metric tensor $g_{ij}$ satisfies the following equations,

$$g_{mm}|_{(z^1,\ldots,z^m)} = 1, \quad g_{jm}|_{(z^1,\ldots,z^m)} = 0, \quad j = 1, \ldots, m-1. \tag{2.18}$$

Let

$$\Omega_s = \{\mathbf{x} \in \mathcal{M} : \ s(\mathbf{x}) = d(\mathbf{x}, \partial\mathcal{M}) = s\}$$

be a surface parallel to $\partial\mathcal{M}$, $\Omega_0 = \partial\mathcal{M}$. Then

$$S_{jk}(\mathbf{z}, s) = \frac{1}{2}\frac{\partial g_{jk}(\mathbf{z}, s)}{\partial s}$$

defines a tensor called the second fundamental form of $\Omega_s$. The corresponding $(1,1)$-tensor, called the shape operator, is given by

$$S^l_k = \frac{1}{2} g^{lj} \frac{\partial g_{jk}}{\partial s}. \tag{2.19}$$

The tensor $S^l_k$ satisfies a Riccati-type equation that is also called *the fundamental equation of Riemannian geometry*,

$$\frac{DS}{ds} + S^2 = \nabla_{\mathbf{v}} S + S^2 = -\widehat{R}. \tag{2.20}$$

Here $\mathbf{v}$ is the velocity vector along the normal geodesics $\gamma_{\mathbf{z},\nu}$, having the form $\mathbf{v} = (0, \ldots, 0, 1)$ in the boundary normal coordinates and $\widehat{R} \in \mathcal{F}^{1,1}\mathcal{N}$,

$$\widehat{R}(\mathbf{w}) = R(\mathbf{w}, \mathbf{v})\mathbf{v}, \quad \mathbf{w} \in T_{\gamma_{\mathbf{z},\nu}(s)}\mathcal{M}.$$

Figure 2.9: Shortest geodesics when $\mathbf{y}$ is conjugate to $\mathbf{z}$

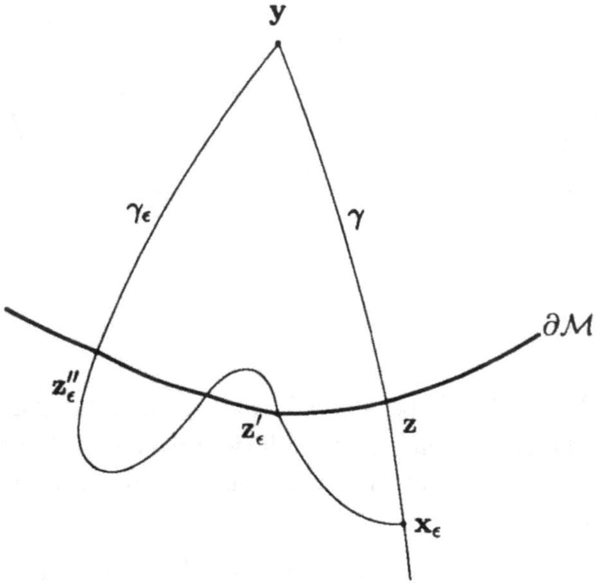


**2.1.19.** Next we will prove that the first conjugate point on a normal geodesic always appears strictly beyond the cut point.

**Lemma 2.13** *For any* $\mathbf{z} \in \partial\mathcal{M}$, $\tau_{\partial\mathcal{M}}(\mathbf{z}) < \tau(\mathbf{z}, \nu)$.

**2.1.20.** ★

**Proof.** Assume the opposite. Then by Theorem 2.12, $\gamma_{\mathbf{z},\nu}(t)$, where $t = \tau(\mathbf{z}, \nu)$, is either the point with more than one shortest geodesic to $\mathbf{z}$, or the first conjugate point along $\gamma_{\mathbf{z},\nu}$. In the first case, let $\gamma_{\mathbf{z},\mathbf{w}}$ be another shortest geodesic to $\gamma_{\mathbf{z},\nu}(t)$ with $\mathbf{w} \neq \nu$. It follows from the definition of $\tau_{\partial\mathcal{M}}$ that, for sufficiently small $s$, there is $\mathbf{z}' \in \partial\mathcal{M}$, such that $d(\mathbf{z}', \gamma_{\mathbf{z},\mathbf{w}}(s)) < s$. Therefore,

$$d(\mathbf{z}', \gamma_{\mathbf{z},\nu}(t)) \leq d(\mathbf{z}', \gamma_{\mathbf{z},\mathbf{w}}(s)) + (t - s) < t = d(\mathbf{z}, \gamma_{\mathbf{z},\nu}(t)) \leq \tau_{\partial\mathcal{M}}(\mathbf{z}),$$

which contradicts the definition of $\tau_{\partial\mathcal{M}}(\mathbf{z})$.

Let now $\mathbf{y} = \gamma_{\mathbf{z},\nu}(t)$ be the first conjugate point along $\gamma_{\mathbf{z},\nu}$. Consider the geodesic $\gamma_{\mathbf{z},\nu}([-\epsilon, t])$ that is an extension of $\gamma_{\mathbf{z},\nu}([0, t])$ to $\mathcal{N}\backslash\mathcal{M}$ (see Fig. 2.9). Denote by $\mathbf{x}_\epsilon = \gamma_{\mathbf{z},\nu}(-\epsilon)$ the endpoint of this geodesic. Since $\mathbf{z}$ is conjugate to $\mathbf{y}$ along the geodesic $\gamma_{\mathbf{z},\nu}([0,t])$, the geodesic $\gamma_{\mathbf{z},\nu}([-\epsilon,t])$ cannot be the shortest one between its endpoints. Thus, by the Klingenberg Lemma 2.7, there exists another shortest geodesic $\gamma_\epsilon$ from $\mathbf{y}$ to $\mathbf{x}_\epsilon$ in $\mathcal{N}$, such that

$$|\gamma_\epsilon| < |\gamma_{\mathbf{z},\nu}([-\epsilon,t])| = t+\epsilon.$$

Consider the intersections of the geodesic $\gamma_\epsilon$ with $\partial\mathcal{M}$ and denote by $\mathbf{z}'_\epsilon$ the first intersection point from $\mathbf{x}_\epsilon$ and by $\mathbf{z}''_\epsilon$ the last one. Then

$$d(\mathbf{y},\mathbf{x}_\epsilon) \geq d(\mathbf{y},\mathbf{z}''_\epsilon) + d(\mathbf{z}'_\epsilon,\mathbf{x}_\epsilon).$$

For sufficiently small $\epsilon$, Theorem 2.12, applied to $\mathcal{N}\backslash\mathcal{M}$, implies that

$$d(\mathbf{z}'_\epsilon,\mathbf{x}_\epsilon) > \epsilon.$$

Since $t \leq \tau_{\partial\mathcal{M}}(\mathbf{z})$, then $d(\mathbf{z}''_\epsilon,\mathbf{y}) \geq t$. Combining these inequalities, we see that

$$|\gamma_\epsilon| = d(\mathbf{x}_\epsilon,\mathbf{y}) > t+\epsilon,$$

which is a contradiction. □

**2.1.21.** The boundary normal coordinates are inappropriate near the cut locus. In this case, we use other coordinates, namely, the boundary distance coordinates.

**Lemma 2.14** *For any $\mathbf{y}\in\mathcal{M}^{int}$ there are points $\mathbf{z}_1,\dots,\mathbf{z}_m\in\partial\mathcal{M}$, such that the functions $(\rho^1(\mathbf{x}),\dots,\rho^m(\mathbf{x}))$, where $\rho^i(\mathbf{x}) = d(\mathbf{x},\mathbf{z}_i)$, are local coordinates in a neighborhood of $\mathbf{y}$.*

Before going to the proof, let us explain how to choose points $\mathbf{z}_i$. For $\mathbf{y}\in\mathcal{M}^{int}$, let $\mathbf{z}\in\partial\mathcal{M}$ be a nearest boundary point, $d(\mathbf{z},\mathbf{y}) = d(\mathbf{y},\partial\mathcal{M})$. Consider any curves $\mathbf{z}_i(t)$, $i=1,\dots,m-1$, in $\partial\mathcal{M}$, such that $\mathbf{z}_i(0)=\mathbf{z}$ and the vectors

$$\eta_i = \frac{d\mathbf{z}_i}{dt}(0), \quad i=1,\dots,m-1,$$

form an orthonormal basis of $T_\mathbf{z}(\partial\mathcal{M})$. Then, for sufficiently small $t>0$, the points $\mathbf{z}_1=\mathbf{z}_1(t),\dots,\mathbf{z}_{m-1}=\mathbf{z}_{m-1}(t),\mathbf{z}_m=\mathbf{z}$ are proper for Lemma 2.14.

Figure 2.10: Boundary distance coordinates

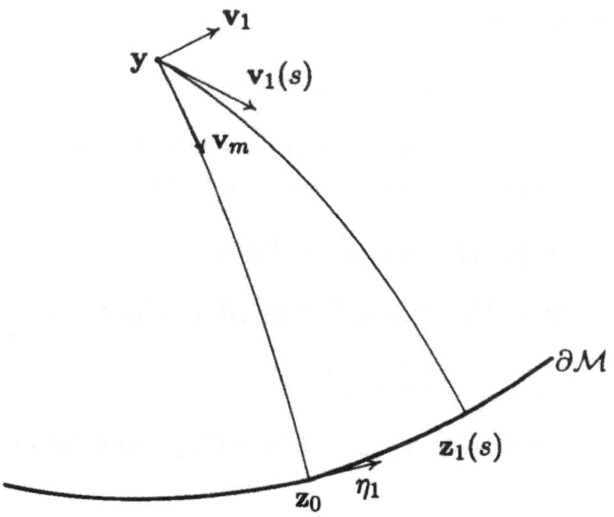


**2.1.22.** ★

**Proof.** As earlier, let $\mathbf{z} \in \partial\mathcal{M}$ be a nearest point to $\mathbf{y}$. Choose $\mathbf{v}$, so that $\exp_{\mathbf{y}}(\mathbf{v}) = \mathbf{z}$. We remind the reader that, due to Lemma 2.13, $\mathbf{z}$ is not conjugate to $\mathbf{y}$. Hence, $d\exp_{\mathbf{y}}|_{\mathbf{v}}$, is non-degenerate, i.e. the mapping $d\exp_{\mathbf{y}}|_{\mathbf{v}} : T_{\mathbf{y}}\mathcal{N} \to T_{\mathbf{z}}\mathcal{N}$ is an isomorphism.

Since $\eta_i$, $i = 1, \ldots, m-1$, and $\nu$ form an orthonormal basis in $T_{\mathbf{z}}\mathcal{M}$, the vectors $\{\mathbf{v}_1, \ldots, \mathbf{v}_{m-1}, \mathbf{v}_m\}$,

$$\mathbf{v}_i = (d\exp_{\mathbf{y}}|_{\mathbf{v}})^{-1}\eta_i, \quad i = 1, \ldots, m-1, \quad \mathbf{v}_m = \mathbf{v},$$

form a basis of $T_{\mathbf{y}}\mathcal{M}$ (see Fig. 2.10).

Moreover, for sufficiently small $s$, there are $\mathbf{v}_i(s) \in T_{\mathbf{y}}\mathcal{M}$ such that,

$$\mathbf{z}_i(s) = \exp_{\mathbf{y}}(\mathbf{v}_i(s)), \quad i = 1, \ldots, m-1,$$

and

$$\mathbf{v}_i(0) = \mathbf{v}, \quad \left.\frac{d\mathbf{v}_i(s)}{ds}\right|_{s=0} = \mathbf{v}_i, \quad i = 1, \ldots, m-1, \quad \mathbf{v}_m = \mathbf{v}.$$

Let $\mathbf{z}_i = \mathbf{z}_i(s_0)$, $i = 1, \dots, m-1$, for sufficiently small $s_0$ and $\mathbf{z}_m = \mathbf{z}$. Then $\mathbf{y} \notin \omega(\mathbf{z}_i)$, $i = 1, \dots, m$, and functions $\rho^i(\mathbf{x})$,

$$\rho^i(\mathbf{x}) = d(\mathbf{y}, \mathbf{z}_i), \quad i = 1, \dots, m-1,$$

are smooth in a neighborhood $U_\mathbf{y}$ of $\mathbf{y}$.

Finally, the gradients of the distance functions

$$\mathrm{Grad}\,\rho^i|_\mathbf{y} = -\frac{\mathbf{v}_i(s_0)}{|\mathbf{v}_i(s_0)|_g}, \quad i = 1, \dots, m,$$

are linearly independent. The claim follows now from the inverse function theorem.

□

The proof of Lemma 2.14 also gives rise to the following result.

**Lemma 2.15** *Let $\mathbf{x}_0 \in \mathcal{M}^{int}$ and $\mathbf{z}_0$ be a nearest point to $\mathbf{x}_0$ on the boundary. Then there are neighborhoods $V \subset \partial\mathcal{M}$ and $U \subset \mathcal{M}^{int}$, $\mathbf{z} \in V$, $\mathbf{x}_0 \in U$, such that*

*i) $d(\mathbf{x}, \mathbf{z}) \in C^\infty(U \times V)$.*

*ii) $\mathrm{Grad}_\mathbf{x} d(\mathbf{x}, \mathbf{z})|_{\mathbf{x}=\mathbf{x}_0}$, considered as a function of $\mathbf{z}$, is a diffeomorphism from $V$ to its image in $S_{\mathbf{x}_0}\mathcal{M}$. In particular, this means that this image is an open set in $S_{\mathbf{x}_0}\mathcal{M}$.*

## 2.2. Elliptic operators on manifolds and gauge transformations

In this section, we will describe second-order elliptic partial differential operators on manifolds. In the first half of this section, we will recall some fundamental results from the general theory of such operators. An interested reader can find a systematic treatment of this theory in the literature cited in Notes to the chapter. The second half of the chapter is devoted to gauge transformations. Here we will give the detailed proofs of the corresponding results.

**2.2.1.** We start with a function $f \in C^\infty(\mathcal{N})$. This means that, in any local coordinates $(U, X)$, $X(\mathbf{x}) = (x^1, \dots, x^m)$, the function

$$f \circ X^{-1}(x^1, \dots, x^m) = f(x^1, \dots, x^m)$$

is infinitely smooth. In local coordinates $(U, X)$ on $\mathcal{N}$, a differential expression $a(\mathbf{x}, D)$ is given by

$$(a(\mathbf{x}, D)f)(\mathbf{x}) = -a^{jk}(x^1, \dots x^m)\partial_j\partial_k f(x^1, \dots, x^m) + \tag{2.21}$$

$$+b^j(x^1, \dots, x^m)\partial_j f(x^1, \dots, x^m) + c(x^1, \dots, x^m)f(x^1, \dots, x^m),$$

where $\partial_j f = \frac{\partial f}{\partial x^j}$. We assume that all coefficients are real and $[a^{jk}]_{j,k=1}^m$ is a symmetric matrix. Local representations (2.21), which are called local differential expressions, are assumed to be defined in such a manner that the value of $a(\mathbf{x}, D)f$ at any point $\mathbf{x} \in \mathcal{N}$ is independent of the choice of local coordinates $(U, X)$ near $\mathbf{x}$. If there is no danger of misunderstanding, we do not distinguish between the point $\mathbf{x}$ and its local coordinates $(x^1, \dots, x^m)$, as well as between the function $f$ on $U \subset \mathcal{N}$ and its representation $f \circ X^{-1}$ on $U' \subset \mathbf{R}^m$.

**Exercise 2.16** *Find the transformation formulae for the local differential expressions in different local coordinates. In particular, show that $[a^{jk}]$ is a tensor.*

The differential expression $a(\mathbf{x}, D)$ is said to be elliptic, if the matrix $[a^{jk}](x^1, \dots, x^m)$ is positive definite in any local coordinates,

$$a^{jk}p_j p_k \geq C \sum_{j=1}^m p_j^2, \quad C > 0. \tag{2.22}$$

The fact that $[a_{jk}] = [a^{jk}]^{-1}$ is positive definite and forms a 2-covariant tensor, implies that we can use $[a_{jk}]$ as a Riemannian metric tensor on $\mathcal{N}$. We call it the metric associated with the differential expression $a(\mathbf{x}, D)$. Differential expressions on a manifold $\mathcal{M}$ with boundary are defined in exactly the same way.

**2.2.2.** In this book we use Sobolev spaces $H^n(\mathcal{M})$ and $H_0^n(\mathcal{M})$, $n = 0, 1, 2, \dots$, which are defined in local coordinates. We remind the reader that, for $U \subset \mathbf{R}^m$, the space $H^n(U)$ consists of all functions $f \in L^2(U)$ such that

$$\|f\|_n^2 = \sum_{|\alpha| \leq n} \int_U |\partial^\alpha f(\mathbf{x})|^2 dx^1 \cdots dx^m < \infty. \tag{2.23}$$

Here $\alpha = (\alpha_1, \dots, \alpha_m)$, $\alpha_j = 0, 1, \dots$, are multi-indexes and

$$\partial^\alpha = \partial_1^{\alpha_1} \partial_2^{\alpha_2} \cdots \partial_m^{\alpha_m}, \quad |\alpha| = \alpha_1 + \cdots + \alpha_m.$$

Clearly, $H^0(U) = L^2(U)$.

When dealing with compact manifolds $\mathcal{M}$ (or $\mathcal{N}$), we use a finite partition of unity $\chi_i$, subordinated to a finite covering $(U_i, X_i)$, $i = 1, \dots, I$, of $\mathcal{M}$. This means that $\chi_i \in C_0^\infty(U_i)$, $\chi_i \geq 0$, and

$$\sum_{i=1}^{I} \chi_i = 1. \tag{2.24}$$

The space $C_0^\infty(U_i)$ consists of smooth functions on $\mathcal{M}$ with compact support, supp $(\chi_i) \subset U_i$. We remind the reader that either $X(U)$ is an open set in $\mathbf{R}^m$ or, in the case of a manifold with boundary, an intersection of an open set in $\mathbf{R}^m$ with $\mathbf{R}^m_+$.

A function $f : \mathcal{M} \to \mathbf{C}$ is in the space $H^n(\mathcal{M})$, if

$$\|f\|_n^2 = \sum_{i=1}^{I} \|\chi_i f\|_n^2 < \infty. \tag{2.25}$$

Here $\chi_i f$ are considered as a function $\chi_i f(x_i^1, \dots, x_i^m)$ in local coordinates $(U_i, X_i)$, $X_i(\mathbf{x}) = (x_i^1, \dots, x_i^m)$. Although the norm $\|f\|_n$ depends on the choice of the covering $(U_i, X_i)$ and partition of unity $\chi_i$, the Sobolev spaces $H^n(\mathcal{M})$ are defined invariantly. Indeed, the norms, which correspond to different coverings, are equivalent.

The space $H_0^n(\mathcal{M})$ is the closure of $C_0^\infty(\mathcal{M})$ in $H^n(\mathcal{M})$.

Occasionally, we will use the Sobolev spaces $H^s(\mathcal{M})$ and $H_0^s(\mathcal{M})$ with non-integer $s \geq 0$. They are defined as interpolation spaces between $H^{[s]}(\mathcal{M})$ and $H^{[s+1]}(\mathcal{M})$ and $H_0^{[s]}(\mathcal{M})$ and $H_0^{[s+1]}(\mathcal{M})$, correspondingly. Here $[s]$ is the integer part of $s$. The space $H_0^s(\mathcal{M})$ is actually the closure of $C_0^\infty(\mathcal{M})$ in $H^s(\mathcal{M})$.

Finally, by trace theorem, the restriction operator $T_{\partial\mathcal{M}} : f \mapsto f|_{\partial\mathcal{M}}$ is a continuous operator, $T_{\partial\mathcal{M}} : H^s(\mathcal{M}) \to H^{s-1}(\partial\mathcal{M})$, or, more precisely, to $H^{s-1/2}(\partial\mathcal{M})$, for $s \neq n + 1/2$. Also, this operator is surjective.

We remark also that $H^s(\mathcal{M})$ and $H_0^s(\mathcal{M})$ are Hilbert spaces.

**2.2.3.** Our next goal is to define second-order self-adjoint operators on $\mathcal{N}$ (or $\mathcal{M}$). To this end, we need a volume element $dV$ on a manifold. In local coordinates $(U, X)$ near $\mathbf{x}$, we have the representation

$$dV = m(\mathbf{x})(g(\mathbf{x}))^{1/2}dx^1dx^2\cdots dx^m = m(\mathbf{x})dV_g,$$

where $m > 0$ is a smooth function on $\mathcal{M}$.

A coordinate invariant norm in $L^2(dV)$ is then

$$\|f\|^2_{L^2(dV)} = \int_{\mathcal{N}} |f|^2 dV = \sum_{i=1}^{I} \int_{U_i'} |\chi_i f(\mathbf{x})|^2 m(\mathbf{x}) dV_g. \tag{2.26}$$

In the same manner, it is possible to define a coordinate invariant norm in $H^1(\mathcal{N})$,

$$\|f\|^2_{H^1(\mathcal{N},dV)} = \int_{\mathcal{N}} (|df|_g^2 + |f|^2)dV = \int_{\mathcal{N}} (|\operatorname{Grad} f|_g^2 + |f|^2)dV.$$

Moreover, using covariant derivatives (see section 2.1.6), it is possible to define coordinate invariant norms in $H^n(\mathcal{N})$, $n \geq 0$. However, in this book we will not use these norms with $n > 1$.

**Exercise 2.17** *Prove that the norms* $\|\cdot\|_{L^2(dV)}$ *and* $\|\cdot\|_{H^1(\mathcal{N})}$ *are equivalent to the norms* $\|\cdot\|_0$ *and* $\|\cdot\|_1$*, correspondingly, and that these norms are independent of the covering* $(U_i, X_i)$ *and partition of unity* $\chi_i$.

In the future, we will denote $\|f\|_{L^2(dV)}$ by $\|f\|$ and the corresponding inner product by $\langle f, h\rangle$,

$$\langle f, h\rangle = \int_{\mathcal{N}} f(\mathbf{x})\overline{h(\mathbf{x})}dV. \tag{2.27}$$

An operator $\mathcal{A}$ in $L^2(\mathcal{N}, dV)$, which corresponds to the differential expression $a(\mathbf{x}, D)$, is defined by the formula

$$\mathcal{A}u(\mathbf{x}) = (a(\mathbf{x}, D)u)(\mathbf{x})$$

on

$$\mathcal{D}(\mathcal{A}) = H^2(\mathcal{N}). \tag{2.28}$$

Since $L^2(\mathcal{N}, dV)$ is a Hilbert space, the adjoint operator $\mathcal{A}^*$ is defined by the formula

$$\langle \mathcal{A}u, v\rangle = \langle u, \mathcal{A}^* v\rangle. \tag{2.29}$$

It turns out that $\mathcal{D}(\mathcal{A}^*) = H^2(\mathcal{N})$ and $\mathcal{A}^*u$ is given by differential expression (2.21) with $a^{ij}, b^j$, and $c$ replaced by $a^{*,ij}, b^{*,j}$, and $c^*$.

**Exercise 2.18** *Find the differential expression $a^*(\mathbf{x}, D)$ of $\mathcal{A}^*$.*

An operator $\mathcal{A}$ is self-adjoint if $\mathcal{A} = \mathcal{A}^*$.

**Exercise 2.19** *Show that an operator $\mathcal{A}$ is self-adjoint if and only if*

$$a(\mathbf{x}, D)f = -m^{-1}g^{-1/2}\partial_i\left(mg^{1/2}g^{ij}\partial_j u\right) + qu, \tag{2.30}$$

*where $g^{ij} = a^{ij}$ (compare with formula (1.4)).*

**2.2.4.** In the case of a manifold $\mathcal{M}$ with boundary, we have to add proper boundary conditions. In this book we deal with the Dirichlet boundary condition,

$$u|_{\partial\mathcal{M}} = 0. \tag{2.31}$$

Then the operator $\mathcal{A}$ defined by differential expression (2.30) on

$$\mathcal{D}(\mathcal{A}) = H^2(\mathcal{M}) \cap H_0^1(\mathcal{M}) = \{u \in H^2(\mathcal{M}) : \ u|_{\partial\mathcal{M}} = 0\} \tag{2.32}$$

is self-adjoint in $L^2(\mathcal{M}, dV)$.

An important object related to any self-adjoint operator $\mathcal{A}$ is its quadratic form (2.32)–2.34),

$$\mathcal{A}[u, w] = \int_{\mathcal{M}} ((du, d\overline{w})_g + qu\overline{w})dV, \tag{2.33}$$

where $(dv, d\overline{w})_g = g^{ij}\partial_i v \partial_j \overline{w}$ (see section 2.1.4). This quadratic form is defined on $H^1(\mathcal{N})$, in the case of a manifold without boundary, and $H_0^1(\mathcal{M})$, in the case of a manifold with boundary. For any $u \in H^1(\mathcal{N})$ or $u \in H_0^1(\mathcal{M})$,

$$|\mathcal{A}[u, u]| \leq C\|u\|^2,$$

and, for sufficiently large $\mu > 0$,

$$\|u\|_1^2 \leq C(\mathcal{A}[u,u] + \mu\|u\|^2).$$

In the future, we will omit references to manifolds $\mathcal{N}$ or $\mathcal{M}$, if the corresponding result is valid for a manifold without boundary as well as for a manifold with boundary, writing e.g., $H^n$, $L^2$, etc., instead of $H^n(\mathcal{N})$, $L^2(\mathcal{M})$, etc.

Summarizing these results, we have the following lemma.

**Lemma 2.20** *Let $L^2(dV)$ be a Hilbert space with a volume element $dV$ and let $\mathcal{A}$ be a second-order differential operator with domain $H^2(\mathcal{N})$, in the case of a manifold without boundary, or with domain $H^2(\mathcal{M}) \bigcap H_0^1(\mathcal{M})$, in the case of a manifold with boundary. The operator $\mathcal{A}$ is self-adjoint, if and only if the corresponding differential expression has form (2.30). Then for $u, w \in \mathcal{D}(\mathcal{A})$,*

$$\mathcal{A}[u,w] = \langle \mathcal{A}u, w\rangle. \tag{2.34}$$

**2.2.5.** A function $\varphi \in \mathcal{D}(\mathcal{A})$, $\varphi \neq 0$, is an eigenfunction of $\mathcal{A}$ if

$$(\mathcal{A} - \lambda)\varphi = 0, \quad \lambda \in \mathbf{C}.$$

In this case $\lambda$ is an eigenvalue of the operator $\mathcal{A}$ and $\varphi$ is an eigenfunction corresponding to $\lambda$. In particular, on a manifold $\mathcal{M}$ with boundary, this means that

$$(a(\mathbf{x}, D) - \lambda)\varphi = 0, \quad \text{in} \quad \mathcal{M},$$

$$\varphi|_{\partial\mathcal{M}} = 0.$$

The set of all eigenfunctions, corresponding to the eigenvalue $\lambda$, together with $\varphi = 0$, form a finite dimensional linear subspace in $L^2(dV)$, which is called the eigenspace of $\mathcal{A}$ corresponding to $\lambda$. The dimension of the eigenspace is the multiplicity of the eigenvalue $\lambda$. When $\mathcal{A}$ is self-adjoint, all eigenvalues are real and all eigenspaces corresponding to the different eigenvalues are orthogonal,

$$\langle\varphi_1, \varphi_2\rangle = 0, \quad \text{if} \quad \mathcal{A}\varphi_1 = \lambda_1\varphi_1, \quad \mathcal{A}\varphi_2 = \lambda_2\varphi_2, \quad \lambda_1 \neq \lambda_2. \tag{2.35}$$

When $\lambda$ is not an eigenvalue, the operator $(\mathcal{A} - \lambda)$ is invertible. Its inverse,

$$\mathcal{R}(\lambda) = (\mathcal{A} - \lambda)^{-1},$$

is called the resolvent of $\mathcal{A}$. This operator is a bounded operator in $L^2(dV)$. The following properties of $\mathcal{A}$ are well known.

**Theorem 2.21** *Let $\mathcal{A}$ be a self-adjoint second-order differential operator of form (2.30) in $L^2(\mathcal{N}, dV)$ and (2.30), (2.32) in $L^2(\mathcal{M}, dV)$. Then its eigenvalues, counted according to their multiplicities, form an increasing sequence $\lambda_1 < \lambda_2 \leq \lambda_3 \leq \ldots$, such that $\lambda_j j^{-2/m} \to C$, for $j \to \infty$. The constant $C > 0$ depends only on the dimension $m$ and the volume $V(\mathcal{M})$ of $\mathcal{M}$. The corresponding eigenfunctions $\varphi_j$ can be chosen so that $\{\varphi_j\}_{j=1}^{\infty}$ form an orthonormal basis in $L^2(dV)$. Moreover, $\varphi_1(\mathbf{x}) \neq 0$ for $\mathbf{x} \in \mathcal{M}^{int}$.*

The above asymptotics of $\lambda_j$ is called Weyl's asymptotics. This theorem implies that any $f \in L^2(dV)$ can be represented by Fourier series,

$$f = \sum_{j=1}^{\infty} f_j \varphi_j, \quad \{f_j\}_{j=1}^{\infty} \in \ell^2.$$

The numbers $f_j$ are the Fourier coefficients of $f$, i.e. $f_j = \langle f, \varphi_j \rangle$, and

$$\|f\| = \|\{f_j\}\|_{\ell^2}.$$

Moreover, if $\lambda$ is not an eigenvalue, then the equation

$$(\mathcal{A} - \lambda)u = f, \tag{2.36}$$

has a unique solution,

$$u = \mathcal{R}(\lambda) f = \sum_{j=1}^{\infty} \frac{1}{\lambda_j - \lambda} \langle f, \varphi_j \rangle \varphi_j. \tag{2.37}$$

In particular, representation (2.37) implies that

$$\|\mathcal{R}(\lambda)\| \leq \min_j |\lambda - \lambda_j|^{-1}, \tag{2.38}$$

where $\|\mathcal{R}(\lambda)\|$ is the operator norm of $\mathcal{R}(\lambda)$ in $L^2(dV)$.

**2.2.6.** In this section, we consider the question of regularity of the solution $u$ of equation (2.36). Since $\mathcal{R}(\lambda) : L^2 \to \mathcal{D}(\mathcal{A}) \subset H^2$,

$$\|u\|_2 \leq C(\|\mathcal{A}u\| + \|u\|).$$

Moreover, if $\mathcal{A}u \in H^s$, $s \geq 0$, the following theorem is valid

**Theorem 2.22 (Gårding inequality)** *Let $u \in \mathcal{D}(\mathcal{A})$ and $\mathcal{A}u \in H^s$, $s \geq 0$. Then $u \in H^{s+2}$ and*

$$\|u\|_{s+2} \leq C(\|\mathcal{A}u\|_s + \|u\|).$$

In particular, this theorem implies that the eigenfunctions $\varphi_j \in \mathcal{D}(\mathcal{A}^s) \subset H^{2s}$, for any $s \geq 0$ and, henceforth, $\varphi_j \in C^\infty$. We consider $\mathcal{D}(\mathcal{A}^s)$ as a Hilbert space with the norm

$$\|u\|^2_{\mathcal{D}(\mathcal{A}^s)} = \|\mathcal{A}^s u\|^2 + \|u\|^2. \tag{2.39}$$

**Lemma 2.23** *The functions $\varphi_j$ form an orthogonal basis in $\mathcal{D}(\mathcal{A}^s)$ and*

$$\mathcal{D}(\mathcal{A}^s) = \{u \in L^2 : \quad \|u\|^2_F = \sum_{j=1}^{\infty}(1 + |\lambda_j|^{2s})|u_j|^2 < \infty\}, \tag{2.40}$$

*where $u_j$ are the Fourier coefficients of $u$. Norms (2.39) and (2.40) are equivalent.*

**Proof.** The proof follows from the fact that $u \in \mathcal{D}(\mathcal{A}^s)$, if and only if

$$\mathcal{A}^s u = \sum_{j=1}^{\infty} \lambda_j^s u_j \varphi_j \in L^2.$$

□

In particular, if $u \in C^\infty(\mathcal{N})$, then the Fourier series converges in $C^\infty(\mathcal{N})$ and, if $u \in C_0^\infty(\mathcal{M})$, then the Fourier series converges in $C^\infty(\mathcal{M})$.

In the case of a manifold $\mathcal{N}$ without boundary, $\mathcal{D}(\mathcal{A}^s) = H^{2s}(\mathcal{N})$. In the case of a manifold $\mathcal{M}$ with boundary and an elliptic second-order operator $\mathcal{A}$ with the Dirichlet boundary condition,

$$H_0^{2s}(\mathcal{M}) \subset \mathcal{D}(\mathcal{A}^s) \subset H^{2s}(\mathcal{M}). \tag{2.41}$$

Moreover, for $0 \le s \le 1/2$,

$$H_0^{2s}(\mathcal{M}) = \mathcal{D}(\mathcal{A}^s) \tag{2.42}$$

and

$$C^{-1}\|u\|_{2s}^2 \le \sum_{j=1}^{\infty}(1+|\lambda_j|)^{2s}|u_j|^2 \le C\|u\|_{2s}^2. \tag{2.43}$$

Negative Sobolev spaces $H^{-s}$ may be defined as spaces dual to $H^s$ in the case of a manifold $\mathcal{N}$ without boundary, and $H_0^s$ in the case of a manifold $\mathcal{M}$ with boundary. Relation (2.43) remains valid and $\mathcal{D}(\mathcal{A}^s) = H^{2s}$ for $-1/2 \le s < 0$.

Next we consider the inhomogeneous Dirichlet boundary value problem on a manifold $\mathcal{M}$ with boundary,

$$a(\mathbf{x}, D)u = \lambda u, \quad u|_{\partial\mathcal{M}} = f. \tag{2.44}$$

The following result is well known.

**Lemma 2.24** *If $\lambda$ is not an eigenvalue of the operator $\mathcal{A}$, then, for any $f \in H^s(\partial\mathcal{M})$, $s \ne n+1/2$, problem (2.44) has a unique solution $u \in H^{s+1/2}(\mathcal{M})$, such that*

$$\|u\|_{s+1/2} \le C(\lambda)\|f\|_{(s,\partial\mathcal{M})}.$$

Here we use the notation, $\|v\|_{(s,\Omega)}$ for the norm of a function $v$, defined on a set $\Omega$, in the Sobolev space $H^s(\Omega)$.

**2.2.7.** As in the one-dimensional case, gauge transformations play an important role in the study of multidimensional inverse problems.

**Definition 2.25** *Let $\kappa \in C^\infty(\mathcal{M})$, $\kappa(\mathbf{x}) > 0$ for $\mathbf{x} \in \mathcal{M}$. The gauge transformation generated by $\kappa$ is the transformation,*

$$S_k : L^2(\mathcal{M}, dV) \to L^2(\mathcal{M}, dV_\kappa), \tag{2.45}$$

*where $dV_\kappa = \kappa^{-2}(\mathbf{x})dV$. It is defined by the formula*

$$S_k u(\mathbf{x}) = \kappa(\mathbf{x})u(\mathbf{x}). \tag{2.46}$$

Each gauge transformation determines the corresponding gauge transformation $\mathcal{A}_\kappa$ of an elliptic second-order operator $\mathcal{A}$. The operator $\mathcal{A}_\kappa$ is defined by

$$\mathcal{A}_\kappa u = \kappa \mathcal{A}(\kappa^{-1} u). \tag{2.47}$$

This operator $\mathcal{A}_\kappa$ is also a second-order elliptic differential operator defined in $L^2(\mathcal{M}, dV_\kappa)$ and

$$\mathcal{D}(\mathcal{A}_\kappa) = H^2(\mathcal{M}) \cap H_0^1(\mathcal{M}). \tag{2.48}$$

Its differential expression is given by the formula

$$a_\kappa(\mathbf{x}, D)u(\mathbf{x}) = \kappa(\mathbf{x}) a(\mathbf{x}, D)(\kappa(\mathbf{x})^{-1} u(\mathbf{x})). \tag{2.49}$$

**Lemma 2.26** *The gauge transformation has the following properties.*

*i) The transformation $S_\kappa$ is a unitary transformation, i.e.*

$$\|S_\kappa u\|^2_{L^2(dV_\kappa)} = \|u\|^2_{L^2(dV)}.$$

*ii) The operator $\mathcal{A}_\kappa$ is self-adjoint in $L^2(\mathcal{M}, dV_\kappa)$, if and only if $\mathcal{A}$ is self-adjoint in $L^2(\mathcal{M}, dV)$, and*

$$\mathcal{A}_\kappa[S_\kappa u, S_\kappa w] = \mathcal{A}[u, w]. \tag{2.50}$$

**Proof.** Statement i) of this lemma is obvious.

The self-adjointness of $\mathcal{A}_\kappa$ in $L^2(\mathcal{M}, dV_\kappa)$ follows from the self-adjointness of $\mathcal{A}$ in $L^2(\mathcal{M}, dV)$, due to the local representation of the differential expression $a_\kappa(\mathbf{x}, D)$ of the operator $\mathcal{A}_\kappa$,

$$a_\kappa(\mathbf{x}, D)u = -m_\kappa^{-1} g_\kappa^{-1/2} \partial_i \left( m_\kappa g_\kappa^{ij} g_\kappa^{1/2} \partial_j u \right) + q_\kappa u. \tag{2.51}$$

Here

$$g_\kappa^{ij} = g^{ij}, \quad m_\kappa = \kappa^{-2} m, \quad q_\kappa = \kappa a(\mathbf{x}, D)(\kappa^{-1}). \tag{2.52}$$

From local representation (2.52) and definition (2.34) of the quadratic form, it is easy to prove formula (2.50). □

**Exercise 2.27** *Prove formulae (2.51), (2.52).*

**2.2.8.** The gauge transformations $S_\kappa : L^2(\mathcal{M}) \to L^2(\mathcal{M})$ form an Abelian group $\mathcal{G}$,

$$\mathcal{G} = \{S_\kappa : \quad \kappa \in C^\infty(\mathcal{M}), \quad \kappa > 0\}, \tag{2.53}$$

with respect to the composition,

$$S_{\kappa_1} \circ S_{\kappa_2} = S_{\kappa_1 \kappa_2}.$$

This group has a representation (or an action) in the set of the second-order elliptic differential operators,

$$S_\kappa(\mathcal{A}) = \mathcal{A}_\kappa = \kappa \mathcal{A} \kappa^{-1}.$$

For any operator $\mathcal{A}$,

$$\sigma(\mathcal{A}) = \{S_\kappa(\mathcal{A}) : \quad S_\kappa \in \mathcal{G}\},$$

is the orbit of the group $\mathcal{G}$ that is generated by $\mathcal{A}$. If $\mathcal{A}$ is self-adjoint, then all operators $\mathcal{A}_\kappa \in \sigma(\mathcal{A})$ are self-adjoint.

**2.2.9.** In this section, we will consider some important invariants of an operator $\mathcal{A}$ with respect to the action of the group $\mathcal{G}$.

First, the metric tensor $[g^{ij}]$ is invariant with respect to gauge transformations, i.e., the metric tensor $g^{ij}$ corresponding to operator $\mathcal{A}$ coincides with the one corresponding to $\mathcal{A}_\kappa$.

Second, due to the unitarity of $S_\kappa$, the eigenvalues are invariant,

$$\lambda_j(\mathcal{A}) = \lambda_j(\mathcal{A}_\kappa), \tag{2.54}$$

while the eigenfunctions are transformed according to the formula

$$\varphi_j^\kappa = S_\kappa \varphi_j = \kappa \varphi_j, \tag{2.55}$$

where $\varphi_j$ and $\varphi_j^\kappa$ are the eigenfunctions of the operators $\mathcal{A}$ and $\mathcal{A}_\kappa$, correspondingly.

Physically, this means that any operator in the orbit $\sigma(\mathcal{A})$ corresponds to the same physical process, but in a different scale of measurements. We remind the reader that this change of the scale is given by the factor $\kappa(\mathbf{x})$, which depends on $\mathbf{x}$.

**2.2.10.** In any orbit $\sigma(\mathcal{A})$, generated by a self-adjoint operator $\mathcal{A}$, there is a distinguished operator. This is the (Riemannian) Schrödinger operator that corresponds to $\mathcal{A}$. To define this operator, let us consider the Riemannian manifold $(\mathcal{M}, g)$, where $g$ is the metric associated with the operator $\mathcal{A}$.

**Definition 2.28** *Consider the operator $\mathcal{A}_g = -\Delta_g + q(\mathbf{x})$ defined by the formula*

$$\mathcal{A}_g u = (-\Delta_g + q)u = -g^{-1/2}\partial_j g^{1/2} g^{ij} \partial_i u + q(\mathbf{x})u, \qquad (2.56)$$

*with*

$$\mathcal{D}(-\Delta_g + q) = H^2(\mathcal{M}) \cap H_0^1(\mathcal{M}).$$

*This operator is called a (Riemannian) Schrödinger operator with the Dirichlet boundary condition. When $q = 0$, this operator is the Laplace-Beltrami operator.*

*We remind the reader that $g = \det(g_{ij})$, $[g^{ij}] = [g_{ij}]^{-1}$, and $q$ is a real-valued smooth function on $\mathcal{M}$.*

Clearly, the Schrödinger operator is of form (2.30) with $m = 1$. Therefore, $-\Delta_g + q$ is self-adjoint in $L^2(dV_g)$,

$$dV_g = g^{1/2} dx^1 \cdots dx^m,$$

which is exactly the Riemannian volume element on $(\mathcal{M}, g)$ (see section 2.1.5).

On the contrary, for a general self-adjoint operator $\mathcal{A}$ of form (2.30), $dV \neq dV_g$, where $g$ is the metric associated with $\mathcal{A}$. Actually, the fact that $dV = dV_g$ is a characterization of a Schrödinger operator. To see this, let us consider the quadratic form associated with this operator,

$$\mathcal{A}_g[u, w] = \int_{\mathcal{M}} [(dv, d\overline{w})_g + q(\mathbf{x}) u \overline{w}] dV_g.$$

Then,

$$\mathcal{A}_g[u, u] = \int_{\mathcal{M}} [|\operatorname{Grad} u|_g^2 + q(\mathbf{x})|u|^2] dV_g,$$

where $dv$ is the differential of $v$ and $\operatorname{Grad} v = I_g dv$ (see section 2.1.4).

**Lemma 2.29** *i) Let $\mathcal{A}$ be a self-adjoint second-order elliptic differential operator in $L^2(dV)$. Then there is a unique Schrödinger operator $-\Delta_g + q$ in the orbit $\sigma(\mathcal{A})$. This means that, for a given operator $\mathcal{A}$, there is a unique $\kappa$ such that $\mathcal{A} = \kappa(-\Delta_g + q)\kappa^{-1}$ and $dV = \kappa^{-2}dV_g$.*

*ii) Let $\mathcal{A}[\cdot,\cdot]$ be quadratic form (2.33) associated with the operator $\mathcal{A}$ of form (2.30). Then $\mathcal{A}$ is a Schrödinger operator, if and only if $dV = dV_g$.*

**Proof.** It is clear from the definition of a Schrödinger operator, that a general operator $\mathcal{A}$ of form (2.30) is a Schrödinger operator, if and only if $m = 1$. Using formulae (2.51), (2.52), which describe the change of $\mathcal{A}$ due to a gauge transformation, we see that $\mathcal{A}_\kappa$ is a Schrödinger operator when

$$\kappa = m^{1/2}.$$

Analogously, comparing the quadratic forms of a Schrödinger and a general operator, we see that they coincide, if and only if $dV = dV_g$. □

To our readers, who undoubtly like beautiful mathematical words, we formulate this result as follows.

The self-adjoint elliptic second-order differential operators form an infinite dimensional manifold which is a fibration with the base of the Schrödinger operators and the fibres being the orbits of the group $\mathcal{G}$.

## 2.3. Initial-boundary value problem for wave equation

In this section we will consider an initial-boundary value problem for the wave equation on a manifold with boundary. This initial-boundary value problem corresponds to an elliptic operator $\mathcal{A}$ of section 2.2. We will develop an invariant approach to prove existence and uniqueness of solutions and to study their regularity properties.

**2.3.1.** Let $\mathcal{A}$ be a second-order self-adjoint elliptic operator in $L^2(\mathcal{M}, dV)$ of form (2.30), (2.32), where $dV = mg^{1/2}dx^1 \cdots dx^m$.

We consider the initial-boundary value problem

$$\partial_t^2 u + a(\mathbf{x}, D)u = F, \quad \text{in} \quad Q^T = \mathcal{M} \times [0,T], \tag{2.57}$$

$$u|_{\Sigma^T} = f, \quad \text{where} \quad \Sigma^T = \partial\mathcal{M} \times [0,T], \tag{2.58}$$

$$u|_{t=0} = \psi_0, \quad \partial_t u|_{t=0} = \psi_1, \tag{2.59}$$

with various assumptions on $F, f, \psi_0$, and $\psi_1$.

Due to the results of sections 2.2.7–2.2.10 and, in particular, Lemma 2.29, we can always transform initial-boundary value problem (2.57)–(2.59) for a general operator $\mathcal{A}$ into an initial-boundary value problem for the Schrödinger operator $a_\kappa(\mathbf{x}, D) = -\Delta_g + q$. Then problem (2.57)–(2.59) takes the form,

$$\partial_t^2 u^\kappa - \Delta_g u^\kappa + q u^\kappa = F^\kappa, \quad \text{in}, \quad Q^T, \tag{2.60}$$

$$u^\kappa|_{\Sigma^T} = \kappa f, \tag{2.61}$$

$$u^\kappa|_{t=0} = \psi_0^\kappa, \quad \partial_t u^\kappa|_{t=0} = \psi_1^\kappa, \tag{2.62}$$

where $u^\kappa = S_\kappa u$, $F^\kappa = S_\kappa F$, etc, and $S_\kappa$ is defined by (2.46). Therefore, without loss of generality, we can assume that $a(\mathbf{x}, D) = -\Delta_g + q$ and consider initial-boundary value problem (2.60)–(2.62) for the Schrödinger operator $-\Delta_g + q$.

Since the coefficients of wave equation (2.60) are real, it is sufficient to consider only the case of real $F$, $f$, $\psi_0$, $\psi_1$ and, henceforth, real $u$. In this connection all spaces in this section are real.

**2.3.2.** For the purposes of this book, our primary interest lies in the study of initial-boundary value problem (2.60)–(2.62), when $u$ is in the energy class. This class consists of the waves $u(t)$, $0 \le t \le T$, such that $u(t) \in C([0,T]; H^1(\mathcal{M})) \cap C^1([0,T]; L^2(\mathcal{M}))$. We remind the reader that $u$ belongs to this class when $u$ is a continuous function of $t$ with values in $H^1(\mathcal{M})$ and is continuously differentiable with respect to $t$ in $L^2(\mathcal{M})$. The following theorem gives the conditions on $F, f, \psi_0, \psi_1$, which guarantee that the solution $u(t)$ of problem (2.60)–(2.62) is in the energy class. To formulate this theorem, we introduce the space $L^1([0,T];\ L^2(\mathcal{M}))$, which consists of all functions in $Q^T$ such that

$$|||F|||_{Q^T} := \int_0^T \|F(t)\|_0 dt < \infty.$$

In the future, if there will be no danger of misunderstanding, we will skip the subscript of the space.

**Theorem 2.30 (Lasiesca, Lions, Triggiani)** *Let*

$$F \in L^1([0,T]; L^2(\mathcal{M})),\ f \in H^1(\Sigma^T),\ \psi_0 \in H^1(\mathcal{M}),\ \psi_1 \in L^2(\mathcal{M}).$$

*Assume, in addition, that the following compatibility condition is valid*

$$f|_{t=0} = \psi_0|_{\partial\mathcal{M}}. \tag{2.63}$$

*Then there is a unique solution $u(t)$ of problem (2.60)–(2.62), such that*

$$u(t) \in C([0,T]; H^1(\mathcal{M})) \cap C^1([0,T]; L^2(\mathcal{M}))$$

*and*

$$\max_{0\le t\le T} \{\|u(t)\|_1 + \|u_t(t)\|_0\} \le$$

$$c(T)\{|||F|||_{Q^T} + \|f\|_{(1,\Sigma^T)} + \|\psi_0\|_1 + \|\psi_1\|_0\}. \tag{2.64}$$

*Moreover, $\partial_\nu u|_{\Sigma^T} \in L^2(\Sigma^T)$ and*

$$\|\partial_\nu u|_{\Sigma^T}\|_{(0,\Sigma^T)} \le c(T)\{|||F|||_{Q^T} + \|f\|_1 + \|\psi_0\|_1 + \|\psi_1\|_0\}. \tag{2.65}$$

We note that the energy class solutions, that is, $u(t) \in C([0,T]; H^1) \cap C^1([0,T]; L^2)$ do not, in general, lie in $H^2(Q^T)$. Thus, the energy class solutions of (2.60)–(2.62) are understood in the weak form, i.e. in the form of the equations,

$$\int_{Q^T} \{-\partial_t u \partial_t v + (du, dv)_g + quv\}\, dV_g dt =$$

$$= \int_{Q^T} Fv\, dV_g dt + \int_{\mathcal{M}} \psi_1 v|_{t=0}\, dV_g, \tag{2.66}$$

$$u|_{t=0} = \psi_0, \quad u|_{\Sigma^T} = f,$$

which should be valid for any $v \in H^1(Q^T)$, $v|_{t=T} = 0$, $v|_{\Sigma^T} = 0$. We remind the reader that $(du, dv)_g = g^{ij}\partial_i u\, \partial_j v$.

More generally, a function $u \in L^2(Q^T)$ is a weak solution of problem (2.60)–(2.62), if, for any $v \in H^2(Q^T)$, $v|_{\Sigma^T} = 0$, and $v|_{t=T} = \partial_t v|_{t=T} = 0$,

$$\int_{Q^T} u(\partial_t^2 v - \Delta_g v + qv) dV_g dt = \tag{2.67}$$

$$= \int_{Q^T} Fv dV_g dt + \int_{\Sigma^T} f \partial_\nu v dS_g dt - \int_{\mathcal{M}} \{\psi_0\, \partial_t v|_{t=0} - \psi_1\, v|_{t=0}\} dV_g.$$

**Exercise 2.31** *Show that, for $u \in H^2(Q^T)$, equations (2.60)–(2.62) are equivalent to (2.66). Moreover, for $u(t) \in C([0,T]; H^1(\mathcal{M})) \cap C^1([0,T]; L^2(\mathcal{M}))$, show that (2.66) is equivalent to (2.67).*

The proof of Theorem 2.30 will be given in a series of lemmas. We start our proof with demonstration of uniqueness.

### 2.3.3.

**Lemma 2.32** *Let $F = 0$, $f = 0$, $\psi_0 = \psi_1 = 0$. Then the only solution $u_0 \in C([0,T], H^1) \cap C^1([0,T], L^2)$ of initial-boundary value problem (2.60)–(2.62) is*

$$u_0 \equiv 0 \quad \textit{for} \quad (\mathbf{x}, t) \in Q^T. \tag{2.68}$$

**Proof.** As $u_0 \in C([0,T], H^1) \cap C^1([0,T], L^2)$ and $f = 0$, it can be represented as Fourier series,

$$u_0(\mathbf{x}, t) = \sum_{k=1}^{\infty} u_k(t)\varphi_k(\mathbf{x}),$$

where $\{\varphi_k\}_{k=1}^{\infty}$ is an orthonormal basis of the eigenfunctions of the Schrödinger operator $\mathcal{A}_0$. Moreover, $u_k(t) \in C^1([0,T])$.

To prove the uniqueness, i.e., to show that for all $k = 1, 2, \ldots$ and $0 < t < T$,

$$u_k(t) = 0,$$

we choose test functions $v_{k,\tau}(\mathbf{x}, t)$, $k = 1, 2, \ldots$, $0 \leq \tau \leq T$, of the form,

$$v_{k,\tau}(\mathbf{x}, t) = \begin{cases} \varphi_k(\mathbf{x}) s_k(\tau - t) & \text{for } 0 \leq t \leq \tau \\ 0 & \text{for } \tau < t \leq T \end{cases}. \tag{2.69}$$

Here

$$s_k(t) = \begin{cases} \frac{\sin\sqrt{\lambda_k}t}{\sqrt{\lambda_k}}, & \lambda_k > 0, \\ t, & \lambda_k = 0, \\ \frac{\sinh\sqrt{|\lambda_k|}t}{\sqrt{|\lambda_k|}}, & \lambda_k < 0. \end{cases}.$$

It follows from definition (2.66) of a weak solution that

$$\int_{Q^T} \{-\partial_t u \partial_t v_{k,\tau} + (du, dv_{k,\tau})_g + q u v_{k,\tau}\} \, dV_g dt = 0.$$

Integrating by parts and using that $f = 0$, $\psi_0 = 0$, we come to the equation

$$0 = -\int_{\mathcal{M}} u_0(\mathbf{x}, \tau) \partial_t v_{k,\tau}(\mathbf{x}, \tau) dV_g +$$

$$+ \int_{Q^\tau} u_0(\mathbf{x}, t) \{\partial_t^2 v_{k,\tau} - \Delta_g v_{k,\tau} + q v_{k,\tau}\}(\mathbf{x}, t) dV_g dt.$$

Due to the special form (2.69) of the test functions $v_{k,\tau}$, we see that, for $0 < t < \tau$,

$$\partial_t^2 v_{k,\tau} = -\lambda_k v_{k,\tau}, \quad -\Delta_g v_{k,\tau} + q v_{k,\tau} = \lambda_k v_{k,\tau},$$

$$\partial_t v_{k,\tau}(\mathbf{x}, \tau) = -\varphi_k(\mathbf{x}).$$

Henceforth,

$$0 = \int_{\mathcal{M}} u_0(\mathbf{x}, \tau) \varphi_k(\mathbf{x}) dV_g = u_k(\tau).$$

□

**Exercise 2.33** *Imitating the proof of Lemma 2.32 and using the test functions* $\widetilde{v}_{k,\tau}$,

$$\widetilde{v}_{k,\tau}=\begin{cases}\varphi_k(\mathbf{x})s_k(\tau-t)(t-\tau) & \text{for } 0\le t\le \tau\\ 0 & \text{for } \tau<t\le T\end{cases},$$

*reduce the uniqueness problem for (2.67) to the uniqueness problem for the Volterra integral equations for the Fourier coefficients* $u_k(t)$ *of the solution* $u(t)$. *Hence, show that, when* $F=0$, $\psi_0=\psi_1=0$, *and* $f=0$, *problem (2.67) has only trivial weak solution,* $u=0$.

**2.3.4.** To obtain the desired energy estimates, we start with the case $f=0$. Then compatibility condition (2.63) implies that $\psi_0\in H^1_0(\mathcal{M})$.

**Lemma 2.34** *Let* $F$, $\psi_1$ *be functions satisfying the conditions of Theorem 2.30,* $\psi_0\in H^1_0(\mathcal{M})$ *and* $f=0$. *Then there is a solution* $u\in C([0,T];H^1(\mathcal{M}))\cap C^1([0,T];L^2(\mathcal{M}))$ *of problem (2.60)–(2.62), which satisfies estimate (2.64).*

**Proof.** *Step 1.* As in the previous lemma we seek the solution in the form of Fourier series,

$$u(t)=\sum_{k=1}^{\infty}u_k(t)\varphi_k. \tag{2.70}$$

Using the same test functions $v_{k,\tau}$ as in Lemma 2.32, we obtain that the Fourier coefficients $u_k$ should satisfy the following equations

$$\begin{aligned}u_k(t)&=\psi_{0,k}s_k'(t)+\psi_{1,k}s_k(t)+\\ &+\int_0^t F_k(\tau)s_k(t-\tau)d\tau=\\ &=w_k^0(t)+w_k^1(t)+w_k^F(t).\end{aligned} \tag{2.71}$$

We remind the reader that $\psi_{0,k}$, $\psi_{1,k}$, and $F_k(\tau)$ are the Fourier coefficients of $\psi_0$, $\psi_1$ and $F(\tau)$, correspondingly.

Representation (2.71) yields that

$$u(t)=w^0(t)+w^1(t)+w^F(t), \tag{2.72}$$

where e.g. $w^0(t) = \sum_{k=1}^{\infty} w_k^0(t)\varphi_k$.

Relation (2.43) implies, in particular, that $\psi_0 \in H_0^1(\mathcal{M})$, if and only if

$$\Sigma(1+|\lambda_k|)|\psi_{0,k}|^2 < \infty,$$

and $\psi_1 \in L^2(\mathcal{M})$, if and only if $\sum |\psi_{1,k}|^2 < \infty$. Hence, it is easy to see that $w^0$ and $w^1$ lie in the energy class and

$$\max_{0\le t\le T}\{\|w^i(t)\|_1 + \|\partial_t w^i(t)\|_0\} \le c(T)\|\psi_i\|_{1-i}, \quad i=0,1. \tag{2.73}$$

*Step 2.* To show that $w^F(t)$ lies in the energy class, we use the representation

$$w^F(t) = \int_0^t w(t;\tau)\,d\tau. \tag{2.74}$$

Here for $0 < t < T$

$$w(t;\tau) = \sum F_k(\tau)s_k(t-\tau)\varphi_k(\mathbf{x}). \tag{2.75}$$

Representation (2.75) implies that $w(t;\tau)$ is the solution of the initial-boundary value problem

$$\partial_t^2 w(\mathbf{x},t;\tau) - \Delta_g w(\mathbf{x},t;\tau) + q(\mathbf{x})w(\mathbf{x},t;\tau) = 0, \quad \text{in} \quad Q^T,$$

$$w(t;\tau)|_{\partial\mathcal{M}\times[0,T]} = 0,$$

$$w(t;\tau)|_{t=\tau} = 0, \quad \partial_t w(t;\tau)|_{t=\tau} = F(\tau). \tag{2.76}$$

By (2.73),

$$\max_{0\le t\le T}\{\|w(t;\tau)\|_1 + \|\partial_t w(t;\tau)\|_0\} \le c(T)\|F(\tau)\|_0. \tag{2.77}$$

As

$$\|w^F(t)\|_1 = \|\int_0^t w(t;\tau)\,d\tau\|_1 \le \int_0^t \|w(t;\tau)\|_1 d\tau,$$

estimate (2.77) implies that

$$\|w^F(t)\|_1 \le c(T)\int_0^t \|F(\tau)\|\,d\tau = c(T)|||F|||_{Q^T}. \tag{2.78}$$

Similar considerations show that $\|\partial_t w^F(t)\|_0 \leq c(T)|||F|||_{Q^T}$.

*Step 3.* Steps 1 and 2 show that, if an energy-class solution $u(\mathbf{x}, t)$ exists, it satisfies estimates (2.64)–(2.65). To prove that the constructed function $u$ is actually a solution of problem (2.57)–(2.59), we should verify integral identity (2.66).

We start with smooth $F^n, \psi_0^n$, and $\psi_1^n$,

$$F^n(\mathbf{x}, t) = \sum_{k=1}^{n} F_k^n(t)\varphi_k(\mathbf{x}),$$

$$\psi_0^n(\mathbf{x}) = \sum_{k=1}^{n} \psi_{0,k}^n \varphi_k(\mathbf{x}), \quad \psi_1^n(\mathbf{x}) = \sum_{k=1}^{n} \psi_{1,k}^n \varphi_k(\mathbf{x}),$$

where $F_k^n(t)$, $k = 1, \ldots, n$, are smooth functions of $t$. We leave as an exercise to prove that $u^n(t)$ is in $H^2(Q^T)$ and is a strong solution of problem (2.57)–(2.59) and, in particular, satisfies identity (2.66) with $F, \psi_0, \psi_1$ replaced by $F^n, \psi_0^n, \psi_1^n$. For general $F, \psi_0$, and $\psi_1$, we approximate them by smooth functions $F^n, \psi_0^n, \psi_1^n$ and use the continuity arguments, based on estimates (2.64)–(2.65), to show that the function $u(t)$ of form (2.72) satisfies identity (2.66), i.e., is an energy class solution of problem (2.57)–(2.59). □

**Exercise 2.35** *Show that the function $u^n(\mathbf{x}, t)$ is a strong solution of problem (2.57)–(2.59), that is, $u^n \in H^2(Q^T)$.*

Considerations of step 3 give an example of the "closure" arguments widely used later in the text. They are based on an approximation of general functions from the considered functional spaces by proper smooth functions, demonstration of the result for these smooth functions and further extension of it to the general functions by the continuity arguments. In particular, integration by parts arguments below are justified by closing arguments.

Later, in Chapter 4, we need solutions of initial-boundary value problem (2.57)–(2.59) that are less regular. To this end, we prove the following result.

**Corollary 2.36** *Let $F \in L^1([0, T]; H^{-1}(\mathcal{M}))$, $\psi_0 \in L^2(\mathcal{M})$, $\psi_1 \in H^{-1}(\mathcal{M})$, and $f = 0$. Then there is a unique solution $u$,*

$$u \in C([0, T]; L^2(\mathcal{M})) \cap C^1([0, T]; H^{-1}(\mathcal{M}))$$

*such that*

$$\max_{0\le t\le T}\{\|u(t)\|_0+\|\partial_t u(t)\|_{-1}\}$$

$$\le c(T)\left\{\int_0^t \|F(\tau)\|_{-1}d\tau+\|\psi_0\|_0+\|\psi_1\|_{-1}\right\}. \tag{2.79}$$

**Proof.** By direct substitution we see that $u(t)$ of form (2.72) is a weak solution of initial-boundary value problem (2.60)–(2.62) with $f=0$.

We leave as an exercise to prove that this function $u(t)$ satisfies estimate (2.79).

**Corollary 2.37** *Let $F\in L^1([0,T];H_0^1(\mathcal{M}))$, $\psi_0\in\mathcal{D}(\mathcal{A})$, $\psi_1\in H_0^1(\mathcal{M})$, and $f=0$. Then there is a unique solution $u$ of problem (2.60)–(2.62), and*

$$u\in C([0,T];\mathcal{D}(\mathcal{A}))\cap C^1([0,T];H_0^1(\mathcal{M}))$$

**Exercise 2.38** *Using equivalences (2.42)–(2.43), prove Corollaries 2.36 and 2.37 by mimicking the proof of Lemma 2.34.*

### 2.3.5.

**Lemma 2.39** *Consider $F\in L^1([0,T];L^2(\mathcal{M}))$, $\psi_0\in H_0^1(\mathcal{M})$, $\psi_1\in L^2(\mathcal{M})$, and $f=0$. Then $\partial_\nu u|_{\Sigma^T}\in L^2(\Sigma^T)$ and*

$$\|\,\partial_\nu u|_{\Sigma^T}\|_0\le c(T)\{|||F|||_{Q^T}+\|\psi_0\|_1+\|\psi_1\|_0\}. \tag{2.80}$$

**Proof.** Continue $F$ by zero onto the interval $T\le t\le 2T$ and denote by $\chi^T(t)$ a smooth cut-off function, $\chi^T(t)=1$ for $t\le T$, $\chi^T(t)=0$ for $t\ge 2T$. Let $\nu(\mathbf{x}),\mathbf{x}\in\mathcal{M}$ be a smooth vector field, such that $\nu(\mathbf{x})=\nu_{\mathbf{x}}\in N_{\mathbf{x}}$ for $\mathbf{x}\in\partial\mathcal{M}$. Multiplying equation (2.60) by $\chi^T(t)\,(du(\mathbf{x},t),\nu(\mathbf{x}))$, and integrating over $Q^T$, we obtain that

$$\int_{Q^T}F(\mathbf{x},t)\chi^T(t)\,(du(\mathbf{x},t),\nu(\mathbf{x}))\,dV_g dt=$$

$$=\int_{Q^{2T}}(\partial_t^2u-\Delta_g u+qu)\,\chi^T(t)\,(du(\mathbf{x},t),\nu(\mathbf{x}))\,dV_g dt.$$

Let us consider separately the first two terms

$$I_1 = \int_{Q^{2T}} \partial_t^2 u(\mathbf{x},t)\, \chi^T(t)(du(\mathbf{x},t), \nu(\mathbf{x}))dV_g\, dt \tag{2.81}$$

and

$$I_2 = -\int_{Q^{2T}} \Delta_g u(\mathbf{x},t)\, \chi^T(t)(du(\mathbf{x},t), \nu(\mathbf{x}))dV_g\, dt \tag{2.82}$$

in the right-hand side of the above equation. Integrating by part with respect to $t$ in the integral $I_1$, we see that

$$I_1 = \left[\int_{\mathcal{M}} \partial_t u \chi^T(du,\nu)dV_g\right]_0^{2T} - \int_{Q^{2T}} \partial_t u\, \chi^T(d(\partial_t u),\nu)\, dV_g\, dt -$$

$$- \int_{Q^{2T}} \partial_t u\, (du,\nu)\partial_t \chi^T dV_g\, dt.$$

Because $\chi^T(2T) = 0$, $\chi^T(0) = 1$, we obtain

$$I_1 = -\int_{\mathcal{M}} \psi_1(d\psi_0,\nu)\, dV_g - \frac{1}{2}\int_{Q^{2T}} \chi^T(d(\partial_t u)^2,\nu)\, dV_g dt -$$

$$- \int_{Q^{2T}} \partial_t \chi^T \partial_t u\, (du,\nu)dV_g dt.$$

In the following, we use the equation,

$$\mathrm{Div}\,(f\nu) = \mathrm{Div}\,(\nu)f + (df,\nu), \tag{2.83}$$

which is valid for any smooth function $f$. Using this equation with $f = (\partial_t u)^2$, we obtain that

$$\int_{Q^{2T}} \chi^T(d(\partial_t u)^2,\nu)\, dV_g dt = -\int_{\Sigma^{2T}} \chi^T(\partial_t u)^2\, dS_g dt -$$

$$- \int_{Q^{2T}} \chi^T\, Div(\nu)\, (\partial_t u)^2\, dV_g dt$$

At last, we take into account that the first term in the right-hand side is equal to 0, since $\partial_t u|_{\partial\mathcal{M}} = 0$ due to $f = 0$.

To analyze $I_2$, we use the following identity, which is valid for any function $u$,

$$(d|w|_g^2, \nu) = 2(\nu, \nabla_w w)_g, \quad w = \operatorname{Grad} u. \tag{2.84}$$

We leave the proof of this identity to an exercise. Applying equations (2.8) and (2.84), we obtain that

$$(d(\nu, w)_g, w) = \frac{1}{2}(d|w|_g^2, \nu) + (w, \nabla_w \nu)_g, \quad w = \operatorname{Grad} u. \tag{2.85}$$

Using equation (2.83) with $f(\mathbf{x}) = |du|_g^2(\mathbf{x})$ and equation (2.85), we obtain that

$$(du, d(du, \nu))_g = \frac{1}{2}\nu(|w|_g^2) + (w, \nabla_w \nu)_g =$$

$$= \frac{1}{2}\operatorname{Div}\{(du, du)_g \nu\} - \frac{1}{2}\operatorname{Div}(\nu)(du, du)_g + (w, \nabla_w \nu)_g.$$

Integrating by parts the right-hand side of (2.82), we get

$$I_2 = -\int_{Q^{2T}} \operatorname{Div}(\operatorname{Grad} u)\, \chi^T (du, \nu)\, dV_g dt =$$

$$= \int_{Q^{2T}} \chi^T (du, d(du, \nu))_g dV_g dt + \int_{\Sigma^{2T}} \chi^T (\partial_\nu u)^2 dS_g dt =$$

$$= -\frac{1}{2}\int_{Q^{2T}} \chi^T \operatorname{Div}(\nu)(du, du)_g\, dV_g dt + \int_{Q^{2T}} \chi^T (w, \nabla_w \nu)_g\, dV_g dt +$$

$$+ \int_{\Sigma^{2T}} \chi^T ((\partial_\nu u)^2 - \frac{1}{2}|du|_g^2)\, dS_g dt, \tag{2.86}$$

where $w = \operatorname{Grad} u$.

As $u|_{\Sigma^T} = f = 0$, we see that

$$(du, du)_g|_{\partial\mathcal{M}} = |\, \partial_\nu u|_{\partial\mathcal{M}}\, |^2.$$

Combining the previous considerations, we obtain that

$$\int_{Q^T} \chi^T F(du, \nu) dV_g dt =$$

$$= -\int_{\mathcal{M}} \psi_1 (d\psi_0, \nu) dV_g + \frac{1}{2} \int_{Q^{2T}} \chi^T \operatorname{Div}(\nu) (\partial_t u)^2 dV_g dt -$$

$$- \int_{Q^{2T}} \partial_t \chi^T \partial_t u (du, \nu) dV_g dt + \frac{1}{2} \int_{\Sigma^{2T}} \chi^T (\partial_\nu u)^2 dS_g dt -$$

$$-\frac{1}{2} \int_{Q^{2T}} \chi^T \operatorname{Div}(\nu) (du, du)_g dV_g dt +$$

$$+ \int_{Q^{2T}} \chi^T (du, \nabla_w \nu) dV_g dt + \int_{Q^{2T}} \chi^T q u (du, \nu) dV_g dt, \qquad (2.87)$$

where $w = \operatorname{Grad} u$. Using equation (2.87), we see that

$$\int_{\Sigma^{2T}} \chi^T (\partial_\nu u)^2 dS_g dt \le c(T) \{ |||F|||_{Q^T}^2 + \|\psi_0\|_1^2 + \|\psi_1\|_0^2 \} +$$

$$+ c(T) \{ \max_{0 \le t \le 2T} \|u(t)\|_1^2 + \max_{0 \le t \le 2T} \|\partial_t u(t)\|_0^2 \}. \qquad (2.88)$$

As $\chi^T(t) = 1$ when $t \le T$, we obtain estimate (2.80) from (2.88) and Lemma 2.34. □

**Exercise 2.40** *Let $u$ be a smooth function and $X$ be a smooth vector field. Show that*

$$(d(w, w)_g, X) = 2(X, \nabla_w w)_g, \quad w = \operatorname{Grad} u.$$

*Since both sides of this equation are invariant, one can use local coordinates to prove the statement.*

**Remark.** For readers who are more familiar with differential geometry, we note that this equation follows from the Cartan formula.

**Corollary 2.41** *Let $F \in L^1([0,T]; H_0^1(\mathcal{M}))$, $\psi_0 \in \mathcal{D}(\mathcal{A})$, $\psi_1 \in H_0^1(\mathcal{M})$ and $f = 0$. Then $\partial_t \partial_\nu u|_{\Sigma^T} \in L^2(\Sigma^T)$ and*

$$\|\partial_t \partial_\nu u|_{\Sigma^T}\| \le \qquad (2.89)$$

$$c(T) \left\{ \int_0^T \|F(\tau)\|_1 d\tau + \|A\psi_0\| + \|\psi_0\| + \|\psi_1\|_1 \right\}.$$

**Proof.** *Step 1.* Analogously to the proof in Lemma 2.34, we represent $u(t)$ in the form,

$$u(t) = w(t) + w^F(t), \quad w(t) = w^0(t) + w^1(t).$$

Consider $e(t) = \partial_t w(t)$. Then $e(t)$ is the solution of initial-boundary value problem (2.60)–(2.62) with $F = 0$, $f = 0$ and

$$e|_{t=0} = \psi_1 \in H_0^1(\mathcal{M}), \quad \partial_t e|_{t=0} = \Delta_g \psi_0 - q\psi_0 \in L^2(\mathcal{M}).$$

Therefore, by Lemma 2.39,

$$\partial_t \partial_\nu w|_{\Sigma^T} = \partial_\nu e|_{\Sigma^T} \in L^2(\Sigma^T),$$

and

$$\| \partial_t \partial_\nu w|_{\Sigma^T} \| \le c(T) \left\{ \|\mathcal{A}\psi_0\| + \|\psi_0\| + \|\psi_1\|_1 \right\}. \tag{2.90}$$

*Step 2.* To analyze $\partial_t w^F$, we use representation (2.74). It implies that

$$\partial_t w^F(t) = \int_0^t \partial_t w(t;\tau) d\tau.$$

Here $w(t;\tau)$ are given by formula (2.75). In particular, $w(\tau;\tau) = 0$.

As $w(t;\tau)$ satisfy equations (2.76), it follows from Step 1 that, for almost all $\tau$, $\partial_t \partial_\nu w(\cdot;\tau)|_{\Sigma^T} \in L^2(\Sigma^T)$ and

$$\| \partial_t \partial_\nu w(\cdot;\tau)|_{\Sigma^T} \| \le c(T) \|F(\tau)\|_1.$$

Using the same considerations as in Lemma 2.39, show that

$$\| \partial_t \partial_\nu w^F|_{\Sigma^T} \|_0 \le \int_0^T \|\partial_t \partial_\nu w(\cdot,\tau)\| d\tau \le$$

$$\le c(T) \int_0^T \|F(\tau)\|_1 d\tau. \tag{2.91}$$

Combining (2.90) and (2.91), we obtain estimate (2.89). □

**2.3.6.** ★ In this section, we start to discuss the case of the inhomogeneous boundary condition, $f \neq 0$.

**Lemma 2.42** *Let $u(x,t)$ be the solution of the initial-boundary value problem (2.60)–(2.62).*
*i) If $\psi_0 \in L^2(\mathcal{M})$, $\psi_1 \in H^{-1}(\mathcal{M})$, $f \in L^2(\Sigma^T)$, $F = 0$, then $u(t) \in C([0,T]; L^2(\mathcal{M})) \cap C^1([0,T]; H^{-1}(\mathcal{M}))$ and*

$$\max_{0\le t\le T}\{\|u(t)\|_0+\|\partial_t u(t)\|_{-1}\}\le c(T)\{\|\psi_0\|+\|\psi_1\|_{-1}+\|f\|_{(0,\Sigma^T)}\}. \quad (2.92)$$

*ii) If $\psi_0 \in H^1(\mathcal{M})$, $\psi_1 \in L^2(\mathcal{M})$, $f, \partial_t f \in L^2(\Sigma^T)$ and $f|_{t=0} = \psi_0|_{\partial\mathcal{M}}$, then $u(t) \in C^1([0,T]; L^2(\mathcal{M})) \cap C^2([0,T]; H^{-1}(\mathcal{M}))$.*

**Proof.** *Step 1.* We will consider only $f \in C_0^\infty(\Sigma^T)$, $\psi_0, \psi_1 \in C_0^\infty(\mathcal{M})$ and prove estimate (2.92) for these functions. The result for the general $f \in L^2(\Sigma^T)$, $\psi_0 \in L^2(\mathcal{M})$, $\psi_1 \in H^{-1}(\mathcal{M})$ will follow by the closure of (2.92) in $L^2(\Sigma^T) \times L^2(\mathcal{M}) \times H^{-1}(\mathcal{M})$.

It follows from Corollary 2.36 that, in case i), it is sufficient to take $\psi_0 = \psi_1 = 0$.

Let $v$ be a solution of the dual problem,

$$\partial_t^2 v - \Delta_g v + qv = H \quad \text{in} \quad Q^T, \quad (2.93)$$

$$v|_{\Sigma^T} = 0, \quad (2.94)$$

$$v|_{t=T} = 0, \quad \partial_t v|_{t=T} = 0, \quad (2.95)$$

where $H \in C_0^\infty(Q^T)$. Then, by Corollary 2.37, $v \in C([0,T]; \mathcal{D}(\mathcal{A})) \cap C^1([0,T]; H_0^1(\mathcal{M}))$. As

$$\partial_t^2 v = \Delta_g v - qv + H,$$

we see that $v \in C^2([0,T]; L^2(\mathcal{M}))$. Using definition (2.67) of the weak solution, where $v$ is a test function, we see that

$$\int_{Q^T} uH\, dV_g dt = \int_{\Sigma^T} f\, \partial_\nu v\, dS_g dt. \quad (2.96)$$

Due to estimate (2.80), the right-hand side of (2.96) depends continuously on $H \in L^1([0,T]; L^2(\mathcal{M}))$. Therefore, equation (2.96) defines a continuous linear functional on $L^1([0,T]; L^2(\mathcal{M}))$. This implies that $u \in L^\infty([0,T]; L^2(\mathcal{M}))$ and depends continuously on $f \in L^2(\Sigma^T)$.

*Step 2.* To show that

$$\|\partial_t u(t)\|_{-1} \le c(T)\|f\|_{(0,\Sigma^T)}, \tag{2.97}$$

we consider the integral $\int_{Q^T} \partial_t u H \, dV_g dt$ for $H \in L^1([0,T]; H_0^1(\mathcal{M}))$. Let $v$ be the solution of (2.93)–(2.95) with such $H$. Then

$$\int_{Q^T} \partial_t u H \, dV_g dt = \int_{\Sigma^T} \partial_t f \, \partial_\nu v \, dS_g dt = -\int_{\Sigma^T} f \, \partial_t \partial_\nu v \, dS_g dt. \tag{2.98}$$

By Corollary 2.41, the right-hand side of (2.98) is defined for any $f \in L^2(\Sigma^T)$. Henceforth, equation (2.98) defines a linear functional on $L^1([0,T]; H_0^1(\mathcal{M}))$, i.e.

$$\partial_t u \in L^\infty([0,T]; H^{-1}(\mathcal{M})),$$

for any $f \in L^2(\Sigma^T)$. Moreover, estimate (2.97) holds true.

*Step 3.* For $f \in C_0^\infty(\Sigma^T)$, let $f^c(\mathbf{x}, t)$ be a smooth continuation of $f$ into $Q^T$ with

$$\mathrm{supp}\,(f^c) \cap \mathcal{M} \times \{t = 0\} = \emptyset.$$

Represent $u$ in the form $u = f^c + w$. Then $w$ is a solution of initial-boundary value problem (2.60)–(2.62) with zero initial and boundary conditions and $F \in C^\infty(Q^T)$. Moreover, $\partial_t w$ is also a solution of initial-boundary value problem (2.60)–(2.62) with $\widehat{\psi_0}$, $\widehat{\psi_1}$ of the form

$$\widehat{\psi_0} = 0, \quad \widehat{\psi_1} = \partial_t^2 w|_{t=0} = F|_{t=0} - a(\mathbf{x}, D)\psi_0 = F|_{t=0} \in L^2(\mathcal{M}).$$

Applying Corollary 2.36, we obtain that

$$u(t) \in C^1([0,T]; L^2(\mathcal{M})) \cap C^2([0,T]; H^{-1}(\mathcal{M})),$$

when $f \in C_0^\infty(\Sigma^T)$. Part i) of the lemma follows now from steps 1 and 2 if we take into account that $C_0^\infty(\Sigma^T)$ is dense in $L^2(\Sigma^T)$ and use the closure arguments.

*Step 4.* Let us consider case ii). We start with $f \in C^\infty(\Sigma^T)$, $\psi_0, \psi_1 \in C^\infty(\mathcal{M})$ and $f|_{t=0} = \psi_0|_{\partial\mathcal{M}}$.

Let $\psi_0^c(\mathbf{x}, t)$ be a smooth continuation of $\psi_0$ into $Q^T$. Then $u(t) = \psi_0^c(t) + w(t)$, where $w$ is the solution of problem (2.60)–(2.62) with

$$F \in C^\infty(Q^T), \quad f \in C^\infty(\Sigma^T), \quad f|_{t=0} = 0$$

and

$$w|_{t=0} = 0, \quad \partial_t w|_{t=0} \in C^\infty(\mathcal{M}).$$

Clearly, $w(t)$ and, henceforth, $u(t) \in C^1([0,T]; L^2(\mathcal{M}))$. Therefore, $\partial_t u$ is a weak solution of initial-boundary value problem (2.60)–(2.62) with

$$F = 0, \quad \partial_t u|_{t=0} = 0, \quad \partial_t^2 u|_{t=0} = -a(\mathbf{x}, D)\psi_0,$$

and $f$ replaced by $\partial_t f$. Using Corollary 2.36 and part i) of the lemma, we see that

$$\partial_t u \in C([0,T]; L^2(\mathcal{M})), \quad \partial_t^2 u(t) \in C([0,T]; H^{-1}(\mathcal{M}))$$

and

$$\max_{0\le t\le T}\{\|\partial_t u(t)\|_0 + \|\partial_t^2 u(t)\|_{-1}\} \le c(T)\{\|\psi_0\|_1 + \|\partial_t f\|_{(0,\Sigma^T)}\}. \tag{2.99}$$

Taking the closure of (2.99) in the space $f, \partial_t f \in L^2(\Sigma^T)$, $\psi_0 \in H_0^1(\mathcal{M})$, $f|_{t=0} = \psi_0|_{\partial\mathcal{M}}$ and using the estimate (2.92), we prove part ii) of the lemma.

□

**Lemma 2.43** *Let $\psi_1 = 0$, $F = 0$, $\psi_0 \in H^1(\mathcal{M})$, $f \in H^1(\Sigma^T)$, and $f|_{t=0} = \psi_0|_{\partial\mathcal{M}}$. Then $u(t) \in C([0,T]; H^1(\mathcal{M})) \cap C^1([0,T]; L^2(\mathcal{M}))$ and $\partial_\nu u|_{\Sigma^T} \in L^2(\Sigma^T)$.*

**Proof.** *Step 1.* Let $f \in H^1(\Sigma^T)$. Since the restriction operator $T_\tau : f \mapsto f|_{t=\tau}$ is an operator from $H^1(\Sigma^T)$ to $H^{1/2}(\partial\mathcal{M})$ depending continuously on $\tau$, we see that $f \in C([0,T]; H^{1/2}(\partial\mathcal{M}))$.

As $f$ satisfies conditions of Lemma 2.42 ii), we have that $u \in C^1([0,T]; L^2(\mathcal{M}))$ and $\partial_t^2 u \in C([0,T]; H^{-1}(\mathcal{M}))$. Then, for any $\sigma \in \mathbf{R}$ and $0 \le t \le T$,

$$a(\mathbf{x}, D)u(t) + \sigma u(t) = -\partial_t^2 u(t) + \sigma u(t) \in C([0,T]; H^{-1}(\mathcal{M})),$$

$$u|_{\partial\mathcal{M}} = f \in C([0,T]; H^{1/2}(\partial\mathcal{M})).$$

Let $\sigma$ not coincide with an eigenvalue of $\mathcal{A}$. Then, due to Lemma 2.24, $u \in C([0,T]; H^1(\mathcal{M}))$ and

$$\max_{0\le t\le T}\|u(t)\|_1 \le c(T)\max_{0\le t\le T}\{\|\partial_t^2 u(t)\|_{-1} + \|u(t)\|_0 + \|f(t)\|_{(1/2,\partial\mathcal{M})}\}$$

$$\leq c(T)\{\|f\|_{(1,\Sigma^T)} + \|\psi_0\|_1\}. \tag{2.100}$$

*Step 2.* To show that $\partial_\nu u|_{\Sigma^T} \in L^2(\Sigma^T)$ and

$$\|\partial_\nu u\|_{(0,\Sigma^T)} \leq c(T)\{\|f\|_{(1,\Sigma^T)} + \|\psi_0\|_1\}, \tag{2.101}$$

we use essentially the same arguments as in the proof of Lemma 2.39. Let $\widetilde{f} \in H^1(\Sigma^{2T})$ be a bounded extension of $f$. Denote by $\widetilde{u}$ the solution of (2.60)–(2.62) in $Q^{2T}$ with $\widetilde{u}|_{\Sigma^T} = \widetilde{f}$, $\widetilde{F} = 0$, and $\widetilde{\psi}_1 = \psi_1 = 0, \widetilde{\psi}_0 = \psi_0$. Multiplying (2.60) by $\chi^T(t)\,(d\widetilde{u},\nu)$ and integrating by parts, we obtain that

$$\int_{\Sigma^{2T}} \chi^T |\partial_\nu \widetilde{u}|^2 dS \leq c(T)\{\|\widetilde{f}\|^2_{(1,\Sigma^{2T})} + \|\widetilde{u}\|^2_{(1,Q^{2T})}\} \leq$$

$$\leq c(T)\{\|f\|^2_{(1,\Sigma^T)} + \|\psi_0\|_1^2\}. \tag{2.102}$$

As $\widetilde{u} = u$ when $t \leq T$, estimate (2.101) follows from (2.102).

□

Summarizing Lemmas 2.34–2.43, we obtain the proof of Theorem 2.30.

**Exercise 2.44** *Following the proof of Lemma 2.39, prove estimate (2.102).*

**2.3.7.** To obtain the higher-order regularity results for $u$, we should make further regularity assumptions on $F, f, \psi_0$ and $\psi_1$. Also, we have to impose higher-order compatibility conditions at $\partial\mathcal{M} \times \{0\}$. To describe these conditions, assume that $u \in C([0,T]; H^{p+1}(\mathcal{M})) \cap C^{p+1}([0,T]; L^2(\mathcal{M}))$. Then we should have compatibility conditions for all derivatives $\partial_t^j f$, $j = 0, 1, .., p$,

$$f|_{t=0} = \psi_0|_{\partial\mathcal{M}}, \quad \partial_t f|_{t=0} = \partial_t u|_{(\partial\mathcal{M}\times\{0\})} = \psi_1|_{\partial\mathcal{M}},$$

$$\partial_t^2 f\big|_{t=0} = \partial_t^2 u\big|_{(\partial\mathcal{M}\times\{0\})} = (-a(\mathbf{x}, D)\psi_0|_{\partial\mathcal{M}} + F|_{(\partial\mathcal{M}\times\{0\})}),$$

etc. Using the considerations similar to those in the demonstration of Lemmas 2.42 ii) and 2.43, we obtain the following theorem.

**Theorem 2.45** *Let $p$ be a non-negative integer. Assume that $F \in L^1([0,T]; H^p(\mathcal{M}))$, $\partial_t^p F \in L^1([0,T]; L^2(\mathcal{M}))$, $\psi_0 \in H^{p+1}(\mathcal{M})$, $\psi_1 \in H^p(\mathcal{M})$, and $f \in H^{p+1}(\Sigma^T)$. If all compatibility conditions up to the order $p$ are satisfied, then*

$$u \in C([0,T]; H^{p+1}(\mathcal{M})) \cap C^{p+1}([0,T]; L^2(\mathcal{M})),$$

$$\partial_\nu u|_{\Sigma^T} \in H^p(\Sigma^T). \tag{2.103}$$

*The dependence of $u$ and $\partial_\nu u|_{\Sigma^T}$ on $F, f, \psi_0, \psi_1$ is continuous in the corresponding spaces, i.e.,*

$$\|u(t)\|_{p+1} + \|\partial_t^{(p+1)} u(t)\|_0 + \|\partial_\nu u\|_{(p,\Sigma^T)} \leq$$

$$\leq c(T)\left\{\|F\|_{L^1([0,T];H^p)} + \|F\|_{L^1([0,T];L^2)} + \|\psi_0\|_{(p+1)} + \right.$$

$$\left. + \|\psi_1\|_p + \|f\|_{(p+1,\Sigma^T)}\right\}, \tag{2.104}$$

*In particular, if $\psi_0, \psi_1 \in C^\infty(\mathcal{M})$, $F \in C^\infty(Q^T)$, $f \in C^\infty(\Sigma^T)$ and all compatibility conditions are satisfied, then $u \in C^\infty(Q^T)$.*

In Chapter 4, we will also need the following global estimate.

**Corollary 2.46** *Let $F \in L^1_{loc}(\mathbf{R}_+; H^p(\mathcal{M}))$ be a function such that $\partial_t^p F \in L^1_{loc}(\mathbf{R}_+; L^2(\mathcal{M}))$. Moreover, let $\psi_0 \in H^{p+1}(\mathcal{M})$, $\psi_1 \in H^p(\mathcal{M})$, and $f \in H^{p+1}_{loc}(\partial\mathcal{M} \times \mathbf{R}_+)$. If these functions satisfy all compatibility conditions up to the order $p$, then*

$$u \in C(\mathbf{R}_+; H^{p+1}(\mathcal{M})) \cap C^{p+1}(\mathbf{R}_+; L^2(\mathcal{M})),$$

$$\partial_\nu u|_{\Sigma^T} \in H^p_{loc}(\partial\mathcal{M} \times \mathbf{R}_+). \tag{2.105}$$

*Moreover, there exist $c_1, c_2 > 0$ such that*

$$\|u(t)\|_{(p+1)} + \|\partial_t^{(p+1)} u(t)\|_0 + \|\partial_\nu u\|_{(p,\Sigma^t)} \leq \tag{2.106}$$

$$\leq c_1 e^{c_2 t}\left\{\|F\|_{L^1([0,t];H^p)} + \|F\|_{L^1([0,t];L^2)} + \|\psi_0\|_{(p+1)} + \right.$$

$$\left. + \|\psi_1\|_p + \|f\|_{(p+1,\Sigma^t)}\right\},$$

*for any $t > 0$.*

We remind the reader that $L^1_{loc}, H^p_{loc}$, etc are defined as classes of functions whose restrictions onto any bounded time interval lie in the proper spaces, e.g.,

$$H^{p+1}_{loc}(\partial\mathcal{M} \times \mathbf{R}_+) = \{f : f|_{\Sigma^T} \in H^{p+1}(\Sigma^T), \text{ for any } T > 0\}.$$

**Proof.** As the coefficients of the wave equation are independent of $t$, Theorem 2.45 and, in particular, estimate (2.104) remain valid for any time cylinder of the form $\mathcal{M} \times [t, t+T]$. Using this observation, we obtain the proof of the corollary by iterations with

$$c_1 = c(T),\ c_2 = \frac{\ln c(T)}{T}.$$

□

**2.3.8.** Another important property of initial-boundary value problem (2.60)–(2.62) is the finite velocity of the wave propagation. To formulate the corresponding result, let $\Omega \subset \mathcal{M}$ be an open set. We remind the reader that, in general, $\partial\mathcal{M} \cap \Omega \neq \emptyset$. Denote by $D^T = D^T(\Omega)$, $T > 0$, the cone

$$D^T = \{(\mathbf{x}, t) \in Q^T : d(\mathbf{x}, \mathcal{M} \setminus \Omega) > |t|\}, \tag{2.107}$$

and by $\Omega^t$ its intersection with $\mathcal{M} \times \{t\}$,

$$\Omega^t = \{\mathbf{x} \in \mathcal{M} : d(\mathbf{x}, M \setminus \Omega) > |t|\}; \quad \Omega = \Omega^0. \tag{2.108}$$

**Theorem 2.47** *Let $u \in C([0,T]; H^1(\mathcal{M})) \cap C^1([0,T]; L^2(\mathcal{M}))$ be a solution of initial-boundary value problem (2.60)–(2.62). Let*

$$f|_{D^T \cap \Sigma^T} = 0, \quad \psi_0 = \psi_1|_\Omega = 0, \quad F|_{D^T} = 0. \tag{2.109}$$

*Then*

$$u|_{D^T} = 0.$$

**Sketch of the proof.** *Step 1.* We remind the reader that, for $\mathbf{x}_0 \in \mathcal{M}^{int}$, $\rho > 0$, we denote by $B_\rho(\mathbf{x}_0)$ the ball of radius $\rho$ with centre in $\mathbf{x}_0$. Let $\rho, \delta > 0$ be sufficiently small, so that $B_{\rho+\delta}(\mathbf{x}_0)$ is a coordinate chart of normal coordinates centered at $\mathbf{x}_0$ and $B_{\rho+\delta}(x_0) \subset \mathcal{M}^{int}$.

Let, in addition, $T \le \delta$. Then, multiplying (2.60) by $\partial_t u$ and integrating over $D^T$, we obtain from (2.109) that

$$\int_{\Omega^T} \{(\partial_t u)^2(T) + (du(T), du(T))_g\}\, dV_g + \sqrt{2}\int_{S^T} (\partial_t u + \partial_\nu u)^2\, d\sigma +$$

$$+2\int_{D^T} qu\partial_t u\, dV_g dt = 0, \tag{2.110}$$

where $S^T$ is the lateral boundary of $D^T$ and $d\sigma$ is the volume element on $S^T$,

$$S^T = \{(\mathbf{x}, t) : d(\mathbf{x}, \mathcal{M} \setminus \Omega) = t,\ 0 \le t \le T\}.$$

As $u(\mathbf{x}, t) = \int_0^t \partial_t u(\mathbf{x}, \xi) d\xi$, it follows from the Cauchy inequality that

$$\int_{\Omega^T} u^2\, dV_g \le c(T) \int_{D^T} \partial_t u^2\, dV_g dt.$$

As the second term in the left-hand side of (2.110) is positive, we see that

$$\int_{\Omega^T} ((\partial_t u)^2(T) + (du(T), du(T))_g) dV_g \le C(T) \int_{D^T} (u^2 + (\partial_t u)^2) dV_g dt.$$

Therefore,

$$H(t) \le c(T) \int_0^t H(\xi) d\xi, \tag{2.111}$$

where

$$H(t) = \frac{1}{2} \int_{\Omega^t} ((\partial_t u)^2 + (du, du)_g + u^2)\, dV_g.$$

As $H(0) = 0$, estimate (2.111) implies that $H(t) = 0$ for $0 \le t \le T$.

*Step 2.* In the general case, we continue $u$ by zero across $D^T \cap \Sigma^T$. The function $\widetilde{u}$ obtained in this way satisfies $\widetilde{u} \in H^1$ in the extended domain due to $f|_{D^T \cap \Sigma^T} = 0$. It is a solution of wave equation (2.60) in this extended domain. We cover $\Omega$ by balls $B_\rho$ and use step 1 to prove that $\widetilde{u} = 0$ in the extended cone $D^\delta$. Repeating this procedure, we see that $u|_{D^T} = 0$.

## 2.4. Gaussian beams

**2.4.1.** In this section, we construct a special class of solutions of the wave equation on a Riemannian manifold without boundary. The corresponding solutions for a manifold with boundary are described in Chapter 3. They are used there for the reconstruction of the Riemannian manifold and the Schrödinger operator on it. The solutions are known as Gaussian beams or quasiphotons. The name "quasiphoton" reflects the fact that any solution of this type is concentrated at time $t$ near a point $\mathbf{x} = \mathbf{x}(t)$. The path $\mu : \mathbf{x} = \mathbf{x}(t)$ turns out to be a geodesic on $(\mathcal{N}, g)$. The quasiphoton moves along this geodesic with unit speed. Moreover, many properties of such solutions (the energy conservation law, reflection from the boundary etc.) are analogous to those of particles. The name "Gaussian beam" reflects the fact that at any time $t$ the energy density of a Gaussian beam coincides with a Gaussian function. The one-dimensional version of such solutions is described in section 1.4. In the multidimensional case, the behaviour of the Gaussian beams is more complicated than in the one-dimensional case. First, Gaussian beams are associated with geodesics on the underlying manifold. In the one-dimensional case there is, in fact, only one geodesic. On the contrary, in the multidimensional case, there is an $(m-1)$-dimensional family of geodesics through any point $\mathbf{x}$. Second, in the multidimensional case, the phase function of a quasiphoton behaves in a more complicated way than in the one-dimensional case.

The construction of the Gaussian beams on a multidimensional manifold without boundary is a rather long procedure and is divided into several steps. First, we construct a special asymptotic solution of the wave equation (2.60) called a formal Gaussian beam. Second, we show that, with the right initial data, the solution of the initial-value problem is close to a formal Gaussian beam. In Chapter 3, we apply these constructions to find a Gaussian beams on a manifold with boundary. This is done by choosing a special boundary data so that the asymptotic expansion of the solution of the wave equation is equal to that of a formal Gaussian beam. This Gaussian beam moves along a geodesic that is transverse to the boundary.

**2.4.2.** We seek an asymptotic solution of the wave equation in the form

$$U_\epsilon^N(\mathbf{x},t) = M_\epsilon \exp\{-(i\epsilon)^{-1}\theta(\mathbf{x},t)\}\sum_{n=0}^{N} u_n(\mathbf{x},t)(i\epsilon)^n, \qquad (2.112)$$

where $M_\epsilon$ is the normalization constant,

$$M_\epsilon = (\pi\epsilon)^{-m/4}. \qquad (2.113)$$

As in the one-dimensional case, the phase function $\theta(\mathbf{x},t)$ and the amplitude functions $u_n(\mathbf{x},t)$, $n = 0,1,\dots,N$, are some smooth complex-valued functions of the variables $(\mathbf{x},t)$. In the following, we construct asymptotic solution (2.112) of the wave equation (2.60) with the phase function $\theta(\mathbf{x},t)$ that satisfies the following conditions

$$\Im\theta(\mathbf{x}(t),t) = 0, \qquad (2.114)$$

$$\Im\theta(\mathbf{x},t) \geq C_0(t)d^2(\mathbf{x},\mathbf{x}(t)), \qquad (2.115)$$

where $C_0(t)$ is a continuous positive function. From (2.112)–(2.115) it follows that $U_\epsilon^N(x,t)$ is concentrated near $\mathbf{x}(t)$ in the sense that for any ball $B = B_\rho(\mathbf{x}(t))$, $\rho > 0$,

$$\|U_\epsilon^N(\cdot\,,t)\| = \mathcal{O}(1), \quad \| U_\epsilon^N(\cdot\,,t)|_{\mathcal{M}\setminus B} \| = \mathcal{O}(\epsilon^p),$$

for any $p > 0$.

Since $\epsilon$ is an asymptotic parameter, we can multiply it by any positive constant $c > 0$, i.e. to replace $\epsilon$ with $\widetilde{\epsilon} = c\epsilon$. As a result, $U_\epsilon^N$ goes to $\widetilde{U}_{\widetilde{\epsilon}}^N$,

$$\widetilde{U}_{\widetilde{\epsilon}}^N(\mathbf{x},t) = M_{\widetilde{\epsilon}} \exp\{-(i\widetilde{\epsilon})^{-1}\widetilde{\theta}(\mathbf{x},t)\}\sum_{n=0}^{N} \widetilde{u}_n(\mathbf{x},t)(i\widetilde{\epsilon})^n,$$

where

$$\widetilde{\epsilon} = \epsilon c, \quad \widetilde{\theta}(\mathbf{x},t) = \theta(\mathbf{x},t)c, \quad \widetilde{u}_n(\mathbf{x},t) = u_n(\mathbf{x},t)c^{-(n-m/2)}.$$

### 2.4.3.

**Definition 2.48** *A formal Gaussian beam of order $N$ is a function $U_\epsilon^N(\mathbf{x},t)$ of form (2.112) with the phase function $\theta(\mathbf{x},t)$ that satisfies conditions (2.114)–(2.115) and that satisfies the inequality*

$$\left|(\partial_t^2 - \Delta_g + q)U_\epsilon^N\right| \le C(T)\, M_\epsilon \epsilon^N, \quad in \quad Q^T \tag{2.116}$$

*with some $C(T)$ independent of $\epsilon$.*

In general, functions satisfying inequality (2.116) are called asymptotic solutions of the wave equation.

Our goal is to construct a phase function $\theta$ and amplitude functions $u^n(\mathbf{x},t)$, $n = 0,\dots,N$, such that $U_\epsilon^N(\mathbf{x},t)$ is a formal Gaussian beam. Condition (2.115) implies that it is sufficient to construct a Gaussian beam in any small neighborhood of the path $(\mathbf{x}(t),t)$. When $t$ is small enough, we can work in a coordinate chart $(U,X)$ around $\mathbf{x}(t)$. Because of this, all further constructions are given in local coordinates, that is, in $\mathbf{R}^m$.

Substitution of (2.112) into equation (2.60) yields that

$$(\partial_t^2 - \Delta_g + q)\, U_\epsilon^N(\mathbf{x},t) =$$

$$= M_\epsilon (i\epsilon)^{-2} \exp\{-(i\epsilon)^{-1}\theta(\mathbf{x},t)\} \sum_{n=0}^{N+2} v_n(\mathbf{x},t)(i\epsilon)^n \tag{2.117}$$

where $v_n$ has the form

$$v_n(\mathbf{x},t) = [(\partial_t\theta)^2 - g^{jl}(\mathbf{x})\partial_j\theta\partial_l\theta]u_n(\mathbf{x},t) - \mathcal{L}_\theta u_{n-1}(\mathbf{x},t) + \tag{2.118}$$

$$+(\partial_t^2 - \Delta_g + q)u_{n-2}(\mathbf{x},t),$$

where $n = 0,\dots,N+2$, and $u_{-2}(\mathbf{x},t) \equiv u_{-1}(\mathbf{x},t) \equiv 0$. Here $\mathcal{L}_\theta$ is the transport operator

$$\mathcal{L}_\theta u = 2\partial_t\theta\partial_t u - 2g^{jl}\partial_j\theta\partial_l u + (\partial_t^2 - \Delta_g)\theta \cdot u. \tag{2.119}$$

**2.4.4.** We start with the following simple lemma.

**Lemma 2.49** *Let $\theta(\mathbf{x},t)$ satisfies conditions (2.114), (2.115) and*

$$|v_n(\mathbf{x},t)| \le Cd(\mathbf{x},\mathbf{x}(t))^{2(N+2-n)}, \quad n = 0,\dots,N+2. \tag{2.120}$$

*Then there is a constant $C(T)$ such that*

$$|(\partial_t^2 - \Delta_g + q)U_\epsilon^N| \le C(T)M_\epsilon \epsilon^N \tag{2.121}$$

*for $(\mathbf{x},t) \in Q^T$.*

**Proof.** In a local coordinates near $\mathbf{x}(t)$ introduce variables

$$\mathbf{y} = \mathbf{x} - \mathbf{x}(t) \tag{2.122}$$

when $|t - t_0|$ is sufficiently small. Then

$$|v_n(\mathbf{x},t)\exp\{-(i\epsilon)^{-1}\theta(\mathbf{x},t)\}| \le C|y|^{2(N+2-n)}\exp(-\frac{C_0(t)|y|^2}{\epsilon})$$

$$\le C'\epsilon^{N+2-n}.$$

Covering the path $\mu : (0,T) \to \mathcal{N}$ by a finite number of coordinate charts and using the fact that in any coordinate chart

$$C^{-1}|\mathbf{x}-\mathbf{y}| \le d(\mathbf{x},\mathbf{y}) \le C|\mathbf{x}-\mathbf{y}|$$

we obtain inequality (2.121). □

It follows from Lemma 2.49 that, to construct $U_\epsilon^N$, it is sufficient to find $\theta$ and $u_n$, $n = 0,1,\dots$, such that corresponding $v_n$, $n = 0,\dots,N+1$, vanish to order $2(N+2-n)$ at $\mathbf{y} = 0$. Note that we do not require that $v_{N+2}$ vanish. In the following we describe a recurrent procedure to construct $\theta$ and $u_n$, so that $\theta$ satisfies (2.114)–(2.115) and the corresponding $v_n$ vanish to an sufficiently large order, say to order $K > 3(N+1)$. (2.120). We use the notation

$$v(\mathbf{y},t) \asymp 0, \tag{2.123}$$

when $\partial_{\mathbf{y}}^\alpha v(\mathbf{y},t)|_{\mathbf{y}=0} = 0$ for any multi-index $\alpha$, $|\alpha| \le K$ at any time $t \in [0,T]$. Notice that this implies

$$(\partial_{\mathbf{x}}^\alpha \partial_t^\beta v)(\mathbf{x}(t),t) = 0$$

for all $|\alpha| + |\beta| \le K$.

**2.4.5.** We start our construction of a formal Gaussian beam with the analysis of the term $v_0$. The corresponding equation $v_0 \asymp 0$ can be written in the form

$$[(\partial_t\theta)^2 - g^{jl}(\mathbf{x})\partial_j\theta\partial_l\theta]u_0(\mathbf{x},t) \asymp 0. \tag{2.124}$$

To satisfy this equation, it is sufficient to find a phase function $\theta(\mathbf{x},t)$ such that

$$(\partial_t\theta)^2 - g^{jl}(\mathbf{x})\partial_j\theta\partial_l\theta \asymp 0. \tag{2.125}$$

Equation (2.125) is called the Hamilton-Jacobi equation for the phase $\theta(\mathbf{x},t)$. Given $\theta$, which satisfies equation (2.125), the other equations $v_{n+1} \asymp 0$, $n = 1,\ldots,N$ take the form of the transport equations

$$\mathcal{L}_\theta u_n \asymp (\partial_t^2 - \Delta_g + q)u_{n-1}, \tag{2.126}$$

$n = 0,\ldots,N$. Equations (2.126) are solved successively, so that the right-hand side is known at any step. In particular, equation (2.126) with $n = 0$ has the form

$$\mathcal{L}_\theta u_0 = 0. \tag{2.127}$$

Because we are interested in Taylor's expansion of $v_n$, it is natural to represent the phase function $\theta$ and the amplitude functions $u_n$ by their Taylor's expansions with respect to $\mathbf{y}$,

$$\theta(\mathbf{x},t) \asymp \sum_{|\gamma|\geq 0} \frac{\theta_\gamma(t)}{\gamma!}\mathbf{y}^\gamma, \tag{2.128}$$

$$u_n(\mathbf{x},t) \asymp \sum_{|\gamma|\geq 0} \frac{u_{n,\gamma}(t)}{\gamma!}\mathbf{y}^\gamma. \tag{2.129}$$

We also use the notations $\theta_l(t)$ and $u_{n,l}(t)$ for the homogeneous terms of order $l$,

$$\theta_l(t) = \sum_{|\gamma|=l} \frac{1}{\gamma!}\theta_\gamma(t)\mathbf{y}^\gamma, \quad u_{n,l}(t) = \sum_{|\gamma|=l} \frac{1}{\gamma!}u_{n,\gamma}(t)\mathbf{y}^\gamma. \tag{2.130}$$

**2.4.6.** The most important part of the construction of the formal Gaussian beam is to find the first three polynomial, $\theta_0$, $\theta_1$, and $\theta_2$ of the phase function $\theta$. Since these polynomials are in the focus of our attention for quite a while, we use special notations for their coefficients. The coefficients of $\theta_1$ are denoted by $p_j$ so that

$$\theta_1(t) = p_j(t)y^j. \tag{2.131}$$

The coefficients of $\theta_2$ are denoted by $H_{jl}$ so that

$$\theta_2(t) = \frac{1}{2}H_{jl}(t)y^j y^l, \tag{2.132}$$

where $H(t)$ is a complex-valued symmetric matrix, $H_{jl} = H_{lj}$.

**2.4.7.** ★A change of variables in the vicinity of $\mathbf{x}(t)$ should not affect $\theta_0(t)$, i.e. $\theta_0(t)$ is a scalar function along the path $\mu : \mathbf{x} = \mathbf{x}(t)$. The components $(p_1(t), \dots, p_m(t))$ form a covariant vector field $\mathbf{p}(t)$ along the path $\mu$. By (2.115), $p_j$ have to be real valued. On the other hand, the complex-valued matrix function $H(t)$ does not represent a tensor field along $\mu$. Indeed, a coordinates transformation $\mathbf{x} \to \widetilde{\mathbf{x}}$ transforms $H(t)$ to $\widetilde{H}(t)$, where

$$H_{ij}(t) = \widetilde{H}_{lk}(t)\frac{\partial \widetilde{x}^l(\mathbf{x}(t))}{\partial x^i}\frac{\partial \widetilde{x}^k(\mathbf{x}(t))}{\partial x^j} + \frac{\partial^2 \widetilde{x}^l(\mathbf{x}(t))}{\partial x^i \partial x^j}\widetilde{p}_l(t). \tag{2.133}$$

However, we see from formula (2.133) that $\Im H(t)$ is a 2-covariant tensor along the path $\mu$. Although $H(t)$ is not a tensor, nevertheless, there is a symmetric tensor $G(t)$ that is closely related with $H(t)$,

$$G_{ij}(t) = H_{ij}(t) - \Gamma^k_{ij}(\mathbf{x}(t))p_k(t), \tag{2.134}$$

where $\Gamma^k_{jl}$ are the Christoffel symbols. Clearly, $\Im H = \Im G$.

**Exercise 2.50** *Using the transformation formulae for the Christoffel symbols (2.7) show that $G$ is 2-covariant tensor.*

**2.4.8.** One of the most important properties of a Gaussian beam is that the corresponding path $\mu : \mathbf{x} = \mathbf{x}(t)$ is a geodesic. To show this, we start with the following lemma

**Lemma 2.51** *Let $U_\epsilon^N(\mathbf{x},t)$ be a formal Gaussian beam of order $N$ that corresponds to wave equation (2.60). Then*
*i) $\theta_0(t) = \theta_0$, i.e. $\theta(\mathbf{x}(t),t)$ is constant.*
*ii) $(\mathbf{x}(t),\mathbf{p}(t))$ ( or $(\mathbf{x}(-t),\mathbf{p}(-t))$) is a bicharacteristic corresponding to the hamiltonian $h(\mathbf{x},\mathbf{p})$,*

$$h(\mathbf{x},\mathbf{p}) = (g^{ij}(\mathbf{x})p_ip_j)^{1/2} = (\mathbf{p},\mathbf{p})_g^{1/2}. \tag{2.135}$$

*We note that condition ii) for the bicharacteristic $(\mathbf{x}(t),\mathbf{p}(t))$ means that $(\mathbf{x}(t),\mathbf{p}(t))$ satisfy the Hamilton system of equations*

$$\frac{dx^i}{dt} = \frac{\partial h}{\partial p_i}, \quad \frac{dp_i}{dt} = -\frac{\partial h}{\partial x^i}. \tag{2.136}$$

*System (2.136) is uniquely solvable if we add proper initial conditions,*

$$\mathbf{x}|_{t=0} = \mathbf{x}_0, \quad \mathbf{p}|_{t=0} = \mathbf{p}_0. \tag{2.137}$$

We remind the reader that the Hamilton-Jacobi equation (2.125) means that all Taylor's coefficients of the function $(\partial_t\theta)^2 - g^{jl}\partial_j\theta\partial_l\theta$ vanish at $\mathbf{y} = 0$.

**Proof.** The proof of the lemma is given under an additional assumption that for any $t$ the instant frequency $\omega(t)$,

$$\omega(t) = -\ \partial_t\theta(\mathbf{x},t)|_{\mathbf{x}=\mathbf{x}(t)} \neq 0. \tag{2.138}$$

Later, we prove that this assumption is satisfied for any Gaussian beam.

*Step 1.* We start our analysis with the terms of degree 0 and 1 in the Taylor expansion of both sides of the Hamilton-Jacobi equation (2.125). These terms give rise to the following equations, which are considered together

$$\left(\frac{d\theta_0}{dt} - p_j\frac{dx^j}{dt}\right)^2 - g^{jl}p_jp_l = 0, \tag{2.139}$$

$$\left(\frac{d\theta_0}{dt} - p_j\frac{dx^j}{dt}\right)\left[2\frac{dp_l}{dt} - (H_{lr} + H_{rl})\frac{dx^r}{dt}\right] - \frac{\partial g^{jr}}{\partial x^l}p_jp_r -$$

$$-g^{jr}p_j(H_{lr} + H_{rl}) = 0, \quad l = 1, \ldots, m. \tag{2.140}$$

Here the metric tensor $g^{jl}$ and its derivatives $\partial_l g^{jr}$ are evaluated at the point $\mathbf{x} = \mathbf{x}(t)$. Consider the imaginary part of equation (2.140). In view of (2.114), (2.115) we have

$$\Im\theta_0(t) = 0, \quad \Im p_j(t) = 0, \quad j = 1, \ldots, m. \tag{2.141}$$

Hence, we obtain the following equation

$$\Im(H_{lr} + H_{rl})\eta^r = 0, \quad l = 1, \ldots, m, \tag{2.142}$$

where $\eta = (\eta^1, \ldots, \eta^m)$ is of the form

$$\eta^r = \left(\frac{d\theta_0}{dt} - p_j \frac{dx^j}{dt}\right)\frac{dx^r}{dt} + g^{jr}p_j = -\omega(t)\frac{dx^r}{dt} + g^{jr}p_j. \tag{2.143}$$

Here we used that from the definition of $\omega(t)$,

$$\omega(t) = p_j \frac{dx^j}{dt} - \frac{d\theta_0}{dt}. \tag{2.144}$$

*Step 2.* Let us show that

$$\frac{d\theta_0}{dt} = 0. \tag{2.145}$$

By condition (2.115), $\Im H(t)$ is positive definite. Hence, equation (2.142) yields that

$$\eta = 0. \tag{2.146}$$

Since $\eta$ is a vector and $\mathbf{p}$ is a covector, we can consider the inner product $(\mathbf{p}, \eta)$,

$$(\mathbf{p}, \eta) = \eta^r p_r = -\omega \frac{dx^r}{dt} p_r + g^{jr} p_j p_r = 0. \tag{2.147}$$

Equation (2.139) implies that

$$g^{jr}p_j p_r = -\omega\left(\frac{d\theta_0}{dt} - p_r \frac{dx^r}{dt}\right) = \tag{2.148}$$

$$= -\omega \frac{d\theta_0}{dt} + \omega p_r \frac{dx^r}{dt} = -\omega \frac{d\theta_0}{dt} + g^{jr} p_j p_r,$$

where the last step is based upon (2.147). Comparing the beginning and the end of the above equation we see that

$$\omega \frac{d\theta_0}{dt} = 0. \tag{2.149}$$

Due to the assumption that $\omega \neq 0$, we obtain that $\frac{d\theta_0}{dt} = 0$.

Statement i) is proven.

*Step 3.* Our next goal is to prove statement ii). Using definition (2.144) of $\omega(t)$, we rewrite equation (2.139) in the form

$$\omega^2(t) = g^{jl}(\mathbf{x}(t))p_j(t)p_l(t).$$

To analyze this equation we consider separately the case $\omega > 0$ and the case $\omega < 0$. When $\omega(t) > 0$, i.e.

$$\omega(t) = (g^{jl}p_jp_l)^{1/2} = h(\mathbf{x}(t), \mathbf{p}(t)), \tag{2.150}$$

equations (2.143) and (2.146) imply that

$$\frac{dx^r}{dt} = [g^{il}p_ip_l]^{-1/2}g^{jr}p_j, \quad r = 1, \ldots, m. \tag{2.151}$$

Now, using definition (2.135) of the Hamiltonian $h(\mathbf{x}, \mathbf{p})$, we rewrite the above equations in the form

$$\frac{dx^r}{dt} = \frac{\partial h}{\partial p_r}, \quad r = 1, \ldots, m,$$

which coincide with the first $m$ equations of the Hamilton system of equations (2.136).

To obtain the last $m$ equations of the Hamilton system, we return to equations (2.140). Taking into account (2.150), we obtain that

$$(H_{lr} + H_{rl})\left(h(\mathbf{x}, \mathbf{p})\frac{dx^r}{dt} - g^{jr}p_j\right) = 2h(\mathbf{x}, \mathbf{p})\frac{dp_l}{dt} + \frac{\partial g^{jr}}{\partial x^l}p_jp_r. \tag{2.152}$$

In view of the first $m$ equations of the Hamilton system, the left-hand side of these equations is equal to zero. Thus,

$$\frac{dp_l}{dt} = -\frac{1}{2h}\frac{\partial g^{jr}}{\partial x^l}p_jp_r = -\frac{\partial h}{\partial x^l}, \quad l = 1, \ldots, m.$$

These equations are just the last $m$ equations of the Hamilton system. Due to (2.150), the fact that the hamiltonian is constant along the bicharacteristic $\mathbf{x} = \mathbf{x}(t)$ is equivalent to the fact that the instant frequency is constant.

$$\omega(t) = \omega(0) = constant.$$

In the case $\omega < 0$ similar considerations as above show that $(\mathbf{x}(-t), \mathbf{p}(-t))$ is a bicharacteristic and $\omega(t) = \omega(0) = const.$

The statement ii) is proven.

*Step 4.* At last, we will prove that $\omega(t) \neq 0$ for any Gaussian beam. Assume that, on the contrary, $\omega(t_0) = 0$ for some $t_0$. Then, due to our previous considerations, $\omega(t) = 0$ for any $t$ near $t_0$. Hence equation (2.139) implies that $\mathbf{p}(t) = 0$. Using (2.139) again, we see that $\frac{d\theta_0}{dt} = 0$ for any $t$. Thus, Taylor's expansion of $\theta(\mathbf{x}, t)$ starts with the term of homogeneity 2 so that the terms of homogeneity 0 and 1 in Hamilton-Jacobi equation (2.125) are equal to zero. Consider the term of homogeneity 2 in this equation. It can be written in the form

$$H_{ki}(t)\left(\frac{dx^i(t)}{dt}\frac{dx^l(t)}{dt} - g^{il}(\mathbf{x}(t))\right)H_{lj}(t)y^k y^j = 0 \tag{2.153}$$

for any $\mathbf{y}$ and $t$.

Denote by $A$ the real-valued symmetric matrix

$$A^{il} = g^{il}(\mathbf{x}(t)) - \frac{dx^i}{dt}\frac{dx^l}{dt}.$$

Then equation (2.153) takes the form

$$HAH = 0, \tag{2.154}$$

which is equivalent to two real equations

$$\Im HA\Re H + \Re HA\Im H = 0, \tag{2.155}$$

$$\Re HA\Re H = \Im HA\Im H.$$

Introduce the matrix $T$,

$$T = \Re H(\Im H)^{-1},$$

where $(\Im H)^{-1}$ exists due to condition (2.115). Thus the equations (2.155) are equivalent to the equations

$$AT + T^*A = 0,$$

$$A = T^*AT, \qquad (2.156)$$

which imply that

$$A(T^2 + I) = 0. \qquad (2.157)$$

It follows from equation (2.157) that $A = 0$, i.e.

$$A^{il}(t) = g^{il}(\mathbf{x}(t)) - \frac{dx^i}{dt}\frac{dx^l}{dt} = 0. \qquad (2.158)$$

However, as matrix $[g^{il}]$ is positive-definite, the matrix $\left[\frac{dx^i(t)}{dt}\frac{dx^l(t)}{dt}\right]$ is degenerate. This contradiction proves that $\omega(t) \neq 0$ for any Gaussian beam.

□

**Exercise 2.52** *Use the fact that $\Re H$ and $\Im H$ are real symmetric matrices to show that $T^2 + I$ is non-degenerate.*

Multiplying $U_\epsilon^N$ by the constant factor $\exp\{i\theta_0/\epsilon\}$ we can always assume that $\theta_0 = 0$.

**2.4.9.** Let us consider an important geometrical consequence of Lemma 2.51. Due to the considerations at the end of section 2.4.2 we can re-normalize $\epsilon$, i.e., take $\widetilde{\epsilon} = c\epsilon$, so that

$$|\omega(0)| = |p_0| = h(\mathbf{x}_0, \mathbf{p}_0) = h_0 = 1,$$

and, henceforth,

$$|\omega(t)| = |p(t)| = h(\mathbf{x}(t), \mathbf{p}(t)) = 1. \qquad (2.159)$$

In this case, the velocity vector $\mathbf{v}(t) = \frac{d\mathbf{x}}{dt}$ is the canonical transformation of $\mathbf{p}(t)$,

$$\frac{d\mathbf{x}(t)}{dt} = I_g\mathbf{p}(t), \ \ i.e. \ \ \frac{dx^j}{dt} = g^{jl}(\mathbf{x}(t))p_j(t). \qquad (2.160)$$

**Lemma 2.53** *Let* $(\mathbf{x}(t), \mathbf{p}(t))$ *be a bicharacteristic of the hamiltonian* $h(\mathbf{x}, \mathbf{p}) = [g^{jl}(\mathbf{x})p_j p_l]^{1/2}$ *with initial data* $(\mathbf{x}_0, \mathbf{p}_0)$, $|\mathbf{p}_0| = 1$. *Then the path* $\mu : \mathbf{x} = \mathbf{x}(t)$ *is the geodesic* $\gamma_{\mathbf{x}_0, \mathbf{v}_0}$, *where* $\mathbf{v}_0 = I_g \mathbf{p}_0$, *and* $t$ *is the arclength along* $\gamma_{\mathbf{x}_0, \mathbf{v}_0}$.

**Exercise 2.54** *Prove Lemma 2.53 by showing that* $\mathbf{x}(t)$ *satisfies the equation of geodesics (2.12). In the proof use the Hamilton system (2.136) and the fact that*

$$\partial_l g^{ij} = -g^{ik}(\partial_l g_{kr})g^{rj}.$$

To conclude this subsection, we discuss the case $|\mathbf{p}_0| = h_0 \neq 1$. Because hamiltonian $h$ is a positive homogeneous function of order 1 with respect to $\mathbf{p}$,

$$h(\mathbf{x}, \lambda\mathbf{p}) = \lambda h(\mathbf{x}, \mathbf{p}), \quad \lambda > 0, \tag{2.161}$$

$\mathbf{x}(t)$ and $\mathbf{p}(t)$ are homogeneous functions of order 0 and 1 with respect to $h_0$. Henceforth, the corresponding path $\mu : \mathbf{x} = \mathbf{x}(t)$ is also a geodesic, with $t$ being its arclength. Using this remark and renormalization of $\epsilon$, we can always consider, without loss of generality, the case $\omega(0) = |\mathbf{p}_0| = 1$.

**2.4.10.** Next, we study the homogeneous term $\theta_2(t)$. To this end we analyze the term of homogeneity 2 with respect to $\mathbf{y}$ in $(\partial_t\theta)^2 - g^{jl}\partial_j\theta\partial_l\theta$. This term which depends on $\theta_2(t)$ and $\theta_3(t)$. However, the terms with $\theta_3$ vanish along the geodesic $\mathbf{x} = \mathbf{x}(t)$ due to the Hamilton system (2.136).

As a result, we obtain a matrix Riccati equation along the bicharacteristic $(\mathbf{x}(t), \mathbf{p}(t))$ for the matrix $H(t)$,

$$\frac{d}{dt}H + D + (BH + HB^t) + HCH = 0. \tag{2.162}$$

The matrices $B$, $C$, and $D$ are $m \times m$ matrices with components given by the second derivatives of the hamiltonian $h(\mathbf{x}, \mathbf{p})$,

$$B^j_l = \frac{\partial^2 h}{\partial x^l \partial p_j}; \quad C^{jl} = \frac{\partial^2 h}{\partial p_j \partial p_l}; \quad D_{jl} = \frac{\partial^2 h}{\partial x^j \partial x^l}, \tag{2.163}$$

where $j, l = 1, \ldots, m$ and the derivatives are evaluated in the point $(\mathbf{x}, \mathbf{p}) = (\mathbf{x}(t), \mathbf{p}(t))$, i.e., on the bicharacteristic. Clearly

$$B^t = B^*, \quad C^t = C^* = C, \quad D^t = D^* = D. \tag{2.164}$$

**Exercise 2.55** *Derive equation (2.162).*

As usual, equation (2.162) is supplemented with initial condition

$$H|_{t=0} = H_0, \tag{2.165}$$

where $H_0$ is any given matrix satisfying

$$H_0 = H_0^t, \tag{2.166}$$

$$\Im H_0 > 0. \tag{2.167}$$

### 2.4.11.

**Lemma 2.56** *Let $H_0$ satisfy conditions (2.166) and (2.167). Then*
*i) The initial-value problem (2.162)–(2.167) has a unique solution $H(t)$, $t \in \mathbf{R}$.*
*ii) The solution $H(t)$, $t \in \mathbf{R}$, is symmetric,*

$$H(t) = H^t(t), \tag{2.168}$$

*and*

$$\Im H(t) > 0. \tag{2.169}$$

*iii) Moreover, for any $Y_0$, $Z_0$, such that*

$$H_0 = Z_0 Y_0^{-1}, \tag{2.170}$$

*the matrix $H(t)$ is represented in the form*

$$H(t) = Z(t)Y(t)^{-1}. \tag{2.171}$$

*The pair of matrices $(Z(t), Y(t))$ is the solution of the initial-value problem,*

$$\frac{d}{dt}Y(t) = B^t \cdot Y + C \cdot Z, \quad Y|_{t=0} = Y_0,$$

$$\frac{d}{dt}Z(t) = -D \cdot Y - B \cdot Z, \quad Z|_{t=0} = Z_0, \tag{2.172}$$

*where the matrix $Y(t)$ is non-degenerate for all $t \in \mathbf{R}$,*

$$\det Y(t) \neq 0. \tag{2.173}$$

**2.4.12.** The proof of the lemma is based on the following statement:

**Lemma 2.57** *Let $(Z(t), Y(t))$ be the solution of initial-value problem (2.172). Then*

$$Z^t(t)Y(t) - Y^t(t)Z(t) = Z_0^t Y_0 - Y_0^t Z_0 = const, \tag{2.174}$$

$$Z^*(t)Y(t) - Y^*(t)Z(t) = Z_0^* Y_0 - Y_0^* Z_0 = const, \tag{2.175}$$

**Proof.** Let us show that the derivative, with respect to $t$ of the left-hand side of (2.175), are equal to zero. Indeed,

$$\frac{d}{dt}(Z^*Y - Y^*Z) = \frac{dZ^*}{dt}Y + Z^*\frac{dY}{dt} - \frac{dY^*}{dt}Z - Y^*\frac{dZ}{dt} = I(Y, Z).$$

Using equations (2.164), (2.172), we rewrite $I(Y, Z)$ in the form

$$I(Y, Z) = -(DY + BZ)^*Y + Z^*(B^tY + CZ) -$$

$$-(B^tY + CZ)^*Z + Y^*(DY + BZ) = -Y^*DY - Z^*B^tY +$$

$$+Z^*B^tY + Z^*CZ - Y^*BZ - Z^*CZ + Y^*DY + Y^*BZ = 0.$$

The proof of relation (2.174) is analogous. □

**2.4.13.** Returning to the proof of Lemma 2.56, we choose

$$Z_0 = H_0, \quad Y_0 = I, \tag{2.176}$$

where $I$ is the identity matrix.

*Step 1.* Due to linearity, the system (2.172)–(2.176) has the unique solution $(Y(t), Z(t))$. Next, we show that $Y(t)$ is non-degenerate for all $t \in \mathbf{R}$. Assume, on the contrary, that there is $t_0 \in \mathbf{R}$ and a vector $\eta \in \mathbf{C}^m$, $\eta \neq 0$, such that

$$Y(t_0)\eta = 0.$$

Clearly,

$$(Y(t_0)\eta, Z(t_0)\eta) - (Z(t_0)\eta, Y(t_0)\eta) = 0.$$

Hence,

$$(Z^*(t_0)Y(t_0) - Y^*(t_0)Z(t_0)\eta, \eta) = 0. \qquad (2.177)$$

In view of (2.175) this equation implies that

$$((Z^*(t_0)Y(t_0) - Y^*(t_0)Z(t_0))\eta, \eta) = ((Z_0^*Y_0 - Y_0^*Z_0)\eta, \eta). \quad (2.178)$$

Introduce the vector $\zeta$ by the formula

$$\zeta = Y_0\eta; \quad \eta = Y_0^{-1}\zeta. \qquad (2.179)$$

Substituting (2.179) into (2.178) and multiplying by $\frac{i}{2}$, we come to the equation

$$\frac{i}{2}((Z_0^*Y_0 - Y_0^*Z_0)Y_0^{-1}\zeta, Y_0^{-1}\zeta) =$$

$$= \frac{i}{2}((H_0^* - H_0)\zeta, \zeta) = (\Im H_0\zeta, \zeta) = 0. \qquad (2.180)$$

In view of (2.167), this equation implies that $\zeta = 0$ so that $\eta = 0$. This contradiction shows that $Y(t)$ is non-degenerate.

*Step 2.* We proceed now to the proof of factorization iii). Differentiating $H(t)$ and using (2.172), we obtain that

$$\frac{dH}{dt} = \frac{dZ}{dt}Y^{-1} + Z\frac{dY^{-1}}{dt} = \frac{dZ}{dt}Y^{-1} - ZY^{-1}\frac{dY}{dt}Y^{-1} =$$

$$= -(DY + BZ)Y^{-1} - ZY^{-1}(B^tY + CZ)Y^{-1} =$$

$$= -D - BZY^{-1} - ZY^{-1}B^t - ZY^{-1}CZY^{-1} =$$

$$= -D - (BH + HB^t) - HCH,$$

which proves iii).

It is known that that the solution of initial-value problem (2.162), (2.165) is locally and, henceforth, globally, unique. Because we have already proven the existence of the solution, this proves part i).

*Step 3.* It remains to prove part ii) of Lemma 2.56. To this end, we use Lemma 2.57, which implies, due to the symmetry of $H_0$, that

$$Y(t)^t Z(t) = Z(t)^t Y(t) \quad \text{for all } t \in \mathbf{R}. \tag{2.181}$$

Therefore, $H^t(t) = H(t)$.

To prove the positive definiteness of $\Im H(t)$, we consider the quadratic form $(\Im H(t)\zeta, \zeta)$, $\zeta \in \mathbf{R}^m$. Then

$$(\Im H(t)\zeta,\zeta) = \frac{i}{2}((Y^*)^{-1}(Z^*Y - Y^*Z)Y^{-1}\zeta,\zeta) = \tag{2.182}$$

$$= \frac{i}{2}((Z^*Y - Y^*Z)Y^{-1}\zeta, Y^{-1}\zeta) = \frac{i}{2}((Z^*(t)Y(t) - Y^*(t)Z(t))\eta,\eta)$$

with

$$\eta = Y^{-1}(t)\zeta.$$

Taking into account equality (2.175), we obtain from (2.182) that

$$(\Im H(t)\zeta,\zeta) = \frac{i}{2}((Z_0^*Y_0 - Y_0^*Z_0)\eta,\eta) = (\Im H_0\widetilde{\zeta},\widetilde{\zeta}),$$

where

$$\widetilde{\zeta} = Y_0\eta = Y_0Y^{-1}(t)\zeta.$$

Due to the positive definiteness of $\Im H_0$, the positive definiteness of $\Im H(t)$ follows from the above equations. This proves Lemma 2.56. □

**Remark.** The representation $H(t) = Z(t)Y^{-1}(t)$ may give an impression that $H(t)$ depends on initial factorization (2.176). However, due to the uniqueness of the solution of the initial-value problem, $H(t)$ does not depend on factorization of $H_0$. Moreover, let $(\widetilde{Z}_0, \widetilde{Y}_0)$ be another factorization of $H_0$. Then, due to the linearity of initial-value problem (2.172)

$$\widetilde{Z}(t) = Z(t)\widetilde{Y}_0, \quad \widetilde{Y}(t) = Y(t)\widetilde{Y}_0,$$

which obviously do not affect $H(t)$.

**2.4.14.** Later, in section 3.6, we need the following result.

**Lemma 2.58** *For any Gaussian beam*

$$det(\Im H(t)) \cdot |det Y(t)|^2 \tag{2.183}$$

*is constant.*

**Proof.** We use the well-known formula

$$\frac{1}{2} tr\left(\frac{dY}{dt} Y^{-1}\right) = \frac{1}{2}\frac{d}{dt}\ln[\det Y(t)], \tag{2.184}$$

where "tr" stands for the trace of matrix

$$trA = tr[A^i_j]^m_{i,j=1} = \sum_{i=1}^{m} A^i_i.$$

The differential equation for matrix function $Y(t)$ implies

$$\frac{d}{dt}\ln(\det Y(t)) = \mathrm{tr}\left(\frac{dY(t)}{dt}Y^{-1}(t)\right) = tr(B^t(t) + C(t)H(t)). \tag{2.185}$$

Thus, we see that

$$|\det Y(t)| = |\det Y(0)| \exp\left\{\int_0^t tr(B^t(\tau) + C(\tau)\Re H(\tau))d\tau\right\}.$$

Analogously,

$$\frac{d}{dt}(\ln(\det(\Im H(t)))) = tr\left(\frac{d\Im H(t)}{dt}(\Im H(t))^{-1}\right)$$

so that

$$\frac{d}{dt}(\ln(\det(\Im H(t)))) =$$

$$= -\mathrm{tr}\{(B(t) + \Re H(t)C(t)) + \Im H(t)(B^t(t) + C(t)\Re H(t))(\Im H(t))^{-1}\}$$

$$= -tr(B(t) + \Re H(t)C(t)) - tr(B^t(t) + C(t)\Re H(t)) =$$

$$= -2tr(B(t) + C(t)\Re H(t)),$$

where we use $C^t = C$, $(\Re H)^t = \Re H$, and $trA = trA^t$, $tr(AB) = tr(BA)$. Hence,

$$\det(\Im H(t)) = \det(\Im H(0)) \exp\left\{-2\int_0^t tr(B(\tau) + C(\tau)\Re H(\tau))d\tau\right\}.$$

Combining this formula with (2.185) we obtain formula (2.183). □

**2.4.15.** ★ In the next two subsections, we will give an invariant geometrical interpretation of the Riccati equation (2.162). Clearly, we should use the tensor field $G(t)$ instead of the matrix function $H(t)$ which is not a tensor. We see that, due to formula (2.134), $G(t)$ is also symmetric and its imaginary part is positive definite,

$$G^t(t) = G(t), \quad \Im G > 0. \tag{2.186}$$

To formulate the analog of Lemma 2.56 for $G(t)$ it is convenient to consider

$$\widetilde{G} = I_g G, \quad \widetilde{G}^i_j = g^{ik} G_{kj}. \tag{2.187}$$

As $\widetilde{G}$ is a $(1,1)$-tensor field along the geodesic $\mathbf{x}(t)$, it can be considered an operator in $T_{\mathbf{x}(t)}\mathcal{N}$. Introduce the operator $\widetilde{C}(t)$ in $T_{\mathbf{x}(t)}\mathcal{N}$,

$$\widetilde{C}(t) = I - P_\gamma(t), \tag{2.188}$$

where $P_\gamma$ is the one-dimensional projector,

$$P_\gamma(t)\mathbf{w} = (\mathbf{w}, \mathbf{v}(t))_g \mathbf{v}(t), \quad \mathbf{w} \in T_{\mathbf{x}(t)}\mathcal{N}. \tag{2.189}$$

We remind the reader that $\mathbf{v}(t) = \frac{d\mathbf{x}}{dt}$ is the unit velocity vector of the geodesic $\mathbf{x}(t)$.

**Exercise 2.59** *Show that*

$$I_g \widetilde{C} = C, \tag{2.190}$$

*where $C$ is given by formula (2.163), and in local coordinates we have the representation*

$$\widetilde{C}^i_j = \delta^i_j - g_{jk} v^i(t) v^k(t). \tag{2.191}$$

Introduce also the operator $\widetilde{R}_\gamma$, which is obtained from the curvature operator $R$ (see section 2.1.7),

$$\widetilde{R}_\gamma(t)\mathbf{w} = R(\mathbf{w}, \mathbf{v})\mathbf{v}, \quad \mathbf{w} \in T_{\mathbf{x}(t)}\mathcal{N}. \tag{2.192}$$

**2.4.16.** ★

**Lemma 2.60** *The* $(1,1)$*-tensor* $\widetilde{G}(t)$ *satisfy the covariant Riccati equation*

$$\frac{D\widetilde{G}}{dt} + \widetilde{G}\widetilde{C}\widetilde{G} + \widetilde{R}_\gamma = 0. \tag{2.193}$$

*where*

$$\frac{D\widetilde{G}}{dt} = \nabla_{\mathbf{v}}\widetilde{G}.$$

**Proof.** Riccati equation (2.162) for $H(t)$ implies the Riccati equation for $G(t)$

$$\frac{dG}{dt} + GCG + (\widehat{B}G + G\widehat{B}^t) + \widehat{D} = 0, \tag{2.194}$$

where

$$\widehat{B}_j^l = B_j^l + \Gamma_{jk}^r p_r C^{kl},$$

$$C^{jl} = \frac{\partial^2 h}{\partial p_j \partial p_l} = g^{jl} - g^{jk} p_k g^{lr} p_r, \tag{2.195}$$

$$\widehat{D}_{jl} = \frac{d}{dt}(\Gamma_{jl}^k p_k) + \Gamma_{jk}^r p_r C^{ks} \Gamma_{sl}^n p_n + (B_j^k \Gamma_{kl}^r + B_l^k \Gamma_{kj}^r) p_r + D_{jl}.$$

Here the matrices $B$, $C$, $D$, $\widehat{B}$, $\widehat{D}$, and the covector $\mathbf{p}$ are functions of $t$, and the metric tensor $g$ and the Christoffel symbols $\Gamma_{jl}^r$ are evaluated in the point $\mathbf{x}(t)$.

The expressions for $\widehat{B}$ and $\widehat{D}$ can be transformed into

$$\widehat{B}_j^l(t) = -\Gamma_{jk}^l(\mathbf{x}(t))\frac{dx^k(t)}{dt}, \tag{2.196}$$

$$\widehat{D}_{jl}(t) = \widehat{R}_{jl}(t) = R_{jklr}(\mathbf{x}(t))\frac{dx^k(t)}{dt}\frac{dx^r(t)}{dt}, \tag{2.197}$$

where $R_{jklr}(\mathbf{x})$ is the curvature tensor of the Riemannian manifold $(\mathcal{M}, g)$,

$$R_{jklr} = g_{jn}\left(\frac{\partial \Gamma_{rk}^n}{\partial x^l} - \frac{\partial \Gamma_{lk}^n}{\partial x^r} + \Gamma_{rk}^s \Gamma_{ls}^n - \Gamma_{lk}^s \Gamma_{rs}^n\right).$$

In particular, (2.197) implies that $\widehat{R}$ is a 2-covariant tensor field along the geodesic $\mathbf{x}(t)$.

Using representations (2.196), (2.197) for $\widehat{B}$ and $\widehat{D}$, Riccati equation (2.194) for $G(t)$ can be rewritten as a covariant Riccati equation for tensor field $G(t)$,

$$\frac{DG}{dt} + GCG + \widehat{R} = 0. \tag{2.198}$$

To write this equation for $\widetilde{G}$ instead of $G$, we note that $I_g \widehat{R} = \widetilde{R}_\gamma$ and $\frac{Dg}{dt} = 0$. Riccati equation (2.193) follows now from equation (2.190).

Concluding our geometrical interpretation of Riccati equation (2.162), we mention that the Riccati equation for $\widetilde{G}$ is a complexification of the fundamental equation of Riemannian geometry for the second fundamental form (see (2.20)).

**2.4.17.** Equations for the homogeneous polynomials $\theta_l$, $l \geq 3$, are obtained by considering the higher-order homogeneous polynomials in the Taylor expansion of the function

$$(\partial_t\theta)^2 - g^{jl}\partial_j\theta\partial_l\theta = (\partial_t\theta)^2 - h^2(\mathbf{x}, \partial\theta), \quad \partial\theta = (\partial_1\theta, \dots, \partial_m\theta).$$

The homogeneous term of order $l$ in this expansion depends on $\theta_k$ with $k \leq l+1$. However, as in section 2.4.10, the terms involving $\theta_{l+1}$, vanish along the bicharacteristic. The resulting differential equations for the homogeneous polynomials $\theta_l$, $l \geq 3$ are linear differential equations,

$$\frac{\partial\theta_l}{\partial t} + N^i_j \frac{\partial\theta_l}{\partial y^i} y^j = \mathcal{F}_l, \quad l = 3, 4, \dots, \tag{2.199}$$

where the right-hand sides $\mathcal{F}_l$ depend upon $\theta_j$, $j < l$. The coefficients $N^i_j = N^i_j(t)$ form an $m \times m$ matrix

$$N^i_j(t) = \frac{\partial^2 h}{\partial x^j \partial p_i} + \frac{\partial^2 h}{\partial p_i \partial p_k} H_{kj} = \tag{2.200}$$

$$= B^i_j(t) + C^{is}(t) H_{sj}(t) = [B^t + CH]^i_j,$$

where the matrices $B$ and $C$ are defined by (2.163). Equations (2.199) are not partial differential equations. They can be considered ordinary differential equations with respect to $t$ for $\theta_\gamma(t)$, $|\gamma| = l$. Therefore, we can consider $\mathbf{y}$ a parameter that the coefficients of these equations depend on. Thus, we can consider these equations also for $\mathbf{y} \in \mathbf{C}^m$.

To simplify the analysis of equations (2.199), $l \geq 3$, we introduce new coordinates $(\widetilde{t}, \widetilde{\mathbf{y}})$,

$$\widetilde{t} = t, \quad \widetilde{\mathbf{y}} = Y^{-1}(t)\mathbf{y}, \tag{2.201}$$

so that

$$\frac{\partial}{\partial \widetilde{t}} = \frac{\partial}{\partial t} + \frac{dY^i_j(t)}{dt}\widetilde{y}^j\frac{\partial}{\partial y^i} = \frac{\partial}{\partial t} + [B^tY + CZ]^i_j\,\widetilde{y}^j\frac{\partial}{\partial y^i}. \tag{2.202}$$

Substitution of $\widetilde{\mathbf{y}}$ of form (2.201) into this equation yields that

$$\frac{\partial}{\partial \widetilde{t}} = \frac{\partial}{\partial t} + [B^t + CH]^i_j\,y^j\frac{\partial}{\partial y^i} = \frac{\partial}{\partial t} + N^i_j y^j\frac{\partial}{\partial y^i}, \tag{2.203}$$

where we make use of the factorization $H(t) = Z(t)Y^{-1}(t)$.

Let $\widetilde{\theta}_l(\widetilde{\mathbf{y}}, \widetilde{t})$ be the representation of the polynomial $\theta_l(\mathbf{y}, t)$ in coordinates $(\widetilde{\mathbf{y}}, \widetilde{t})$. Then equations (2.199) take the form

$$\frac{\partial}{\partial \widetilde{t}}\widetilde{\theta}_l = \widetilde{\mathcal{F}}_l, \quad l = 3, 4, \ldots. \tag{2.204}$$

We can supplement this equation with initial data,

$$\widetilde{\theta}_l(\widetilde{\mathbf{y}}, \widetilde{t})\Big|_{\widetilde{t}=0} = \theta_l(\mathbf{y}, t)|_{t=0} = \widetilde{\theta}^0_l(\widetilde{\mathbf{y}}) = \theta^0_l(\mathbf{y}). \tag{2.205}$$

We recall that $Y(0) = I$. With these initial conditions, the equations (2.204) determine $\widetilde{\theta}(\widetilde{\mathbf{y}}, \widetilde{t})$ for any $\widetilde{t}$. Using transformation (2.201), we find $\theta_l(\mathbf{y}, t)$ for any $t$.

### 2.4.18.

**Lemma 2.61** *Let $\Theta(\mathbf{x})$ be a smooth function near $\mathbf{x}_0$ with Taylor's expansion*

$$\Theta(\mathbf{x}) \asymp \sum_{l \geq 1}\Theta_l = \sum_{|\gamma| \geq 1}\frac{1}{\gamma!}\Theta_\gamma \mathbf{y}^\gamma, \quad \mathbf{y} = \mathbf{x} - \mathbf{x}_0.$$

*Let, in addition,*

$$\Theta_1 = (\mathbf{p}_0, \mathbf{y}), \qquad |\mathbf{p}_0| = 1,$$

*be real and*

$$\Theta_2 = \frac{1}{2}(H_0\mathbf{y}, \mathbf{y}), \quad \Im H_0 > 0.$$

*Then, for any integer $K > 1$, there exists a function $\theta(\mathbf{x}, t) = \theta_K(\mathbf{x}, t)$, which satisfies conditions (2.114), (2.115), and the estimate*

$$|(\partial_t\theta)^2 - g^{ij}(\mathbf{x})\partial_i\theta\partial_j\theta| \le C_K|\mathbf{x} - \mathbf{x}(t)|^K, \tag{2.206}$$

*where $\mathbf{x}(t)$ is the geodesic $\gamma_{\mathbf{x}_0,\mathbf{v}_0}$, $\mathbf{v}_0 = I_g\mathbf{p}_0$.*

**Proof.** Using $(\mathbf{x}_0, \mathbf{p}_0)$ as initial conditions in Lemma 2.51, we construct $(\mathbf{x}(t), \mathbf{p}(t))$ and $\theta_1(t)$,

$$\theta_1(t) = (\mathbf{p}(t), \mathbf{y}).$$

Having constructed the bicharacteristic $(\mathbf{x}(t), \mathbf{p}(t))$ and using $H_0$ as initial data in Lemma 2.56, we construct $\theta_2(t)$,

$$\theta_2(t) = \frac{1}{2}(H(t)\mathbf{y}, \mathbf{y}).$$

At last, using the procedure described in section 2.4.17 and taking $\Theta_l$, $3 \le l \le K-1$ as initial data, we construct $\theta_l(t)$.

The desired function $\theta_K(\mathbf{x}, t)$ is then given by the formula

$$\theta_K(\mathbf{x}, t) = \sum_{l=1}^{K-1} \theta_l(t) = \sum_{|\gamma|=1}^{K-1} \frac{1}{\gamma!}\theta_\gamma(t)(\mathbf{x} - \mathbf{x}(t))^\gamma. \tag{2.207}$$

□

**2.4.19.** Having constructed the phase $\theta$, we focus next on amplitude functions $u_n(x, t)$. The analysis of the transport equations,

$$\mathcal{L}_\theta u_n - (\partial_t^2 - \Delta_g + q)u_{n-1} \asymp v_n \asymp 0, \tag{2.208}$$

at $\mathbf{x} = \mathbf{x}(t)$ is also based upon Taylor's expansion. Near $\mathbf{x} = \mathbf{x}(t)$ of the amplitudes $u_n$ are represented as

$$u_n(\mathbf{x}, t) \asymp \sum_{l\geq 0} u_{n,l}(\mathbf{y}, t) \asymp \sum_{l\geq 0} \widetilde{u}_{n,l}(\widetilde{\mathbf{y}}, t),$$

where $u_{n,l}$ and $\widetilde{u}_{n,l}$ are homogeneous polynomials of order $l$, $l = 0, 1, \ldots$, with respect to $\mathbf{y}$ and $\widetilde{\mathbf{y}}$. Since the operator $\mathcal{L}_\theta$,

$$\mathcal{L}_\theta u = 2\partial_t\theta\partial_t u - 2g^{jl}\partial_j\theta\partial_l u + (\partial_t^2 - \Delta_g)\theta \cdot u,$$

is a first order linear differential operator, the homogeneous polynomials of order $l$, $l \geq 0$ in equation (2.208) depend only upon the coefficients $\widetilde{u}_{n,k}$ for $k \leq l+1$. However, the same considerations as in section 2.4.10 show that the terms involving $u_{n,l+1}$ vanish along the bicharacteristic.

We consider $\widetilde{u}_{n,l}$ as a function of $t$ with values in the space of the homogeneous polynomials of order $l$ with respect to $\widetilde{\mathbf{y}}$. Then the resulting differential equations for $\widetilde{u}_{n,l}$ take the form

$$\frac{d}{dt}\widetilde{u}_{n,l}(t) + r(t)\widetilde{u}_{n,l}(t) = \widetilde{\mathcal{F}}_{n,l}(t), \quad n, l = 0, 1, \ldots. \tag{2.209}$$

The right-hand sides $\widetilde{\mathcal{F}}_{n,l}(t)$ are homogeneous polynomials of order $l$ that depend on $\widetilde{u}_{r,k}(t)$ and $\widetilde{\theta}_k$ with $k \leq l+2$, $r < n$. The factor $r(t)$ in this equation is equal to

$$r(t) = -\frac{1}{2}(\partial_t^2 - \Delta_g)\theta|_{\mathbf{x}=\mathbf{x}(t)}$$

and can be written also in the form

$$r(t) = \frac{1}{2}tr(B^t + CH) + \frac{1}{4}\frac{d}{dt}\ln g(t). \tag{2.210}$$

Using factorization $H(t) = Z(t)Y^{-1}(t)$ and the differential equation (2.172) for $Y(t)$, we see that

$$\frac{1}{2}tr(B^t + CH) = \frac{1}{2}tr[(B^tY + CZ)Y^{-1}] = \frac{1}{2}tr\left(\frac{dY}{dt}Y^{-1}\right).$$

By formula (2.184) we obtain

$$\frac{1}{2}tr(B^t + CH) = \frac{1}{2}\frac{d}{dt}\ln[\det Y(t)].$$

Using this representation, we rewrite equation (2.209) in the form

$$\frac{d}{dt}\widetilde{u}_{n,l}(t) + \left(\frac{1}{4}\frac{d}{dt}\ln[(\det(Y(t))^2 g(t)]\right)\widetilde{u}_{n,l}(t) = \widetilde{\mathcal{F}}_{n,l}(t). \quad (2.211)$$

The solutions to these equation are given by the formula

$$\widetilde{u}_{n,l}(t) = \varrho(t)\left(\widetilde{u}_{n,l}(0) + \int_0^t \varrho^{-1}(t')\widetilde{\mathcal{F}}_{n,l}(t')\,dt'\right), \quad (2.212)$$

where $\widetilde{u}_{n,l}(0)$ are initial data and

$$\varrho(t) = \sqrt{\frac{\det Y(0)}{\det Y(t)}}\sqrt[4]{\frac{g(0)}{g(t)}}. \quad (2.213)$$

In the future, we are specially interested in the behaviour of Gaussian beams on the corresponding path $\mu : \mathbf{x} = \mathbf{x}(t)$. Moreover, it is sufficient for our purposes to analyze only the main terms $u_0(\mathbf{x}(t), t) = u_{0,0}(t)$ and $u_1(\mathbf{x}(t), t) = u_{1,0}(t)$. Since $\widetilde{\mathcal{F}}_{0,0} = 0$,

$$\widetilde{u}_{0,0}(t) = \widetilde{u}_{0,0}(0)\varrho(t). \quad (2.214)$$

Considering equation (2.211) at $\mathbf{y} = 0$, we see that

$$\widetilde{\mathcal{F}}_{1,0} = -\frac{1}{2}\left(\frac{\partial^2}{\partial t} - \Delta_g + q(\mathbf{x})\right) u_0\bigg|_{x=x(t)} = \quad (2.215)$$

$$= -\frac{1}{2}q(\mathbf{x}(t))\widetilde{u}_{0,0}(t) + \widetilde{\mathcal{F}}^1_{0,0}(t).$$

The term $\widetilde{\mathcal{F}}^1_{0,0}(t)$ depends on $\widetilde{u}_{0,0}(t)$, $\widetilde{u}_{0,1}(t)$, $\widetilde{u}_{0,2}(t)$ but does not depend on $q$. Putting together (2.212), (2.215) and (2.214) we see that

$$\widetilde{u}_{1,0}(t) = -\frac{1}{2}\widetilde{u}_{0,0}(t)\int_0^t q(\mathbf{x}(t'))\,dt' + \widetilde{u}^1_{1,0}(t). \quad (2.216)$$

The integration in this formula is, in fact, the integration along the geodesic $\gamma_{\mathbf{x}_0,\mathbf{v}_0}$, $\mathbf{v}_0 = I_g\mathbf{p}_0$. The function $\widetilde{u}^1_{1,0}(t)$ depends on $\widetilde{u}_{0,0}, \widetilde{u}_{0,1}$, $\widetilde{u}_{0,2}$, $\mathbf{p}$, and $Y$, but not q.

**2.4.20.**

**Theorem 2.62** *Consider wave equation (2.60) on a compact manifold $(\mathcal{N}, g)$. Let $\Theta(\mathbf{x})$, $U_n(\mathbf{x})$, $n = 0, \ldots, N$, be smooth complex valued functions given in a neighborhood $V \subset \mathcal{N}$ of point $\mathbf{x}_0$. Assume that $\Theta$ and $U_0$ satisfy the following conditions*
*i)* $\Theta(\mathbf{x}_0) = 0$,
*ii)* $\Im\Theta(\mathbf{x}) \geq cd^2(\mathbf{x}, \mathbf{x}_0)$, $c > 0$,
*iii)* $d\Theta(\mathbf{x}_0) = \mathbf{p}_0$, $|\mathbf{p}_0|_g = 1$,
*iv)* $U_0(\mathbf{x}_0) \neq 0$.
*Let $\gamma = \gamma_{\mathbf{x}_0, \mathbf{v}_0}$ be the geodesic on $\mathcal{N}$ where $\mathbf{v}_0 = I_g \mathbf{p}_0$. Let for any $T > 0$, $(\gamma(t), t)$ be the corresponding trajectory in $\mathcal{N} \times \mathbf{R}$, $t \in [0, T]$. Then there exist smooth complex-valued functions $\theta(\mathbf{x}, t)$, $u_n(\mathbf{x}, t)$, $n = 0, \ldots, N$, in a neighborhood $W$ of this trajectory such that*

*i)* $\theta(\mathbf{x}(t), t) = 0$,
*ii)* $\Im\theta(\mathbf{x}, t) \geq cd^2(\mathbf{x}, \mathbf{x}(t))$,
*iii)* $d_{\mathbf{x}}\theta(\mathbf{x}(t), t) = \mathbf{p}(t)$, $|\mathbf{p}(t)|_g = 1$,
*iv)* $u_0(\mathbf{x}(t), t) \neq 0$.

*Then the function $U_\epsilon^N(\mathbf{x}, t)$, $(\mathbf{x}, t) \in W$,*

$$U_\epsilon^N(\mathbf{x}, t) = M_\epsilon \exp\{-(i\epsilon)^{-1}\theta(\mathbf{x}, t)\} \sum_{n=0}^{N} u_n(\mathbf{x}, t)(i\epsilon)^n,$$

*is a formal Gaussian beam in $W$ such that*

*v) For $\mathbf{x} \in V' \subset V$, $\mathbf{x}_0 \in V'$,*

$$|U_\epsilon^N(\mathbf{x}, 0) - M_\epsilon \exp\{-(i\epsilon)^{-1}\Theta(\mathbf{x})\} \sum_{n=0}^{N} U_n(\mathbf{x})(i\epsilon)^n| \leq C\epsilon^{N+1},$$

*vi)* $|(\partial_t^2 - \Delta_g + q)U_\epsilon^N| \leq CM_\epsilon \epsilon^N$ *for* $(\mathbf{x}, t) \in W$.

**2.4.21. Proof.** *Step 1.* If $T$ is sufficiently small so that $\gamma_{\mathbf{x}_0, \mathbf{v}_0} : [0, T] \to \mathcal{N}$ lies in a coordinate chart $(U, X)$, the statement of the Theorem follows immediately from Lemma 2.61 and considerations of sections 2.4.17-2.4.19. Indeed, we just take

$$\theta(\mathbf{x}, t) = \sum_{l \leq 2(N+2)} \theta_l(t),$$

and

$$u_n(\mathbf{x},t) = \sum_{l\leq 2(N+1-n)} u_{n,l}(t)$$

where $\theta_l(t)$ and $u_{n,l}(t)$ are polynomials of $\mathbf{x} - \mathbf{x}(t)$.

*Step 2.* In the general case, our considerations are valid for $t \in (-\delta, t_1+\delta)$ in a coordinate chart $(U_{(1)}, X_{(1)})$. As a result, we obtain a formal Gaussian beam $U^N_{\epsilon(1)}$ with the corresponding phase function $\theta_{(1)}$ and amplitude functions $u_{n(1)}$. Let $(U_{(2)}, X_{(2)})$ be another coordinate chart near $\mathbf{x}(t_1)$. We use $\theta_{(1)}(\mathbf{x}, t_1)$ and $u_{n(1)}(\mathbf{x}, t_1)$ as initial data to construct a formal Gaussian beam $U^N_{\epsilon(2)}$ for $t \in (t_1-\delta, t_2+\delta)$ in the coordinate chart $(U_{(2)}, X_{(2)})$. Thus $U^N_{\epsilon(2)}$ is determined by the phase function $\theta_{(2)}$ and amplitude functions $u_{n(2)}$.

When $\mathbf{x}(t) \in U_{(1)} \cap U_{(2)}$, then Taylor's expansions of $\theta_{(1)}$ and $\theta_{(2)}$ coincide up to the order $2(N+2)$ near $\mathbf{x} = \mathbf{x}(t)$. Similarly, Taylor's expansions $u_{n(1)}$, $u_{n(2)}$ coincide up to the order $2(N+1-n)$. Hence,

$$|U^N_{\epsilon(1)}(\mathbf{x},t) - U^N_{\epsilon(2)}(\mathbf{x},t)| \leq C\epsilon^{N+1}, \tag{2.217}$$

$$|\partial_t(U^N_{\epsilon(1)}(\mathbf{x},t) - U^N_{\epsilon(2)}(\mathbf{x},t))| \leq C\epsilon^{N+1/2}.$$

*Step 3.* For any $T > 0$ we need only a finite number $J$ of coordinate charts to cover the geodesic $\gamma_{\mathbf{x}_0,\mathbf{v}_0} : [0,T] \to \mathcal{N}$. Hence, we can use the procedure described in step 2 to construct $U^N_{\epsilon(1)}, \dots, U^N_{\epsilon(J)}$, $U^N_{\epsilon(j)}$ defined on the time interval $(t_{j-1}-\delta, t_j+\delta)$, $t_0 = 0$, $t_J = T$. Let $\chi_j(t)$ $j = 1, \dots, J$, be a partition of unity, $\chi_j(t) \in C^\infty(t_{j-1} - \delta, t_j + \delta)$, $\sum_{j=1}^J \chi_j(t) = 1$, for $t \in [0,T]$. Then function

$$U^N_\epsilon(\mathbf{x},t) = \sum_{j=1}^J \chi_j(t) U^N_{\epsilon(j)}(\mathbf{x},t)$$

is well defined in a neighborhood $W$ of the trajectory $(\gamma(t), t)$, $t \in [0,T]$. Inequality (2.217) and the fact that $U^N_{\epsilon(j)}$ are locally formal Gaussian beams imply that $U^N_\epsilon(\mathbf{x},t)$ is a formal Gaussian beam in the set $W$. □

**2.4.22.**

**Corollary 2.63** *Let $U_\epsilon^N(\mathbf{x},t)$ be a formal Gaussian beam constructed in Theorem 2.62. Then, for any $j \geq 0$ and multi-index $\alpha$, $|\alpha|+j < N$*

$$|\partial_t^j \partial_{\mathbf{x}}^\alpha (\partial_t^2 - \Delta_g + q) U_\epsilon^N| \leq C_{j\alpha} M_\epsilon \epsilon^{N-j-|\alpha|}.$$

**Proof.** The statement follows from equation (2.117) and Lemma 2.49 if we take into account that $v_n(\mathbf{x},t)$ are smooth functions satisfying

$$|v_n(\mathbf{x},t)| \leq C d(\mathbf{x}, \mathbf{x}(t))^{2(N+2-n)}.$$

□

**2.4.23.**

**Theorem 2.64** *Let $\Theta(\mathbf{x})$ and $U_n(\mathbf{x})$, $n = 0, \ldots, N$, be defined as in Theorem 2.62. Then there is a solution $u_\epsilon(\mathbf{x},t)$ of the equation*

$$(\partial_t^2 - \Delta_g + q) u_\epsilon(\mathbf{x},t) = 0, \quad (\mathbf{x},t) \in [0,T] \times \mathcal{N}, \tag{2.218}$$

*such that, for any $j \geq 0$ and multi-index $\alpha$,*

$$|\partial_t^j \partial_{\mathbf{x}}^\alpha (u_\epsilon(\mathbf{x},t) - \chi(\mathbf{x},t) U_\epsilon^N(\mathbf{x},t))| \leq C M_\epsilon \epsilon^{N-(j+|\alpha|)}. \tag{2.219}$$

*Here $U_\epsilon^N(\mathbf{x},t)$ is the formal Gaussian beam of Theorem 2.62 and $\chi$ is a smooth cut-off function, $\chi = 1$ near the trajectory $(\gamma(t), t)$, $t \in [0,T]$.*

**Proof.** The proof follows from Theorem 2.45 in view of Theorem 2.62 and Corollary 2.63. □

**Definition 2.65** *A Gaussian beam of order $N$ is a function $u_\epsilon(x,t)$ satisfying conditions (2.218) and (2.219).*

We note that the solution of the initial-value problem

$$(\partial_t^2 - \Delta_g + q) u = 0, \quad (\mathbf{x},t) \in [0,T] \times \mathcal{N},$$

$$u(\mathbf{x},t)|_{t=0} = U_\epsilon^N(\mathbf{x},t)|_{t=0}, \quad \partial_t u(\mathbf{x},t)|_{t=0} = \partial_t U_\epsilon^N(t,\mathbf{x})|_{t=0}$$

defines a Gaussian beam $u = u_\epsilon(\mathbf{x},t)$.

**2.4.24.** The reader might wonder why, dealing with the second-order wave equation, we use only one initial condition for $U_\epsilon^N(\mathbf{x},0)$.

$$U_\epsilon^N\big|_{t=0} = M_\epsilon \exp\{-(i\epsilon)^{-1}\Theta(\mathbf{x})\} \sum_{n=0}^{N} U_n(\mathbf{x})(i\epsilon)^n$$

The answer to the question lies in the observation that these data generate two Gaussian beams that correspond to $\omega > 0$ and $\omega < 0$. Because in the construction of the Gaussian beam, we require that there be one path $\mathbf{x} = \mathbf{x}(t)$ such that the solution is at any time localized near $\mathbf{x}(t)$, this requirement forces us to choose one of these beams.

## 2.5. Carleman estimates and unique continuation

In this section we will present a local Hölmgren-John unique continuation theorem, which will later play a crucial role in proving necessary controllability results.

**2.5.1.** We consider a solution $u = u(\mathbf{x},t)$ of the hyperbolic equation

$$P(\mathbf{x},D)u = \partial_t^2 u + a(\mathbf{x},D)u = 0 \text{ in } U' \times (-\delta,\delta), \tag{2.220}$$

where $U' \subset \mathcal{N}$ is an open set on a manifold $\mathcal{N}$. Because in this section we consider local results, $U'$ can be taken as a chart of local coordinates. Hence, we can assume that $U' = B'_\delta \subset \mathbf{R}^m$, where $B'_\delta$ is the ball of radius $\delta$ with centre at $\mathbf{0}'$ and study equation (2.220) in $\mathbf{R}^m \times \mathbf{R}$. We also consider the time- and space-variables simultaneously, and introduce the coordinates

$$\mathbf{y} = (\mathbf{x},t) = (\mathbf{y}',y^0) \in \mathbf{R}^{m+1}, \quad \mathbf{y}' = (y^1,\ldots,y^m),$$

We will use the index $'$ for the objects that lie in $\mathbf{R}^m$, writing, e.g., $\mathbf{0}', B'_\delta$, and not use this index when we deal with objects in $\mathbf{R}^{m+1}$, writing, e.g., $\mathbf{0}, B_\delta$, etc. We denote the operator $P(\mathbf{x},D)$ in these coordinates by $P(\mathbf{y},D)$. Our goal is to show that, if a solution $u$ of wave equation (2.220) vanishes on one side of a non-characteristic surface $\Gamma$, i.e.,

$$\text{supp }(u) \subset \{\mathbf{y} : \ \psi(\mathbf{y}) \le 0\}, \tag{2.221}$$

where

$$\Gamma = \{\mathbf{y} : \ \psi(\mathbf{y}) = 0\} \tag{2.222}$$

then $u$ vanishes near $\Gamma$ and, in particular, supp $(u) \cap \Gamma = \emptyset$.

**2.5.2.** As in section 2.2,

$$a(\mathbf{x}, D) = -a^{jk}(\mathbf{x})\partial_j\partial_k + b^j(\mathbf{x})\partial_j + c(\mathbf{x}), \tag{2.223}$$

where

$$D = (D', D_0) = (D_1, \ldots, D_m, D_0), \quad D_\alpha = -i\partial_\alpha, \ \alpha = 0, 1, \ldots, m,$$

and we denote by Greek indices, $\alpha$, $\beta = 0, 1, \ldots, m$, the indices that correspond to $\mathbf{y}$, and by Latin indices, $j, k = 1, \ldots, m$, – those that correspond to $\mathbf{y}' = \mathbf{x}$. The matrix $[a^{jk}(\mathbf{x})]$ is a real, symmetric, positive definite matrix. As we know from sections 2.2.1–2.2.3, $[a^{jk}]$ determines a Riemannian metric $g$ on $\mathcal{N}$. In $\mathbf{y}$-coordinates, we write $P(\mathbf{y}, D) = P(\mathbf{y}', D)$ as a sum of the leading, second-order terms $p(\mathbf{y}', D)$ and the lower-order terms $P_1(\mathbf{y}', D)$. In terms of symbols, $P(\mathbf{y}, \xi) = p(\mathbf{y}, \xi) + P_1(\mathbf{y}, \xi)$, where

$$p(\mathbf{y}, \xi) = -\xi_0^2 + a^{jk}(\mathbf{y}')\xi_j\xi_k, \tag{2.224}$$

and

$$P_1(\mathbf{y}, \xi) = ib^j(\mathbf{y}')\xi^j + c(\mathbf{y}').$$

Remember that, by definition, a surface $\Gamma$ is non-characteristic at a point $\mathbf{y}$ on $\Gamma$, if

$$p(\mathbf{y}, \nu) \neq 0.$$

Here $\nu = (\nu', \nu^0) = (\nu^1, \ldots, \nu^m, \nu^0)$ is a normal vector to $\Gamma$ with respect to the metric $g_{ij}dx^i dx^j + dt^2$ on $\mathcal{N} \times \mathbf{R}$, where $dt^2$ is the canonical Euclidean metric of $\mathbf{R}$.

The main result of this section is:

**Theorem 2.66 (Tataru)** *Assume that $a^{jk}$, $b^j$, $c \in C^\infty(B'_\delta)$ and that $[a^{jk}(x)]$ is a real, symmetric, positive definite matrix. Assume, in addition, that the surface $\Gamma \in B_\delta$ is non-characteristic. Then if $u \in H^1(B_\delta)$ is a solution of wave equation (2.220), which is equal to 0 on one side of $\Gamma$, i.e., (2.221) is satisfied, then supp $(u) \cap \Gamma = \emptyset$.*

**2.5.3.** We will prove Theorem 2.66 for a general elliptic operator of form (2.223). However, for our purposes, it is sufficient to consider the special case $a(\mathbf{x}, D) = -\Delta_g + q$, where $\Delta_g$ is the Laplace-Beltrami operator (see (2.56)).

**2.5.4.** ★ The proof of Theorem 2.66 is rather long and it will be given in a series of lemmas. Besides the formal proofs of the statements, we will try to explain the ideas behind these proofs.

We start with the following observation. Let $v$ and $\phi$ be smooth real valued functions such that, for any $\tau > \tau_0 > 0$,

$$\int e^{\tau\phi(\mathbf{y})} |v(\mathbf{y})|^2 d\mathbf{y} \le C,$$

where $C$ does not depend on $\tau$. Then, clearly, supp $(v) \subset \{\mathbf{y} : \phi(\mathbf{y}) \le 0\}$. We are going to prove a generalization of this statement. We start with some definitions.

Let $O$ be a point on $\Gamma$, $O \in \Gamma$, and assume that coordinates are chosen so that $\mathbf{y}(O) = \mathbf{0}$ and $(y^1, \ldots, y^m)$ are normal coordinates near $\mathbf{0}'$ with respect to the metric $g^{ij} = a^{ij}$. Moreover, we assume that $y^m$-unit vector is normal to $\Gamma \cap \{y^0 = 0\}$ at $\mathbf{0}$. We denote

$$\mathbf{y} = (\mathbf{y}', y^0) = (\mathbf{y}'', y^m, y^0)$$

and use the corresponding notations for the Fourier variable $\xi = (\xi', \xi_0) = (\xi'', \xi_m, \xi_0)$.

Let again $\phi \in C^\infty(\mathbf{R}^{m+1})$ be a real valued function. We define the operator,

$$Q^{\phi}_{\epsilon,\tau} u = \exp\left(-\frac{\epsilon}{2\tau} D_0^2\right) \left(e^{\tau\phi(\mathbf{y})} u(\mathbf{y})\right), \quad \epsilon, \tau > 0, \tag{2.225}$$

where the operator $\exp(-\frac{\epsilon}{2\tau} D_0^2)$ is determined in terms of the Fourier transform,

$$\exp\left(-\frac{\epsilon}{2\tau} D_0^2\right) w(\mathbf{y}) = \frac{1}{2\pi} \int_{\mathbf{R}} e^{iy^0\xi_0} \exp\left(-\frac{\epsilon}{2\tau} \xi_0^2\right) \mathcal{F}_0 w(\mathbf{y}', \xi_0)\, d\xi_0.$$

Here

$$\mathcal{F}_0 u(\mathbf{y}', \xi_0) = \int_{\mathbf{R}} e^{-i\xi_0 y^0} u(\mathbf{y}', y^0) dy^0$$

is the Fourier transform with respect to the $y^0$-variable. Similarly, we denote by $\mathcal{F}$ the Fourier transform with respect to all variables.

### 2.5.5. ★

**Lemma 2.67** *Let $u \in L^2_{\rm C}(\mathbf{R}^{m+1})$, where $L^2_{\rm C}$ is the space of $L^2$-functions with compact support. Let $\epsilon > 0$. Assume that*

$$\|Q^{\phi}_{\epsilon,\tau} u\|_0 \leq C$$

*for $\tau > \tau_0 > 0$, where $C$ is independent of $\tau$. Then supp $(u) \subset \{\mathbf{y} : \phi(\mathbf{y}) \leq 0\}$.*

**Proof.** Let $f$ be a function in $\mathbf{R}^{m+1}$ such that its Fourier transform $\mathcal{F}f$ is a compactly supported smooth function. Introduce a distribution $h = \phi_*(fu)$ by the formula,

$$\langle h, g\rangle_{L^2(\mathbf{R})} = \langle fu, \phi^* g\rangle_{L^2(\mathbf{R}^{m+1})}, \quad g \in C_0^\infty(\mathbf{R}), \tag{2.226}$$

where $\phi^*$ is the pull-back operator given by formula (2.4). In particular, this implies that supp $(h) \subset \phi($supp $(u))$ is compact. Therefore, formula (2.226) can be generalized to $g \in C^\infty(\mathbf{R})$ and, for any $\tau \in \mathbf{C}$, the Fourier transform $\widehat{h}(\tau)$ of $h$ is given by

$$\widehat{h}(\tau) = \langle fu, e^{i\tau\phi}\rangle = \langle u, \overline{f} e^{i\tau\phi}\rangle.$$

It is then clear that $|\widehat{h}(\tau)|$ is uniformly bounded for $\tau \in \mathbf{R}$. Moreover, since $\mathcal{F}f \in C_0^\infty(\mathbf{R}^{m+1})$ is compactly supported, then, for any $\tau > 1$,

$$|\widehat{h}(i\tau)| = |\langle ue^{\tau\phi}, \overline{f}\rangle| = |\langle \mathcal{F}(ue^{\tau\phi}), \mathcal{F}\overline{f}\rangle| =$$

$$= |\langle e^{-\epsilon\xi_0^2/2\tau}\mathcal{F}(ue^{\tau\phi}), e^{\epsilon\xi_0^2/2\tau}\mathcal{F}\overline{f}\rangle| = |\langle \mathcal{F}(Q^{\phi}_{\epsilon,\tau}u), e^{\epsilon\xi_0^2/2\tau}\mathcal{F}\overline{f}\rangle| \leq$$

$$\leq \|Q^{\phi}_{\epsilon,\tau}u\| \|e^{\epsilon\xi_0^2/2\tau}\mathcal{F}\overline{f}\| \leq C. \tag{2.227}$$

On the other hand, $u \in L^2_{\rm C}(\mathbf{R}^{m+1})$ and $f$ is a smooth function. Thus by applying formula (2.226) with $g \in C_0^\infty(\mathbf{R})$, $g(s) = e^{i\tau s}$ in supp $(h)$ we see that

$$|\widehat{h}(\tau)| \leq C' e^{c|\tau|}, \tag{2.228}$$

for any complex $\tau \in \mathbf{C}$. We use the following Phragmén-Lindelöf principle:

**Theorem 2.68** *Let $S \subset \mathbf{C}$ be a sector of an opening $\frac{\pi}{\alpha}$ and let an analytic function $\widehat{h}(\tau)$ be bounded on $\partial S$. If, in the interior of $S$,*

$$|\widehat{h}(\tau)| \leq C' \exp(c|\tau|^{\alpha'}), \quad \alpha' < \alpha, \tag{2.229}$$

*then $\widehat{h}$ is bounded in $S$.*

In our case, we actually have two sectors $S_1$ and $S_2$, being, respectively, the first and the second quadrants of the complex plane, so that $\alpha = 2$. Conditions (2.228) and (2.227) together with the uniform boundedness of $\widehat{h}(\tau)$ for $\tau \in \mathbf{R}$ show that $\widehat{h}(\tau)$ satisfies estimate (2.229) with $\alpha' = 1$. Hence, it follows from the Phragmén-Lindelöf principle that $\widehat{h}$ is a bounded function in the upper half plane. Thus, by the Paley-Wiener theorem, supp $(h) \subset \overline{\mathbf{R}}_-$. Therefore, we have proven that, for any $g \in C_0^\infty(\mathbf{R}_+)$,

$$\int_{\mathbf{R}^{m+1}} f u \, \overline{(\phi^* g)} \, dy = \int_{\mathbf{R}} h(s) \, \overline{g(s)} \, ds = 0.$$

Because the set of functions $f$ with compactly supported $\mathcal{F}f$ is dense in $L^2(\mathbf{R}^{m+1})$, we see that $g(\phi(\mathbf{y}))u(\mathbf{y}) = 0$. At last, choosing $g$ such that $g = 1$ on any closed interval $I \subset \mathbf{R}_+$, we see that $u(\mathbf{y}) = 0$ when $\psi(\mathbf{y}) > 0$. □

**2.5.6.** ★ Our further considerations are based on the following idea. Consider a solution $u(\mathbf{y})$ of equation (2.220) in a vicinity of the point $\mathbf{0} \in \Gamma$. If $\delta > 0$ is sufficiently small, we will construct a second-order polynomial $\phi(\mathbf{y})$, which satisfies the following conditions,

$$\text{supp }(u) \subset \{\mathbf{y} : \psi(\mathbf{y}) \leq 0\} \subset \{\mathbf{y} : \phi(\mathbf{y}) \leq 0\}$$

in $B_\delta$, and $\mathbf{0}$ is the only point that lies on both surfaces $\phi = 0$ and $\psi = 0$, i.e.,

$$\{\mathbf{y} : \phi(\mathbf{y}) = 0\} \cap \{\mathbf{y} : \psi(\mathbf{y}) = 0\} = \{\mathbf{0}\}.$$

We will prove then that there are constants $C$, $c > 0$, such that

$$\|Q^{\phi}_{\epsilon,\tau}(\chi u)\| \leq C e^{-c\tau} \|u\|_1, \tag{2.230}$$

where $\chi \in C_0^\infty(B_\delta)$ is a cut-off function, which is equal to 1 near the point $\mathbf{0}$. We remind the reader that $\|u\|_1$ is the norm of $u$ in $H^1(B_\delta)$. Combining estimate (2.230) with Lemma 2.67, we obtain that

$$\text{supp }(\chi u) \subset \{\mathbf{y} : \phi(\mathbf{y}) \leq -c\}$$

(see Fig. 2.11). Because $\chi(\mathbf{0}) = 1$ this implies that $\mathbf{0} \notin \text{supp }(u)$.

Figure 2.11: Pseudo-convex surfaces and the support of $U$

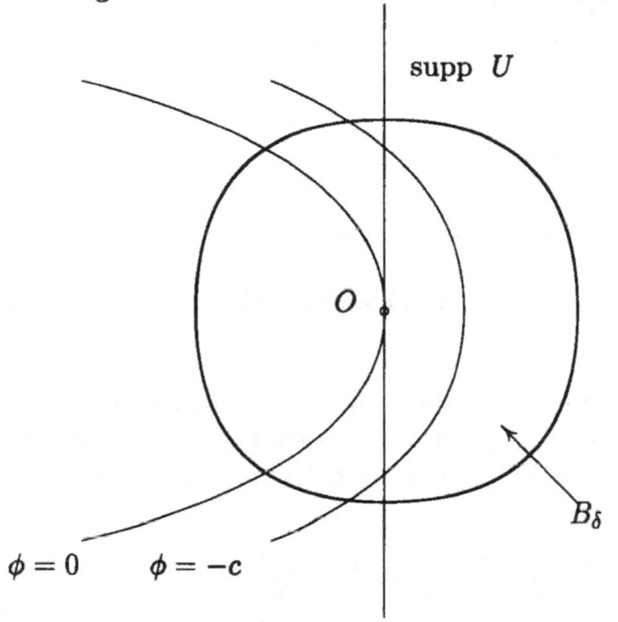


**2.5.7.** ★ We remind the reader that, due to the choice of coordinates,

$$g^{ij}(\mathbf{0}) = \delta^{ij}, \quad \Gamma^k_{ij}(\mathbf{0}) = 0,$$

and

$$d\psi(\mathbf{0}) = (0, \ldots, 0, \partial_m\psi(\mathbf{0}), \partial_0\psi(\mathbf{0})), \quad d = (\partial_1 \ldots, \partial_m, \partial_0).$$

Hence, in these coordinates,

$$a^{ij}(\mathbf{0}) = \delta^{ij}, \quad \partial_l a^{ij}(\mathbf{0}) = 0.$$

The fact that $\Gamma$ is not characteristic at $\mathbf{0}$ means that

$$(\partial_0\psi(\mathbf{0}))^2 - (\partial_m\psi(\mathbf{0}))^2 \neq 0.$$

Next, we use the function $\psi$ to define a function $\widetilde{\psi}$,

$$\widetilde{\psi}(y) = \partial_0\psi(\mathbf{0})y^0 + \partial_m\psi(\mathbf{0})y^m - \rho|\mathbf{y}|^2, \tag{2.231}$$

where $\rho \geq 0$ is sufficiently large so that, in the vicinity of $\mathbf{0}$,

$$\widetilde{\psi}(\mathbf{y}) < \psi(\mathbf{y}) \quad \text{if} \quad \mathbf{y} \neq 0. \tag{2.232}$$

Obviously,

$$\widetilde{\psi}(\mathbf{0}) = \psi(\mathbf{0}) = 0, \tag{2.233}$$

and, decreasing $\delta$ if necessary, we see that

$$\text{supp}\,(u) \subset \{\mathbf{y} \,:\, \widetilde{\psi}(\mathbf{y}) < 0\} \cup \{0\} \tag{2.234}$$

in $B_\delta$ (see Fig. 2.11).

Next, we will introduce a second-order polynomial $\phi$, which satisfies conditions (2.233) and (2.234) and is appropriate for estimate (2.230). To this end, we will look for a polynomial $\phi(\mathbf{y})$, such that the condition

$$\left.\frac{\{p(\mathbf{y},\xi + i\tau d\phi(\mathbf{y})), \overline{p(\mathbf{y},\xi + i\tau d\phi(\mathbf{y}))}\}}{8i\tau}\right|_{\mathbf{y}=\mathbf{0}} > 0 \tag{2.235}$$

is satisfied, whenever

$$\xi_0 = 0, \quad p(\mathbf{0},\xi + i\tau d\phi(\mathbf{0})) = 0, \quad \tau > 0. \tag{2.236}$$

Here $d\phi$ is the differential of $\phi$ defined in section 2.1.3 and $\{\cdot,\cdot\}$ is the Poisson brackets given by

$$\{p_1, p_2\} = \partial_{y^\alpha} p_1 \cdot \partial_{\xi_\alpha} p_2 - \partial_{\xi_\alpha} p_1 \cdot \partial_{y^\alpha} p_2. \tag{2.237}$$

**2.5.8.** ★

**Definition 2.69** *A function $\phi(\mathbf{y})$, which satisfies condition (2.235) on set (2.236), is pseudo-convex with respect to $p(\mathbf{y},\xi)$ at the point* $\mathbf{0}$.

**2.5.9.** ★ The pseudo-convexity conditions (2.235), (2.236) involve $d\phi(\mathbf{0})$ and the Hessian $H_\phi(\mathbf{0})$,

$$H_\phi(\mathbf{0}) = [\partial_\alpha \partial_\beta \phi(\mathbf{0})]_{\alpha,\beta=0,1,\ldots,m}.$$

Simple calculations show that

$$\{p(\mathbf{y},\xi+i\tau d\phi(\mathbf{y})),\overline{p(\mathbf{y},\xi+i\tau d\phi(\mathbf{y}))}\}/8i\tau\big|_{\mathbf{y}=\mathbf{0}} =$$

$$-\{\Re p(\mathbf{y},\xi+i\tau d\phi(\mathbf{y}),\Im p(\mathbf{y},\xi+i\tau d\phi(\mathbf{y})\}/4\tau\big|_{\mathbf{y}=\mathbf{0}} = \qquad (2.238)$$

$$= (JH_\phi(\mathbf{0})J\xi,\xi)+\tau^2(JH_\phi(\mathbf{0})Jd\phi(\mathbf{0}),d\phi(\mathbf{0})) = \Pi_\phi(\xi,\tau),$$

where

$$J = \operatorname{diag}\{1,\ldots,1,-1\}$$

and $(\cdot,\cdot)$ is the usual inner product in $\mathbf{R}^{m+1}$.

**2.5.10.** ★

**Exercise 2.70** *Prove equation (2.238).*

**2.5.11.** ★ To analyze pseudo-convexity conditions (2.235), (2.236), we need to distinguish between two cases.

When the surface $\phi(\mathbf{y}) = 0$ is space-like at $\mathbf{y} = \mathbf{0}$, i.e.

$$\sigma = (Jd\phi(\mathbf{0}),d\phi(\mathbf{0})) < 0, \qquad (2.239)$$

the pseudo-convexity conditions are void. Indeed, the set of points defined by (2.236) is empty.

Therefore, we need to consider only the case when the surface $\phi(\mathbf{y})$ is time-like at $\mathbf{y} = \mathbf{0}$, i.e.

$$\sigma = (Jd\phi(\mathbf{0}),d\phi(\mathbf{0})) > 0. \qquad (2.240)$$

When $\xi$ and $\tau$ satisfy (2.236), we see that $\tau^2 = |\xi|^2/\sigma$. Introduce the notations,

$$\eta = (JH_\phi(\mathbf{0})Jd\phi(\mathbf{0}),d\phi(\mathbf{0})) \qquad (2.241)$$

and

$$\Pi_\phi(\xi) = \Pi_\phi(\xi,\tau), \quad \tau^2 = |\xi|^2/\sigma. \qquad (2.242)$$

Then, the condition (2.235), considered on the set (2.236), becomes the inequality,

$$\Pi_\phi(\xi) = (JH_\phi(\mathbf{0})J\xi, \xi) + \frac{|\xi|^2\eta}{\sigma} \geq c_0|\xi|^2, \tag{2.243}$$

which should be valid for all

$$\xi = (\xi'', 0, 0). \tag{2.244}$$

The pseudo-convexity conditions, written in form (2.243)–(2.244), mean that the surface $\Pi_\phi(\xi) = 1$, $\xi = (\xi'', 0, 0)$, is actually an $(m-2)$-dimensional ellipsoid in variables $\xi''$. It is the convexity of this surface that explains the term pseudo-convexity.

At last, let

$$\gamma = \min_{\xi=(\xi'',0,0),|\xi|=1} (JH_\phi(\mathbf{0})J\xi, \xi). \tag{2.245}$$

Then, in terms of $\sigma$, $\eta$, and $\gamma$ the pseudo-convexity condition can be rewritten as

$$\gamma + \frac{\eta}{\sigma} > 0. \tag{2.246}$$

**2.5.12.** ★ To construct the polynomial $\phi$, we start with a function $\widetilde{\phi}$, $\widetilde{\phi} = \lambda\widetilde{\psi} + \lambda^2\widetilde{\psi}^2$, where $\lambda > 0$. We are going to choose $\lambda$ so large that $\phi$ satisfies pseudo-convexity condition (2.243) when $\xi = (\xi'', 0, 0)$. By direct calculations,

$$\Pi_{\widetilde{\phi}}(\xi) = \lambda\Pi_{\widetilde{\psi}}(\xi) + 2\lambda^2|\xi''|^2(Jd\widetilde{\psi}(\mathbf{0}), d\widetilde{\psi}(\mathbf{0})). \tag{2.247}$$

**Exercise 2.71** *Prove formula (2.247).*

As $\Gamma$ is time-like at $\mathbf{y} = \mathbf{0}$ and, due to (2.231), $d\widetilde{\psi}(\mathbf{0}) = d\psi(\mathbf{0})$, then $(Jd\widetilde{\psi}(\mathbf{0}), d\widetilde{\psi}(\mathbf{0})) > 0$. Hence, it is possible to choose $\lambda > 0$ large enough to obtain that

$$\Pi_{\widetilde{\phi}}(\xi) \geq c_0|\xi|^2, \quad \xi = (\xi'', 0, 0),$$

for some $c_0 > 0$. Obviously, in a sufficiently small neighborhood of $\mathbf{0}$, the surface $\widetilde{\phi}(\mathbf{y}) = 0$ coincides with the surface $\widetilde{\psi}(\mathbf{y}) = 0$, and $\widetilde{\phi}(\mathbf{y}) \leq 0$ if and only if $\widetilde{\psi}(\mathbf{y}) \leq 0$. This implies, in particular, that, decreasing if necessary $\delta$,

$$\text{supp}\,(u) \subset \{\mathbf{y} : \widetilde{\phi}(\mathbf{y}) < 0\} \cup \{\mathbf{0}\} \tag{2.248}$$

in $B_\delta$.

We will now construct a second-order polynomial $\phi(\mathbf{y})$ from $\widetilde{\phi}(\mathbf{y})$ by using the second-order Taylor approximation of $\widetilde{\phi}(\mathbf{y})$. We define

$$\phi(\mathbf{y}) = \sum_{|\alpha|\leq 2} \frac{1}{\alpha!}\partial^\alpha \widetilde{\phi}(\mathbf{0})y^\alpha - \mu|\mathbf{y}|^2, \quad \mu > 0. \qquad (2.249)$$

When $\mu$ is small enough,

$$\Pi_\phi(\xi) \geq \frac{c_0}{2}|\xi|^2 \quad \text{for} \quad \xi = (\xi'', 0, 0), \qquad (2.250)$$

and, decreasing if necessary $\delta$, we see that $\phi(\mathbf{y}) \leq \widetilde{\phi}(\mathbf{y})$ in $B_\delta$.

**2.5.13.** ★ Hence, we have proven the following lemma.

**Lemma 2.72** *Let supp $u \subset \{\mathbf{y} : \psi(\mathbf{y}) \leq 0\}$ and assume that $\mathbf{0}$ is a non-characteristic point of $\Gamma$. Then, for sufficiently small $\delta$, there exists a pseudo-convex second-order polynomial $\phi(\mathbf{y})$, $\phi(\mathbf{0}) = 0$, such that*

$$\text{supp} \;\; u \subset \{\mathbf{y} : \phi(\mathbf{y}) < 0\} \cup \{0\} \qquad (2.251)$$

*in $B_\delta$.*

If $\Gamma$ is space-like, then any second-order polynomial that satisfies (2.248), is appropriate. Assume now that $\phi$ is pseudo-convex with respect to $p(\mathbf{y}, \xi)$. If we make a sufficiently small $C^1$-perturbation of the coefficients of $p(\mathbf{y}, \xi)$, then $\phi$ remains pseudo-convex with respect to this new symbol. Thus, we obtain the following corollary.

**Corollary 2.73** *Let $\phi(\mathbf{y})$ be a pseudo-convex polynomial of Lemma 2.72. Then, for sufficiently small $|\epsilon|$, $\phi(\mathbf{y})$ is also pseudo-convex with respect to the symbol $p(\mathbf{y}, \xi - \epsilon H_\phi(\mathbf{0})(0, \ldots, 0, \xi_0))$.*

**2.5.14.** ★ Our next goal is to prove estimate (2.230) with the function $\phi(\mathbf{y})$ constructed in the previous section. To this end, we will first show that

$$\|Q^\phi_{\epsilon,\tau} v\|_{(1,\tau)} \leq c\tau^{-1/2}\|Q^\phi_{\epsilon,\tau} P(\mathbf{y}, D)v\| + ce^{-c\tau}\|e^{\tau\phi}v\|_{(1,\tau)}, \quad (2.252)$$

for any $v \in H^1(\mathbf{R}^{m+1})$, supp $(v) \in B_\delta$. Here

$$\|v\|^2_{(1,\tau)} = \|dv\|^2 + \tau^2\|v\|^2, \qquad (2.253)$$

and $Q^{\phi}_{\varepsilon,\tau}$ is determined in (2.225). The proof of inequality (2.252) is also given in a series of lemmas.

In what follows, we use the notation

$$P(\mathbf{y}, D, \tau) = P(\mathbf{y}, D + i\tau d\phi). \tag{2.254}$$

Then

$$e^{\tau\phi} P(\mathbf{y}, D)v = P(\mathbf{y}, D, \tau)w, \quad w = e^{\tau\phi}v. \tag{2.255}$$

**Exercise 2.74** *Prove formula (2.255).*

Next we write $P(\mathbf{y}, D + i\tau d\phi)$ in the form

$$P(\mathbf{y}, D + i\tau d\phi) = p(\mathbf{y}, D, \tau) + P_1(\mathbf{y}, D, \tau). \tag{2.256}$$

where $P_1(\mathbf{y}, D, \tau)$ is a first order operator respect of $D$ and $\tau$ and

$$p(\mathbf{y}, D, \tau) = p_R(\mathbf{y}, D, \tau) + i p_I(\mathbf{y}, D, \tau).$$

Here $p_R$, $p_I$ are given by formulae

$$p_R(\mathbf{y}, D, \tau)w = \Box_a w - \tau^2 (J_a d\phi, d\phi)w, \tag{2.257}$$

$$p_I(\mathbf{y}, D, \tau)w = \tau\big((J_a d\phi, D) + (D, J_a d\phi)\big)w,$$

and

$$\Box_a = -D_0^2 + \sum_{j,k=1}^{m} D_j a^{jk} D_k,$$

$$(J_a \xi, \zeta) = -\xi_0 \zeta_0 + \sum_{j,k=1}^{m} a^{jk} \xi_j \zeta_k, \tag{2.258}$$

and, for $B = (B^1, \ldots, B^m, B^0)$,

$$(B, D)u = -i \sum_{\alpha=0}^{m} B^\alpha \partial_\alpha u, \quad (D, B)u = -i \sum_{\alpha=0}^{m} \partial_\alpha (B^\alpha u).$$

Operators $p_R$ and $p_I$ are formally self-adjoint and $P_1(\mathbf{y}, D, \tau)$ contains only terms of order 0 and 1 with respect to $(D, \tau)$. This makes it possible to treat the term $P_1(y, D, \tau)$ as a perturbation.

**2.5.15.** ★

**Lemma 2.75** *Let $\phi$ be a pseudo-convex second-order polynomial at* 0. *Then, there exist $C > 0$, $\delta > 0$, and $\tau_0 > 0$, such that*

$$\|w\|^2_{(1,\tau)} \le C\left[\frac{\|p(\mathbf{y}, D, \tau)w\|^2}{\tau} + \|D_0 w\|^2\right] \tag{2.259}$$

*for $w \in C_0^\infty(B_\delta)$, and $\tau > \tau_0$.*

**Remark.** Estimate (2.259) is closely related to the Gårding inequality. However, the operator $p(\mathbf{y}, D)$ in (2.259) is not elliptic. This necessitates the use of an auxiliary parameter $\tau$.

**2.5.16.** ★ **Proof.** To prove estimate (2.259), we will show that

$$\|w\|^2_{(1,\tau)} \le C_1\frac{\|p(\mathbf{y}, D, \tau)w\|^2}{\tau} + C_2\frac{\|p(\mathbf{y}, D, \tau)w\|^2}{\tau^2} +$$

$$+C_3\|D_0 w\|^2, \tag{2.260}$$

when $\tau$ is large enough. We have that

$$\|p(\mathbf{y}, D, \tau)w\|^2 = \|p_R w\|^2 + \|p_I w\|^2 + \langle i[p_R, p_I]w, w\rangle, \tag{2.261}$$

where $[p_R, p_I] = p_R p_I - p_I p_R$ is the commutator of the operators $p_R(\mathbf{y}, D, \tau)$ and $p_I(\mathbf{y}, D, \tau)$.

**2.5.17.** ★We start our considerations by analysis of the last term, $\langle i[p_R, p_I]w, w\rangle$, in the right-hand side of (2.261). We approximate $p_R$ and $p_I$ with $p_R^0$ and $p_I^0$,

$$p_R^0 = (JD, D) - \tau^2\Big((Jd\phi(0), d\phi(0)) + 2(JH_\phi(0)\mathbf{y}, d\phi(0))\Big),$$

$$p_I^0 = 2\tau(Jd\phi, D) = 2\tau(Jd\phi(0), D) + 2\tau(JH_\phi(0)\mathbf{y}, D). \tag{2.262}$$

To analyze the difference $p_R - p_R^0$ and $p_I - p_I^0$, we introduce the classes of operators $\mathcal{R}_k^{(l)}$, where $k, l$ are non-negative integers. The class $\mathcal{R}_k^{(l)}$ consists of the operators of order $k$ with respect to $(D, \tau)$. The coefficients of these operators are smooth functions of order $O(|\mathbf{y}|^l)$

near $\mathbf{y} = \mathbf{0}$, which means that their Taylor expansions start with the terms of homogeneity $l$. Later, we will use these classes to estimate terms of the form $\langle Rw, w\rangle$, $R \in \mathcal{R}_k^{(l)}$.

In this notations,

$$p_R, p_R^0 \in \mathcal{R}_2^{(0)}, \quad \frac{p_I}{\tau}, \frac{p_I^0}{\tau} \in \mathcal{R}_1^{(0)}, \tag{2.263}$$

$$p_R - p_R^0 \in \mathcal{R}_2^{(2)}, \quad \frac{p_I - p_I^0}{\tau} \in \mathcal{R}_1^{(2)} + \mathcal{R}_0^{(0)}. \tag{2.264}$$

Direct calculations show that

$$i[p_R^0, p_I^0] = 4\tau\Pi_\phi(D, \tau) + \tau^3 R_0, \tag{2.265}$$

where $R_0 \in \mathcal{R}_0^{(1)}$. Then, the remainder term $\tau^3 R_0$ in formula (2.265) can be estimated as

$$\tau^3 |\langle R_0 w, w\rangle| \le c\delta\tau^3 \|w\|^2, \tag{2.266}$$

when supp $(w) \subset B_\delta$.

**Exercise 2.76** *Prove formula (2.265) and find the remainder $R_0$.*

As

$$[p_R, p_I] - [p_R^0, p_I^0] = \tau\left(\left[p_R - p_R^0, \frac{p_I}{\tau}\right] + \left[p_R^0, \frac{p_I}{\tau} - \frac{p_I^0}{\tau}\right]\right), \tag{2.267}$$

it follows from (2.263), (2.264), that

$$[p_R, p_I] - [p_R^0, p_I^0] = \tau(R_1 + R_2), \quad R_1 \in \mathcal{R}_1^{(0)}, \ R_2 \in \mathcal{R}_2^{(1)}.$$

Indeed, the commutators of the operators in the right-hand side of (2.267) contain derivatives of the coefficients of these operators. Each differentiation of coefficients decreases the order of the resulting operator by 1. On the other hand, differentiation of a function of order $O(|\mathbf{y}|^l)$, $l \ge 1$, produces a function of order $O(|\mathbf{y}|^{l-1})$. Returning to the expression in the right-hand side of (2.267) and taking into account (2.263)–(2.264), we see that, when we differentiate coefficients only once, we obtain an operator of order 2 with respect to $(D, \tau)$ with coefficients of order $O(|\mathbf{y}|)$, i.e. an operator from $\mathcal{R}_2^{(1)}$.

When we differentiate coefficients twice, we obtain an operator from $\mathcal{R}_1^{(0)}$.

Consider the term $\langle R_1 w, w\rangle$. Since $R_1 \in \mathcal{R}_1^{(0)}$,

$$|\langle R_1 w, w\rangle| \leq c(\tau\|w\|^2 + \|dw\|\|w\|), \quad \tau \geq 1.$$

Here, the first term in the right-hand side is due to the terms in $R_1$ of order 1 with respect to $\tau$ and of order 0 with respect to $D$, while the second term is due to the terms of order 0 with respect to $\tau$ and of order 1 with respect to $D$. Possible terms of order 0 with respect to $(D,\tau)$ are included in $\tau\|w\|^2$, since $\tau \geq 1$.

At last, by means of the Cauchy-Schwartz inequality,

$$2\|w\|\,\|dw\| \leq \frac{1}{\tau}\|dw\|^2 + \tau\|w\|^2,$$

we see that

$$|\langle R_1 w, w\rangle| \leq C\left(\tau\|w\|^2 + \frac{1}{\tau}\|dw\|^2\right) \leq \frac{C}{\tau}\|w\|^2_{(1,\tau)}. \tag{2.268}$$

If we take into account that supp $(w) \subset B_\delta$, i.e. $|\mathbf{y}| \leq c\delta$, then, by similar considerations, we obtain that

$$|\langle R_2 w, w\rangle| \leq C\delta(\|dw\|^2 + \tau^2\|w\|^2) + C\|w\|\,\|dw\| \leq \tag{2.269}$$

$$\leq C(\delta + \frac{1}{\tau})\|w\|^2_{(1,\tau)}.$$

Combining (2.268) and (2.269), we see that

$$|\langle([p_R, p_I] - [p_R^0, p_I^0])w, w\rangle| \leq C\tau\left(\delta + \frac{1}{\tau}\right)\|w\|^2_{(1,\tau)}. \tag{2.270}$$

Finally, we need also a rough estimate for the commutator $[p_R, p_I]$. Using (2.263), we see that

$$\left|\langle[p_R, p_I]w, w\rangle\right| \leq c\tau\|w\|^2_{(1,\tau)}. \tag{2.271}$$

**2.5.18.** ★ To analyze $\|p_I w\|^2$ in formula (2.261), we use the operator $p_I^c$ of the form

$$p_I^c = 2\tau(Jd\phi(0), D). \tag{2.272}$$

This operator is obtained from $p_I$ by freezing coefficients at $\mathbf{y} = 0$ and keeping only the terms of order 2 with respect to $(D, \tau)$. We remind the reader that

$$d\phi(0) = (0, \ldots 0, \partial_m\phi(0), \partial_0\phi(0)), \quad (\partial_m\phi(0))^2 - (\partial_0\phi(0))^2 \neq 0.$$

Then

$$\|p_I w\|^2 - \|p_I^c w\|^2 = \tau^2 \left( \left\langle \frac{(p_I - p_I^c)w}{\tau}, \frac{p_I w}{\tau} \right\rangle + \left\langle \frac{p_I^c w}{\tau}, \frac{(p_I - p_I^c)w}{\tau} \right\rangle \right),$$

where

$$\frac{p_I}{\tau}, \frac{p_I^c}{\tau} \in \mathcal{R}_1^{(0)}, \quad \frac{p_I - p_I^c}{\tau} \in \mathcal{R}_1^{(1)} + \mathcal{R}_0^{(0)}. \tag{2.273}$$

Estimates similar to those that yield formula (2.268) show that

$$\left| \|p_I w\|^2 - \|p_I^c w\|^2 \right| \leq c\tau^2 \left( \delta + \frac{1}{\tau} \right) \|w\|^2_{(1,\tau)}. \tag{2.274}$$

**2.5.19.** ★ To estimate

$$\frac{1}{\tau^2}\|p_R w\|^2 = \frac{1}{\tau^2}\|\Box_a w - \tau^2 (J_a d\phi, d\phi) w\|^2 \tag{2.275}$$

in formula (2.261), we start with the representation,

$$\frac{1}{\tau^2}\|p_R w\|^2 = \frac{1}{\tau^2}\|\Box_a w\|^2 - 2\Re\langle \Box_a w, (J_a d\phi, d\phi) w\rangle + \tag{2.276}$$

$$+\tau^2 \|(J_a d\phi, d\phi) w\|^2.$$

Next, we consider the homogeneous terms of order 2 with respect to $(D, \tau)$ in the second and third members in the right-hand side of (2.276). Freezing coefficients at $\mathbf{y} = 0$ and integrating by parts in $\langle \Box_a w, (J_a d\phi, d\phi) w\rangle$, we obtain that

$$\frac{1}{\tau^2}\|p_R w\|^2 = \frac{1}{\tau^2}\|\Box_a w\|^2 - 2\sigma\langle JDw, Dw\rangle +$$

$$+\tau^2\sigma^2\|w\|^2 + \langle (R_3 + R_4)w, w\rangle, \tag{2.277}$$

where

$$R_3 \in \mathcal{R}_1^{(0)}, \quad R_4 \in \mathcal{R}_2^{(1)}, \tag{2.278}$$

so that

$$|\langle (R_3 + R_4)w, w\rangle| \le c\left(\delta + \frac{1}{\tau}\right)\|w\|^2_{(1,\tau)}. \tag{2.279}$$

**2.5.20.** ★ After these technical preparations, we are ready to prove the fundamental estimate (2.260). The cases of the time-like and space-like surfaces $\phi(\mathbf{y}) = 0$ are different, and will be considered separately.

**2.5.21.** ★ The case of the time-like surface is more difficult and is considered first. We will use a number of constants $C_1, C_2$, etc. and a generic constant $c_0$, which depends on $C_1, C_2, \ldots$, but not on $\tau$ or $\delta$. In the proof, we will impose different conditions on these constants. Appropriate values of these constants are chosen at the end of the proof.

It follows from (2.261) that, for $p = p(\mathbf{y}, D, \tau)$,

$$C_1\frac{\|pw\|^2}{\tau} + C_2\frac{\|pw\|^2}{\tau^2} + C_3\|\partial_0 w\|^2 \ge$$

$$\ge C_1\frac{\langle i[p_R, p_I]w, w\rangle}{\tau} + C_2\left(\frac{\|p_R w\|^2}{\tau^2} + \frac{\|p_I w\|^2}{\tau^2}\right) + C_3\|\partial_0 w\|^2 +$$

$$+C_2\frac{\langle i[p_R, p_I]w, w\rangle}{\tau^2} \ge \tag{2.280}$$

$$\ge 4C_1\langle \Pi_\phi(D,\tau)w, w\rangle + C_4\left(\frac{\|\Box_a w\|^2 - 2\sigma\tau^2\langle JDw, Dw\rangle}{\tau^2} + \right.$$

$$\left. +\sigma^2\tau^2\|w\|^2\right) + C_2\frac{\|p_I^c w\|^2}{\tau^2} + C_3\|\partial_0 w\|^2 + \langle R_5 w, w\rangle,$$

where $C_4$ is a constant which satisfies inequality $C_4 \le C_2$. The remainder term $\langle R_5 w, w\rangle$ is a sum of the remainder terms described in sections 2.5.17–2.5.19 and the term $C_2\langle i[p_R, p_I]w, w\rangle/\tau^2$.

Combining estimates, which are similar to (2.268), with (2.271), we obtain that

$$|\langle R_5 w, w\rangle| \le c_0 \left(\delta + \frac{1}{\tau}\right) \|w\|^2_{(1,\tau)}. \tag{2.281}$$

**2.5.22.** ★ Denote by $I_1$ the expression

$$I_1 = C_2 \frac{\|p_I^c w\|^2}{\tau^2} + C_3\|\partial_0 w\|^2 =$$

$$= 4C_2\|\partial_0\phi(0)\,\partial_0 w - \partial_m\phi(0)\,\partial_m w\|^2 + C_3\|\partial_0 w\|^2.$$

Since $\partial_m\phi(0) \ne 0$, there is a constant $C_5 > 0$, such that

$$C_5 \le \min_{s^2+t^2=1} \left\{4C_2|t\partial_0\phi(0) - s\partial_m\phi(0)|^2 + C_3 t^2\right\}, \tag{2.282}$$

so that

$$I_1 \ge C_5(\|\partial_m w\|^2 + \|\partial_0 w\|^2). \tag{2.283}$$

**2.5.23.** ★ Denote by $C_6$ the constant, $C_6 = 4C_1/C_4$ and let $I_2$ contains the members in the right-hand side of (2.280), which do not appear in $I_1$ and $\langle R_5 w, w\rangle$,

$$I_2 = C_4\Bigg(C_6\langle JH_\phi(0)JDw, Dw\rangle + C_6\tau^2\eta\|w\|^2 +$$

$$+\frac{\|\Box_a w\|^2}{\tau^2} - 2\sigma\langle JDw, Dw\rangle + \sigma^2\tau^2\|w\|^2\Bigg), \tag{2.284}$$

where we have used definition (2.242) of $\Pi_\phi(D,\tau)$. To analyze the term $\langle JH_\phi(0)JDw, Dw\rangle$, we take into account that it contains a term $\langle JH_\phi(0)JD''w, D''w\rangle$, where $D''$ is interpreted as $(D'', 0, 0)$, and some other terms, which contain either second-order derivatives with respect to $(y^0, y^m)$ or mixed derivatives with respect to $(y^0, y^m)$

and $\mathbf{y}''$. Using the Cauchy-Schwartz inequality for the terms with mixed derivatives, we obtain that

$$\langle JH_\phi(0)JDw, Dw\rangle \geq \langle JH_\phi(0)JD''w, D''w\rangle -$$

$$-\epsilon\|D''w\|^2 - K_\epsilon(\|D_0w\|^2 + \|D_mw\|^2), \tag{2.285}$$

where $\epsilon > 0$ is arbitrary. Since, by definition (2.245),

$$\langle JH_\phi(0)J\xi'', \xi''\rangle \geq \gamma|\xi''|^2,$$

we come to the inequality

$$I_2 \geq C_4\Big(C_6(\gamma - \epsilon)\|D''w\|^2 + \tau^{-2}\|\Box_a w\|^2 - 2\sigma\langle JDw, Dw\rangle +$$

$$+(\sigma^2 + C_6\eta)\tau^2\|w\|^2 - K_\epsilon(\|D_0w\|^2 + \|D_mw\|^2)\Big) =$$

$$= C_4\Big(\tau^{-2}\|[\Box_a - \tau^2(\sigma - \frac{C_6\gamma}{2} + \lambda)]w\|^2 + (2\sigma - C_6\gamma + 2\lambda)\langle\Box_a w, w\rangle$$

$$+\tau^2\left[\sigma^2 + C_6\eta - (\sigma - C_6\gamma/2 + \lambda)^2\right]\|w\|^2 - 2\sigma\langle JDw, Dw\rangle +$$

$$+C_6(\gamma - \epsilon)\|D''w\|^2 - K_\epsilon(\|D_0w\|^2 + \|D_mw\|^2)\Big), \tag{2.286}$$

where $\lambda > 0$ is arbitrary. To proceed further, we will consider different members in the right-hand side of (2.286). We can write

$$\langle\Box_a w, w\rangle = \langle JDw, Dw\rangle + \langle Rw, w\rangle, \tag{2.287}$$

where $R \in \mathcal{R}_2^{(2)} + \mathcal{R}_1^{(1)}$, so that

$$|\langle Rw, w\rangle| \leq c_0\left(\delta + \frac{1}{\tau}\right)\|w\|^2_{(1,\tau)}. \tag{2.288}$$

Therefore,

$$I_2 \geq C_4\Big((-C_6\gamma + 2\lambda)\langle JDw, Dw\rangle + C_6(\gamma - \epsilon)\|D''w\|^2 +$$

$$+\tau^2\left[C_6(\eta+\sigma\gamma)-(C_6\gamma/2)^2-\lambda^2-2\sigma\lambda+C_6\gamma\lambda\right]\|w\|^2-$$

$$\left.-K_\epsilon(\|D_0w\|^2+\|D_mw\|^2)-c_0(\delta+\tau^{-1})\|w\|^2_{(1,\tau)}\right). \tag{2.289}$$

Because

$$\langle JDw, Dw\rangle=\|D''w\|^2+\|D_mw\|^2-\|D_0w\|^2, \tag{2.290}$$

we have

$$I_2\geq C_4\Big((2\lambda-C_6\epsilon)\|D''w\|^2-K'_\epsilon(\|D_0w\|^2+\|D_mw\|^2)+$$

$$+\tau^2\left[C_6(\eta+\sigma\gamma)-(C_6\gamma/2)^2-\lambda^2-2\sigma\lambda+C_6\gamma\lambda\right]\|w\|^2-$$

$$\left.-c_0(\delta+\tau^{-1})\|w\|^2_{(1,\tau)}\right)=\widetilde{I}_2, \tag{2.291}$$

where $K'_\epsilon=K_\epsilon+C_6|\gamma|+2\lambda$. Since $\Gamma$ is a time-like surface, $\sigma>0$, so that pseudo-convexity condition (2.246) implies that $\eta+\sigma\gamma>0$. Choose $C_6\leq 2(\eta+\sigma\gamma)/\gamma^2$. Then, for sufficiently small $\lambda$ and $\epsilon\leq\lambda/C_6$,

$$\widetilde{I}_2\geq C_4c_1(\|D''w\|^2+\tau^2\|w\|^2)- \tag{2.292}$$

$$-C_4K'_\epsilon(\|D_0w\|^2+\|D_mw\|^2)-C_4c_0(\delta+\tau^{-1})\|w\|^2_{(1,\tau)},$$

with some constant $c_1>0$ depending on $C_6$ and $\lambda$.

**2.5.24.** ★ Combining equations (2.281), (2.283), (2.291) and (2.292) of sections 2.5.21–2.5.23, we obtain from (2.280) that

$$C_1\frac{\|pw\|^2}{\tau}+C_2\frac{\|pw\|^2}{\tau^2}+C_3\|D_0w\|^2\geq$$

$$\geq C_4c_1(\|D''w\|^2+\tau^2\|w\|^2)-C_4K'_\epsilon(\|D_0w\|^2+\|D_mw\|^2)+$$

$$+C_5(\|D_0w\|^2+\|D_mw\|^2)-c_0(\delta+\tau^{-1})\|w\|^2_{(1,\tau)}. \tag{2.293}$$

Now we are ready to complete the proof in the time-like case. First, we choose $C_6$, $\lambda$, $\epsilon$ as above, so that inequality (2.292) is satisfied. These parameters determine also $c_1$. Next, we choose large parameters $C_4$ and then $C_5$, such that the right-hand side of (2.293) is more than $(2-c_0(\delta+\tau^{-1}))\|w\|^2_{(1,\tau)}$. Therefore, the parameter $C_1 = 4C_4C_6$ is also determined. Finally, we choose parameters $C_2 \geq C_4$ and $C_3$, so that (2.282) is valid.

At last, taking $\delta$ sufficiently small and $\tau_0$ sufficiently large, we see that

$$C_1\frac{\|pw\|^2}{\tau} + C_2\frac{\|pw\|^2}{\tau^2} + C_3\|D_0w\|^2 \geq \|w\|^2_{(1,\tau)},$$

for $\tau > \tau_0$. This yields the desired inequality (2.260), when the surface $\{\mathbf{y} : \phi(\mathbf{y}) = 0\}$ is time-like.

**2.5.25.** ★ When the surface $\{\mathbf{y} : \phi(\mathbf{y}) = 0\}$ is space-like, the proof is easier. It is based on the inequalities,

$$I_1 = C_2\frac{\|p_I^c w\|^2}{\tau^2} + C_3\|\partial_0 w\|^2 \geq C_5\|\partial_0 w\|^2,$$

and

$$\frac{\|p_R w\|^2}{\tau^2} \geq c_1(\|D'w\|^2 + \tau^2\|w\|^2) - K\|D_0w\|^2, \qquad (2.294)$$

which is valid for the space-like surfaces when $\tau$ is sufficiently large.

We would like to point out that, if $\partial_m\phi(\mathbf{0}) \neq 0$, then stronger inequality (2.283) is valid for $I_1$.

**Exercise 2.77** *Prove inequality (2.294) in the case of space-like surface and derive that, for some $C_1, C_2 > 0$,*

$$\|w\|^2_{(1,\tau)} \leq C_1\frac{\|pw\|^2}{\tau^2} + C_2\|D_0w\|^2. \qquad (2.295)$$

This completes the proof of Lemma 2.75. □

**2.5.26.** ★ To proceed further with the proof of Carleman estimate (2.252), we need some lemmas about the operator $Q^\phi_{\epsilon,\tau}$. The idea is to use the pseudo-convexity and smoothing property in the $y^0$-direction of the operator $Q^\phi_{\epsilon,\tau}$. We start with some one-dimensional estimates in the $y^0$-space. Let $\chi \in C_0^\infty(\mathbf{R})$, $0 \le \chi(y^0) \le 1$, be a cut-off function, such that $\chi(y^0) = 1$ for $|y^0| < 1$ and supp $(\chi) \subset (-2,2)$. For $0 < \kappa < 1$, we define

$$\chi_\kappa(y^0) = \chi(y^0/\kappa), \quad \widetilde{\chi}_\kappa = 1 - \chi_\kappa. \tag{2.296}$$

**Lemma 2.78** *Let $\epsilon, \kappa > 0$. Then, for any sufficiently large $\tau$ and any $w \in C_0^\infty(\mathbf{R})$,*

$$\left\|(\epsilon D_0/\tau)\exp(-\epsilon D_0^2/2\tau)w\right\| \le$$

$$\le 2\kappa\left\|\exp(-\epsilon D_0^2/2\tau)w\right\| + \exp(-\tau\kappa^2/4\epsilon)\|w\|_{(1,\tau)}.$$

**Proof.** Since $|\epsilon\xi_0/\tau| < 2\kappa$ when $\chi_\kappa(\epsilon\xi_0/\tau) \ne 0$, the Plancherel theorem implies that

$$\left\|(\epsilon D_0/\tau)\,\chi_\kappa(\epsilon D_0/\tau)\,\exp(-\epsilon D_0^2/2\tau)w\right\| \le$$

$$\le 2\kappa\left\|\exp(-\epsilon D_0^2/2\tau)w\right\|. \tag{2.297}$$

Here

$$\chi_\kappa(\epsilon D_0/\tau)u = \mathcal{F}_0^{-1}\left(\chi_\kappa\left(\epsilon\xi_0/\tau\right)\mathcal{F}_0 u\right)$$

is the cut-off operator in the Fourier variable.

Next we will show that, for large $\tau$,

$$\left\|(\epsilon D_0/\tau)\widetilde{\chi}_\kappa(\epsilon\, D_0/\tau)\exp(-\epsilon D_0^2/2\tau)w\right\| \le$$

$$\le \exp(-\tau\kappa^2/4\epsilon)\,\|w\|_{(1,\tau)}. \tag{2.298}$$

By the Plancherel theorem, this inequality is equivalent to

$$\left|\frac{\epsilon\xi_0}{\tau}\right|\,\widetilde{\chi}_\kappa\left(\epsilon\xi_0/\tau\right)\,(\xi_0^2+\tau^2)^{-1/2} \le \exp\left(\frac{\epsilon\xi_0^2}{2\tau} - \frac{\tau\kappa^2}{4\epsilon}\right). \tag{2.299}$$

The left-hand side of (2.299) vanishes if $|\epsilon\xi_0| < |\tau\kappa|$. When $|\epsilon\xi_0| \ge |\tau\kappa|$,

$$\frac{\epsilon\xi_0^2}{2\tau} - \frac{\tau\kappa^2}{4\epsilon} \ge \kappa|\xi_0|\left(\frac{1}{2}-\frac{1}{4}\right) \ge \frac{\kappa^2\tau}{4\epsilon}.$$

For fixed $\epsilon$ and $\kappa$, the left-hand side of (2.299) is uniformly bounded when $\tau > 1$. Therefore, the above estimate yields estimate (2.299), when $\tau$ is sufficiently large. Combining inequalities (2.297) and (2.298), which follows from (2.299), we obtain the claim of the lemma.

□

**2.5.27.** ★

**Lemma 2.79** *Let $\epsilon, \kappa > 0$. Then, for sufficiently large $\tau$ and any $w \in C_0^\infty((-\kappa/4, \kappa/4))$,*

$$\|\widetilde{\chi}_\kappa(y^0) \exp(-\epsilon D_0^2/2\tau) w\| \leq c \exp(-\tau\kappa^2/4\epsilon) \|w\|.$$

**Proof.** We will prove the estimate,

$$\left\| \left(\frac{y^0}{\kappa}\right)^{1/2} \widetilde{\chi}_\kappa(y^0) \exp(-\epsilon D_0^2/2\tau) w \right\| \leq c \exp(-\tau\kappa^2/4\epsilon) \|w\|,$$

which implies the lemma because $|y^0/\kappa| \geq 1$ on supp $(\widetilde{\chi}_\kappa)$. The operator $\exp(-\epsilon D_0^2/2\tau)$ is an integral operator with the kernel

$$k(y^0, z^0) = \sqrt{\frac{\tau}{2\pi\epsilon}} \exp[-\tau(y^0 - z^0)^2/2\epsilon], \quad y^0, z^0 \in \mathbf{R}.$$

By the Cauchy-Schwartz inequality,

$$\begin{aligned} \left|\exp(-\epsilon D_0^2/2\tau)\, w(y^0)\right| &\leq c\sqrt{\frac{\tau}{\epsilon}} \|w\| \left( \int_{|z^0|<\kappa/4} e^{-\tau(y^0-z^0)^2/\epsilon}\, dz^0 \right)^{\frac{1}{2}} \\ &\leq c\sqrt{\frac{\kappa\tau}{\epsilon}}\, e^{-\tau(y^0)^2/4\epsilon}\, \|w\|, \end{aligned}$$

when $|y^0| > \kappa$. Hence,

$$\left\| \left(\frac{y^0}{\kappa}\right)^{1/2} \widetilde{\chi}_\kappa(y^0) \exp(-\epsilon D_0^2/2\tau) w \right\|^2 \leq$$

$$\leq c\frac{\tau}{\epsilon} \|w\|^2 \int_{|y^0|>\kappa} e^{-\tau|y^0|^2/2\epsilon} y^0\, dy^0 \leq$$

$$\leq c e^{-\tau\kappa^2/2\epsilon}\, \|w\|^2.$$

This proves the claim for large enough $\tau$. □

**2.5.28.** ★ We return to the proof of Carleman estimate (2.252).
We will first show that

$$\exp(-\epsilon D_0^2/2\tau)\, P(\mathbf{y}, D + i\tau d\phi(\mathbf{y}))u = \tag{2.300}$$

$$= P(\mathbf{y}, D - \epsilon H_\phi(\mathbf{0})(0, \dots, 0, D_0) + i\tau d\phi(\mathbf{y}))\exp(-\epsilon D_0^2/2\tau)u.$$

Indeed,

$$\exp(-\epsilon\xi_0^2/2\tau) i\partial_{\xi_0} u = \tag{2.301}$$

$$= i\partial_{\xi_0}\left(\exp(-\epsilon\xi_0^2/2\tau)u\right) + \frac{i\epsilon\xi_0}{\tau}\exp(-\epsilon\xi_0^2/2\tau)u,$$

and, applying the inverse Fourier transform to (2.301), we obtain that

$$\exp(-\epsilon D_0^2/2\tau)(y^0 u) = (y^0 + i\epsilon D_0/\tau)\exp(-\epsilon D_0^2/2\tau)u.$$

Iterating this equation, we have

$$\exp(-\epsilon D_0^2/2\tau)[((y^0)^k u)] = (y^0 + i\epsilon D_0/\tau)^k \exp(-\epsilon D_0^2/2\tau)u. \tag{2.302}$$

Since $\phi$ is a second-order polynomial, its differential $d\phi$ depends on $\mathbf{y}$ linearly,

$$d\phi(\mathbf{y}) = H_\phi(\mathbf{0})\,\mathbf{y} + d\phi(\mathbf{0}).$$

Thus,

$$exp(-\epsilon D_0^2/2\tau)\left(D_0 + i\tau d\phi(\mathbf{y}', y^0)\right) =$$

$$= \left(D_0 - \epsilon H_\phi(0)(0, \dots, 0, D_0) + i\tau d\phi(\mathbf{y})\right) exp(-\epsilon D_0^2/2\tau). \tag{2.303}$$

The coefficients of the operator $P(\mathbf{y}, D)$ are independent of the variable $y^0$. Therefore, for a given $\mathbf{y}$, the coefficients of $P(\mathbf{y}, D+i\tau d\phi(\mathbf{y}))$ depend on $y^0$ polynomially. Equation (2.300) follows now from equations (2.302) and (2.303).

**2.5.29.** ★ Now we are ready to prove Carleman estimate (2.252).

**Theorem 2.80** *Let $\kappa$ and $\epsilon$ be sufficiently small positive numbers. Let $u \in H_0^1(B_{\kappa/16}(0))$ and $P(\mathbf{y}, D)u \in L^2(B_{\kappa/16}(0))$. Then, there is $\tau_1(\delta,\kappa)$, such that for $\tau > \tau_1(\delta,\kappa)$,*

$$\|Q^{\phi}_{\epsilon,\tau}u\|_{(1,\tau)} \leq c\tau^{-\frac{1}{2}}\|Q^{\phi}_{\epsilon,\tau}P(\mathbf{y},D)u\| + ce^{-\tau\kappa^2/16\epsilon}\|e^{\tau\phi}u\|_{(1,\tau)}, \tag{2.304}$$

*where $c$ depends only on $\epsilon$ and $\kappa$.*

**2.5.30.** ★ **Proof.** We start with $u \in C_0^\infty(B_{\kappa/8}(0))$. Let $w = e^{\tau\phi}u$. By formula (2.255),

$$e^{\tau\phi}P(\mathbf{y},D)u = P(\mathbf{y}, D + i\tau d\phi)w.$$

Thus, relation (2.300) implies that

$$Q^{\phi}_{\epsilon,\tau}P(\mathbf{y},D)u = P(\mathbf{y}, D - \epsilon H_\phi(0)(0,\ldots,0,D_0) + i\tau d\phi(\mathbf{y}))Q^{\phi}_{\epsilon,\tau}u. \tag{2.305}$$

We intend to apply Lemma 2.75 to the main term, due to $p(\mathbf{y},D)$, in the right-hand side of this equation. However, supp $(Q^{\phi}_{\epsilon,\tau}u)$ is no more compact, because $Q^{\phi}_{\epsilon,\tau}$ is not a local operator in the $y^0$-variable. Since supp $(u) \subset B_{\kappa/8}(0)$,

$$\text{supp } (Q^{\phi}_{\epsilon,\tau}u) \subset \{(\mathbf{y}', y^0) : |\mathbf{y}'| < \kappa/8\}.$$

Represent $Q^{\phi}_{\epsilon,\tau}u$ in the form

$$Q^{\phi}_{\epsilon,\tau}u = \chi_\kappa(y^0)\,Q^{\phi}_{\epsilon,\tau}u + \widetilde{\chi}_\kappa(y^0)\,Q^{\phi}_{\epsilon,\tau}u, \tag{2.306}$$

where $\chi_\kappa$, $\widetilde{\chi}_\kappa$ are of form (2.296). By Lemma 2.79,

$$\|\widetilde{\chi}_\kappa(y^0)Q^{\phi}_{\epsilon,\tau}u\| = \|\widetilde{\chi}_\kappa(y^0)\exp(-\epsilon D_0^2/2\tau)\,w\| \leq$$

$$\leq ce^{-\tau\kappa^2/4\epsilon}\|e^{\tau\phi}u\|, \tag{2.307}$$

if $\tau$ is sufficiently large. Furthermore, Lemma 2.79 implies also that

$$\|D_\alpha(\widetilde{\chi}_\kappa Q^{\phi}_{\epsilon,\tau}u)\| \leq \|D_\alpha\widetilde{\chi}_\kappa\|_\infty\,\|\widetilde{\chi}_{\kappa/2}Q^{\phi}_{\epsilon,\tau}u\| +$$

$$+\|\widetilde{\chi}_\kappa\exp(-\epsilon D_0^2/2\tau)D_\alpha(e^{\tau\phi}u)\| \leq$$

$$\leq c\kappa^{-1}e^{-\tau\kappa^2/16\epsilon}\|e^{\tau\phi}u\| + ce^{-\tau\kappa^2/4\epsilon}\|e^{\tau\phi}u\|_{(1,\tau)}. \tag{2.308}$$

Therefore, for sufficiently large $\tau > \tau_1(\kappa,\epsilon)$,

$$\|\widetilde{\chi}_\kappa Q^{\phi}_{\epsilon,\tau}u\|_{(1,\tau)} \leq c\kappa^{-1}e^{-\tau\kappa^2/16\epsilon}\|e^{\tau\phi}u\|_{(1,\tau)}. \tag{2.309}$$

**2.5.31.** ★ In view of inequality (2.309), it remains to estimate $\chi_\kappa Q^\phi_{\epsilon,\tau}u$.

By Corollary 2.73, if $\epsilon > 0$ is sufficiently small, then the function $\phi$ is pseudo-convex at 0 with respect to the symbol $p(\mathbf{y}, \xi - \epsilon H_\phi(\mathbf{0})(0,\dots,0,\xi_0))$. Hence, Lemma 2.75 remains valid for $p(\mathbf{y}, D - \epsilon H_\phi(\mathbf{0})(0,\dots,0,D_0),\tau)$ and we have the inequality,

$$\|\chi_\kappa Q^\phi_{\epsilon,\tau}u\|^2_{(1,\tau)} \le c\| D_0(\chi_\kappa Q^\phi_{\epsilon,\tau}u)\|^2 + \tag{2.310}$$

$$+c\tau^{-1}\|p(\mathbf{y}, D - \epsilon H_\phi(\mathbf{0})(0,\dots,0,D_0) + i\tau d\phi(\mathbf{y}))(\chi_\kappa Q^\phi_{\epsilon,\tau}u)\|^2.$$

Clearly,

$$\|P_1(\mathbf{y}, D - \epsilon H_\phi(\mathbf{0})(0,\dots,0,D_0),\tau)\chi_\kappa Q^\phi_{\epsilon,\tau}u\| \le c\|\chi_\kappa Q^\phi_{\epsilon,\tau}u\|_{(1,\tau)},$$

since $P_1(\mathbf{y}, D,\tau) = P(\mathbf{y}, D,\tau) - p(\mathbf{y}, D,\tau)$ is a first-order operator in $(D,\tau)$. Hence, inequality (2.310) implies that

$$\|\chi_\kappa Q^\phi_{\epsilon,\tau}u\|^2_{(1,\tau)} \le c\tau^{-1}\|\chi_\kappa Q^\phi_{\epsilon,\tau}u\|^2_{(1,\tau)} + c\| D_0(\chi_\kappa Q^\phi_{\epsilon,\tau}u)\|^2 +$$

$$+c\tau^{-1}\|P(\mathbf{y}, D - \epsilon H_\phi(\mathbf{0})(0,\dots,0,D_0) + i\tau d\phi(\mathbf{y}))(\chi_\kappa Q^\phi_{\epsilon,\tau}u)\|^2.$$

Thus,

$$\|\chi_\kappa Q^\phi_{\epsilon,\tau}u\|^2_{(1,\tau)} \le c\| D_0(\chi_\kappa Q^\phi_{\epsilon,\tau}u)\|^2 + \tag{2.311}$$

$$+c\tau^{-1}\|P(\mathbf{y}, D - \epsilon H_\phi(\mathbf{0})(0,\dots,0,D_0) + i\tau d\phi(\mathbf{y}))(\chi_\kappa Q^\phi_{\epsilon,\tau}u)\|^2,$$

when $\tau$ is large enough.

Consider the first term in the right-hand side of (2.311). Then,

$$P(\mathbf{y}, D - \epsilon H_\phi(\mathbf{0})(0,\dots,0,D_0) + i\tau d\phi(\mathbf{y}))\,(\chi_\kappa Q^\phi_{\epsilon,\tau}u) = \tag{2.312}$$

$$= \chi_\kappa P(\mathbf{y}, D - \epsilon H_\phi(\mathbf{0})(0,\dots,0,D_0) + i\tau d\phi(\mathbf{y}))\, Q^\phi_{\epsilon,\tau}u + N(Q^\phi_{\epsilon,\tau}u),$$

where $N$ is the commutator

$$N = \left[P(\mathbf{y}, D - \epsilon H_\phi(\mathbf{0})(0,\dots,0,D_0) + i\tau d\phi(\mathbf{y}))\,,\ \chi_\kappa(y^0)\right].$$

However, $N$ is a first-order operator in $(D,\tau)$ and, therefore, for some $c_3$ depending on $\kappa$,

$$\|N(Q^{\phi}_{\epsilon,\tau}u)\| \leq c_3\left(\|\chi_{\kappa}Q^{\phi}_{\epsilon,\tau}u\|_{(1,\tau)} + \|\widetilde{\chi}_{\kappa}Q^{\phi}_{\epsilon,\tau}u\|_{(1,\tau)}\right) \leq$$

$$\leq c_3\|\chi_{\kappa}Q^{\phi}_{\epsilon,\tau}u\|_{(1,\tau)} + c\kappa^{-1}e^{-\tau\kappa^2/16\epsilon}\|e^{\tau\phi}u\|_{(1,\tau)}, \tag{2.313}$$

where we have used formula (2.309). Analogously, with some $c_4$ depending on on $\kappa$

$$\|\,D_0\chi_{\kappa}Q^{\phi}_{\epsilon,\tau}u\,\| \leq \|\,D_0Q^{\phi}_{\epsilon,\tau}u\,\| + c\kappa^{-1}e^{-\tau\kappa^2/16\epsilon}\|e^{\tau\phi}u\|_{(1,\tau)}| + \tag{2.314}$$

$$+\frac{c_4}{\tau}\|\chi_{\kappa}Q^{\phi}_{\epsilon,\tau}u\|_{(1,\tau)}.$$

Combining equations (2.312) and (2.305) with inequalities (2.311) and (2.313), we see that

$$\|\chi_{\kappa}Q^{\phi}_{\epsilon,\tau}u\|_{(1,\tau)} \leq c\tau^{-1/2}(\|Q^{\phi}_{\epsilon,\tau}P(\mathbf{y},D)u\| + c_3\|\chi_{\kappa}Q^{\phi}_{\epsilon,\tau}u\|_{(1,\tau)}) +$$

$$+c\|\,D_0Q^{\phi}_{\epsilon,\tau}u\,\| + c\kappa^{-1}e^{-\tau\kappa^2/16\epsilon}\|e^{\tau\phi}u\|_{(1,\tau)}. \tag{2.315}$$

For sufficiently large $\tau$ this inequality implies that

$$\|\chi_{\kappa}Q^{\phi}_{\epsilon,\tau}u\|_{(1,\tau)} \leq c\tau^{-1/2}\|Q^{\phi}_{\epsilon,\tau}P(\mathbf{y},D)u\| +$$

$$+c\kappa^{-1}e^{-\tau\kappa^2/16\epsilon}\|e^{\tau\phi}u\|_{(1,\tau)} + c_5\|\,D_0Q^{\phi}_{\epsilon,\tau}u\|. \tag{2.316}$$

Employing Lemma 2.78, we see that

$$\|\,D_0Q^{\phi}_{\epsilon,\tau}u\,\| \leq \frac{c\kappa\tau}{\epsilon}\|Q^{\phi}_{\epsilon,\tau}u\| + \frac{c\tau}{\epsilon}e^{-\tau\kappa^2/4\epsilon}\|e^{\tau\phi}u\|_{(1,\tau)}. \tag{2.317}$$

Decreasing $\kappa$ if necessary, inequalities (2.316) and (2.317) yield that

$$\|\chi_{\kappa}Q^{\phi}_{\epsilon,\tau}u\|_{(1,\tau)} \leq c\tau^{-1/2}\|Q^{\phi}_{\epsilon,\tau}P(\mathbf{y},D)u\| + \frac{\tau}{2}\|\,Q^{\phi}_{\epsilon,\tau}u\| +$$

$$+\frac{c}{\kappa}e^{-\tau\kappa^2/16\epsilon}\|e^{\tau\phi}u\|_{(1,\tau)} \tag{2.318}$$

for sufficiently large $\tau$. Combining inequalities (2.309) and (2.318), we obtain estimate (2.304) for $u \in C_0^{\infty}(B_{\kappa/8}(0))$.

**2.5.32.** ★ In this section, we will complete the proof of Theorem 2.80. Let $u \in H_0^1(B_\rho(0))$, $P(\mathbf{y}, D)u \in L^2(B_\rho(0))$, where $\rho < \kappa/8$. Consider a mollification $u_s$ of the function $u$,

$$u_s(\mathbf{y}', y^0) = u * \psi_s = \int_{\mathbf{R}} u(\mathbf{y}', y^0 - t)\psi_s(t)\, dt,$$

where $\psi_s(t) = s^{-1}\psi(t/s)$ and $\psi \in C_0^\infty(\mathbf{R})$ satisfies $\int \psi(t)dt = 1$. Clearly,

$$u_s \in C_0^\infty(\mathbf{R}; H_0^1(B'_\rho(0'))). \tag{2.319}$$

As the coefficients of $P(\mathbf{y}, D)$ are independent of $y^0$,

$$P(\mathbf{y}, D)u_s = (P(\mathbf{y}, D)u) * \psi_s \to P(\mathbf{y}, D)u \quad \text{when } s \to 0 \tag{2.320}$$

in $L^2(B_{\kappa/8}(0))$. Moreover, relation (2.320) shows that $P(\mathbf{y}, D)u_s \in C_0^\infty(\mathbf{R}; L^2(B'_\rho(0')))$. Therefore,

$$a(\mathbf{y}, D)u_s = P(\mathbf{y}, D)u_s - \partial_0^2 u_s \in C_0^\infty(\mathbf{R}; L^2(B'_\rho(0'))). \tag{2.321}$$

Combining (2.319) and (2.321) and using Gårding Theorem 2.22, we see that $u_s \in C_0^\infty(\mathbf{R}; H^2(B'_\rho(0')))$. As supp $(u_s) \subset B_{\kappa/8}(0)$ for sufficiently small $s$, we can approximate $u_s$ in $C_0^\infty(\mathbf{R}; H^2(B'_\rho(0))$ by smooth functions from $C_0^\infty(B_{\kappa/8}(0))$. These smooth approximations satisfy estimate (2.304) and, by closure arguments, $u_s$ also satisfy this estimate. As $u_s \to u$ in $H_0^1$ and, by relation (2.320), $P(\mathbf{y}, D)u_s \to P(\mathbf{y}, D)u_s$ in $L^2$, estimate (2.304) holds true for $u \in H_0^1(B'_\rho(0))$ with $P(\mathbf{y}, D)u \in L^2(B_\rho(0))$, for any $\rho < \kappa/8$, in particular, for $\rho = \kappa/16$. □

**2.5.33.** ★ Using Carleman estimate (2.252), we can easily complete the proof of Theorem 2.66.

Let $u$ be a solution of hyperbolic equation (2.220). Let $\chi \in C_0^\infty(B_{\kappa/16}(0))$ be a cut-off function such that $\chi(\mathbf{y}) = 1$ in a ball $B_\rho(0)$, $\rho < \kappa/16$. By Lemma 2.72,

$$\text{supp } (u) \subset \{\mathbf{y} : \ \phi(\mathbf{y}) < 0\} \cup \{0\}.$$

(see Fig. 2.11). Since

$$\text{supp } (P(\mathbf{y}, D)(\chi u)) \subset \text{supp } (u) \backslash B_\rho(0),$$

there is $c_0 > 0$, such that $\phi(\mathbf{y}) \leq -c_0$ on supp $(P(\mathbf{y}, D)(\chi u))$. However, $P(\mathbf{y}, D)u = 0$, so that

$$\|e^{\tau\phi} P(\mathbf{y}, D)(\chi u)\| \leq c e^{-c_0\tau} \|u\|_1.$$

Clearly,

$$\|e^{-\epsilon D_0^2/2\tau} v\| \leq \|v\|,$$

for any $v \in L^2$. Taking $v = e^{\tau\phi} P(\mathbf{y}, D)(\chi u)$, we see that Theorem 2.80, with $u$ replaced by $\chi u$, implies that

$$\|Q^\phi_{\epsilon,\tau}(\chi u)\|_{(1,\tau)} \leq c\tau^{-1/2} \|Q^\phi_{\epsilon,\tau} P(\mathbf{y}, D)(\chi u)\| + c\tau e^{-\tau\kappa^2/16\epsilon} \|u\|_1 \leq$$

$$\leq c e^{-c'\tau} \|u\|_1,$$

for sufficiently large $\tau$. Hence,

$$\|e^{-\epsilon D_0^2/\tau} e^{\tau(\phi + c')} (\chi u)\| \leq C \|u\|_1,$$

when $\tau$ is large enough. By Lemma 2.67, this implies that $u = 0$ in $\{\mathbf{y} : \phi(\mathbf{y}) > -c'\}$, which is an open neighborhood of $\mathbf{y} = 0$ (see Fig. 2.11). □

**Notes.** Section 2.1 contains rather standard material on differential and Riemannian geometries. A complete exposition, including proofs, of this material can be found in numerous textbooks on the subject, e.g., [BuZa], [Cv], [GrKlMe], [Kl], [GaHuLa], [Sp]. Our presentation pays special attention to manifolds with boundary and different phenomena related to boundary, e.g., the boundary normal coordinates, cut locus with respect to the boundary, etc. This material is not traditional in the textbooks on Riemannian geometry. However, in many cases, structures on manifolds with boundary have direct analogy to manifolds without boundaries. For instance, construction of the normal and boundary normal coordinates and the demonstration of the Klingenberg lemma are completely parallel.

Our main references for section 2.2 are [Ev], [Ld], [EgSb]. Namely, second-order elliptic differential operators are considered in [Ev], Chapter 6, [Ld], Chapter 2, [EgSb], Chapters 3-4. Hyperbolic initial-boundary values problems are discussed in [Ev], section 7.2, and [Ld],

Chapter 4. However, some of the results, e.g., the asymptotic properties of the eigenfunctions and the Fourier expansions in different Sobolev spaces, cannot be found in these textbooks. More advanced material on elliptic differential operators is contained in [Ag2], [Ho2], and [LiMg]. For instance, the interpolation result used in section 2.3 can be found in [LiMg].

The basic references for section 2.3 are the same as for section 2.2. However, the case of the inhomogeneous boundary conditions and the behaviour of the normal derivatives of the waves on the boundary, cannot be found in these textbooks. For these and related results, we refer to the original papers [LsLiTr] (see also [LsTr1]–[LsTr2]).

The Gaussian beams, which are considered in section 2.4 are, representatives of a new class of the WKB-type asymptotic solutions, which have a complex phase. There are no textbooks devoted to this subject (see, however, [BaBuMo]. In the time-harmonic case, an analog of the Gaussian beams was developed in [BaPa]. The nonstationary case appeared later [BaUl], [Ka], [Rl]. Different types of Gaussian beams were used for constructing the parametrics for the wave equations [Rl], for developing efficient numeric algorithms to compute high-frequency wave fields [KaPo1], [KaPo2], etc. This type of solution was developed also for the Maxwell and elasticity system of equations [Po], [KaPo2], [KaPo3], [Nm].

Unique continuation results for solutions of partial differential equations with analytic coefficients are based, in general, on the Cauchy-Kovalevskaya theorem, see, e.g., [Ru], [Ho1]. When coefficients are non-analytic, the existing methods to prove the unique continuation go back to Carleman [Cr]. They were developed to prove numerous unique continuation results for elliptic and parabolic systems, see, e.g., [Ho1] and literature cited here. In the hyperbolic case, the method developed in [Ho1] is suitable to prove the unique continuation across a space-like surface. The problem of unique continuation across a time-like surface turns out to be more complicated. First results in this direction were obtained in [Rb], [Ho3], see also [RuTy]. The complete solution of the unique continuation problem for the wave equation with time-analytic coefficients that involve a number of important new ideas was given by Tataru [Ta1], see also further developments in [Ta2], [Ho3], [RbZu]. The scheme used in section 2.5 is a modification of the scheme in [Ho4]. However, in our exposition, we use only the energy-type estimates, avoiding the pseudodifferential calculus. In recent years, the technique of Carleman estimates was extended to boundary value problems [Ta3] and certain hyperbolic systems [EIsNkTa]. We mention also that the technique of the Carleman estimates is now widely used to study multidimensional inverse problems, see, e.g., [Is2], [Is3], [Kb], [Ya], [ImYa]. More detailed discussions of the Carleman-type technique for inverse problems and further references are given in [Is1], [Is4].

The most commonly used versions of Phragmen-Lindelöf principle and Paley-Wiener theorem in section 4.5 can be found in [Hi], [Rd].

# Chapter 3

# Gel'fand inverse boundary spectral problem for manifolds

## 3.1. Formulation of the problem and the main result

**3.1.1.** This is the main chapter of the book. Here, we generalize for the multidimensional case the method described in Chapter 1 for the one-dimensional case. We consider a multidimensional inverse boundary spectral problem for the general second-order self-adjoint operator on a manifold with boundary. This problem is the direct multidimensional analog of Problem 1 of section 1.1.3. On the one hand, this problem is quite general and embraces a large number of specific inverse problems that occur in practical applications, e.g., the inverse problem for a Schrödinger operator in a domain of an Euclidean space, or for an anisotropic conductivity operator, see section 4.5. On the other hand, the unified approach to all these different inverse problems elucidates the essential features of these problems.

**3.1.2.** As observed in Chapter 1, inverse problems should be considered taking into account all admissible transformations that preserve the structure of the problem, i.e., from the invariant point of view. Physically, this means that, although the same physical process can be described by different models, all these models should possess some characteristic features that are invariant with respect to the admissible transformations. The goal of inverse problems is to find these characteristic features. There are two types of transformations that should be taken into account, namely, changes of coordinates and gauge transformations. Both have fundamental physical meaning. Indeed, changes of coordinates correspond to re-scaling of the independent variables $\mathbf{x}$, i.e., the space variables, while gauge transformations correspond to re-scaling of the dependent variable, i.e., the field $u$. As we will show, they also have an adequate mathematical meaning. In the one-dimensional case, changes of coordinates are irrelevant, as the geometry remains essentially the same. Indeed, we always have some finite interval of the real axis. In the multi-dimensional case, the underlying geometry becomes much more complicated, for instance, a domain can contain holes or it may lie on a curved surface. It is the concept of a manifold that covers all these cases. That is why we consider inverse problems on manifolds. We understand that a manifold is a rather abstract concept that does not look natural in practical problems used to find physical parameters in specific domains. However, on the one hand, in a number of applications, the reconstruction is not carried out globally in Euclidean coordinates, but locally in some special coordinates, e.g., travel-time coordinates. Mathematically speaking, this corresponds to the reconstruction of the underlying manifold structure in some distance coordinates, e.g., the boundary normal coordinates. On the other hand, when an actual domain lies in an Euclidean space, the reconstruction of a manifold and an operator on it is only the first step in the solution of the inverse problem. The second step consists of the reconstruction of the unknown operator in the Euclidean domain and is based on the results obtained at the first step (see, e.g., section 4.5).

What does it mean to reconstruct a Riemannian manifold? As we know from section 2.1.5, isometric Riemannian manifolds are identical from the point of view of Riemannian geometry. Therefore, to reconstruct a Riemannian manifold means to construct some Riemannian manifold that is isometric to the original one. This is exactly what we will do in this chapter. We will develop a procedure to construct a particular Riemannian manifold from the boundary spectral data and then show that this manifold is isometric to the considered one.

The role of the gauge transformations is related to re-scaling of the solution. This re-scaling may vary from point to point so that the values of the field are multiplied by a smooth positive function. When the structure of the operator is known *a priori*, this may decrease the set of admissible gauge transformations. For instance, these re-scalings may be non-trivial only in the interior of the manifold, i.e., the corresponding re-scaling function is equal to 1 on $\partial\mathcal{M}$. Thus, the values of the solutions do not change on the boundary.

In many practical applications, we know both the structure of the operator and the domain in the Euclidean space. Sometimes this additional information is sufficient to find the operator uniquely. This takes place, for instance, for a Schrödinger operator, $-\Delta+q(x)$. This, and other examples of operators in $\mathbf{R}^m$, will be considered in section 4.5.

The above discussion implies that we look at inverse problems from an invariant point of view. Therefore, although the inverse boundary spectral problem for a general anisotropic operator does not have a unique solution, we can still construct the underlying manifold and the class of operators, which are gauge equivalent to the original one. To stress the invariant nature of the approach to inverse problems, which is developed later in this chapter, we will introduce and study inverse problems with the boundary spectral data given in the gauge invariant form.

**3.1.3.** This section is devoted to the rigorous formulation of the problems that we consider in this chapter.

Let $\mathcal{M}$ be a manifold with boundary $\partial\mathcal{M}$ and $\mathcal{A}$ be a second-order elliptic differential operator with the Dirichlet boundary condition. We assume that $\mathcal{A}$ is self-adjoint in $L^2(\mathcal{M}, dV)$. As shown in sections 2.2.3–2.2.4, $\mathcal{A}$ has the form

$$\mathcal{A}u = a(\mathrm{x}, D)u = -g^{-1/2}m^{-1}\partial_j(g^{1/2}mg^{jk}\partial_k u) + qu, \qquad (3.1)$$

$$\mathcal{D}(\mathcal{A}) = H^2(\mathcal{M}) \cap H_0^1(\mathcal{M}), \qquad (3.2)$$

$$dV = m dV_g = mg^{1/2}dx^1 \wedge \cdots \wedge dx^m. \qquad (3.3)$$

Denoting the eigenvalues and the orthonormal eigenfunctions of $\mathcal{A}$ by $\lambda_j$ and $\varphi_j$, $j = 1, 2, \ldots$, we come to the following definition.

**Definition 3.1** *The collection*

$$\{\partial\mathcal{M},\ \lambda_j,\ \partial_\nu\varphi_j|_{\partial\mathcal{M}},\ j=1,2,\dots\}$$

*is called the boundary spectral data of* $(\mathcal{M},\mathcal{A})$.

Here $\partial_\nu$ is the normal derivative with respect to the metric associated with operator $\mathcal{A}$, i.e,.

$$\partial_\nu u = \nu^j\partial_j u,$$

where $\nu = \nu^j\frac{\partial}{\partial x^j}$ is the unit inward normal vector to $\partial\mathcal{M}$.

**3.1.4.** As we have shown in sections 2.2.7–2.2.9, a gauge transformation $S_\kappa$ does not change the eigenvalues, while the eigenfunctions are multiplied by $\kappa$. Therefore, if $\mathcal{A}_\kappa = S_\kappa(\mathcal{A})$, then the boundary spectral data of $\mathcal{A}_\kappa$ is the collection

$$\{\partial\mathcal{M},\ \lambda_j,\ \partial_\nu\varphi_j^\kappa|_{\partial\mathcal{M}},\ j=1,2,\dots\} =$$

$$= \{\partial\mathcal{M},\ \lambda_j,\ \kappa_0\partial_\nu\varphi_j|_{\partial\mathcal{M}},\ j=1,2,\dots\},$$

where $\kappa_0 = \kappa|_{\partial\mathcal{M}}$.

**Definition 3.2** *The boundary spectral data* $\{\partial\mathcal{M},\ \lambda_j,\ \partial_\nu\varphi_j|_{\partial\mathcal{M}} : j = 1,2,\dots\}$ *and* $\{\partial\widetilde{\mathcal{M}},\ \widetilde{\lambda}_j,\ \partial_\nu\widetilde{\varphi}_j|_{\partial\mathcal{M}} :\ j=1,2,\dots\}$ *are gauge equivalent if* $\partial\mathcal{M} = \partial\widetilde{\mathcal{M}}$, $\lambda_j = \widetilde{\lambda}_j$ *and there is a smooth positive function* $\kappa_0$ *on* $\partial\mathcal{M}$ *such that*

$$\partial_\nu\widetilde{\varphi}_j|_{\partial\mathcal{M}} = \kappa_0\partial_\nu\varphi_j|_{\partial\mathcal{M}},\ j=1,2,\dots$$

**3.1.5.** We have observed that, if two operators on the same manifold are gauge equivalent, their boundary spectral data are gauge equivalent. The natural inverse boundary spectral problem is the following.

**Problem 3** *Assume that two pairs* $(\mathcal{M},\mathcal{A})$ *and* $(\widetilde{\mathcal{M}},\widetilde{\mathcal{A}})$ *have the gauge equivalent boundary spectral data. Does it follow that manifolds* $\mathcal{M}$ *and* $\widetilde{\mathcal{M}}$ *coincide and that* $\mathcal{A}$ *and* $\widetilde{\mathcal{A}}$ *are gauge equivalent, i.e., lie in the same orbit of the group of gauge transformations?*

The answer to this question is positive:

**Theorem 3.3** *Let*

$$\{\partial\mathcal{M},\ \lambda_j,\ \partial_\nu\varphi_j|_{\partial\mathcal{M}},\ j=1,2,\dots\}$$

*and*

$$\{\partial\widetilde{\mathcal{M}},\ \widetilde{\lambda}_j,\ \partial_\nu\widetilde{\varphi}_j|_{\partial\mathcal{M}},\ j=1,2,\dots\}$$

*be the boundary spectral data of pairs $(\mathcal{M},\mathcal{A})$ and $(\widetilde{\mathcal{M}},\widetilde{\mathcal{A}})$. If these boundary spectral data are gauge equivalent, then $\mathcal{M}=\widetilde{\mathcal{M}}$ and $\mathcal{A}$ and $\widetilde{\mathcal{A}}$ are gauge equivalent, i.e., lie in the same orbit of the group of gauge transformations, $\widetilde{\mathcal{A}}\in\sigma(\mathcal{A})$.*

**3.1.6.** Theorem 3.3 is the main result of this chapter. To prove this theorem, we use the observation made in section 2.2.10. It shows that, in any orbit of the group of gauge transformations, there is a unique Schrödinger operator. This observation makes it possible to reduce Problem 3, which is posed for a general operator, to the corresponding inverse problem for a Schrödinger operator.

**Problem 4** *Let*

$$\{\partial\mathcal{M},\ \lambda_j,\ \partial_\nu\varphi_j|_{\partial\mathcal{M}},\ j=1,2,\dots\}$$

*be the boundary spectral data of a Schrödinger operator $-\Delta_g+q$. Do these data determine uniquely the Riemannian manifold $(\mathcal{M},g)$ and the potential $q$?*

Moreover, we give an answer to a more general question.

**Problem 5** *Assume that we are given a collection,*

$$\{\partial\mathcal{M},\ \lambda_j,\ \kappa_0\partial_\nu\varphi_j|_{\partial\mathcal{M}},\ j=1,2,\dots\},$$

*where $\lambda_j$ and $\partial_\nu\varphi_j|_{\partial\mathcal{M}}$ are the eigenvalues and normal derivatives on $\partial\mathcal{M}$ of the orthonormal eigenfunctions of a Schrödinger operator $-\Delta_g+q$ and $\kappa_0$ is an unknown smooth positive function on $\partial\mathcal{M}$. Do these data, which we call the gauge equivalent boundary spectral data, determine uniquely the Riemannian manifold $(\mathcal{M},g)$ and the potential $q$?*

The answers to both these questions are positive. Actually, we do not only prove the uniqueness but also develop a procedure to construct the Riemannian manifold and the Schrödinger operator on it.

Let us explain how the proof of Theorem 3.3 follows from the positive answer to Problem 5. Assume that the boundary spectral data that correspond to the operators $\mathcal{A}$ and $\widetilde{\mathcal{A}}$ are gauge equivalent. Then the Schrödinger operators, which correspond to $\mathcal{A}$ and $\widetilde{\mathcal{A}}$, have the gauge equivalent boundary spectral data. The positive answer for Problem 5 implies that these Riemannian Schrödinger operators coincide. Hence, $\mathcal{M} = \widetilde{\mathcal{M}}$ and $\mathcal{A}$ and $\widetilde{\mathcal{A}}$ lie in the orbit of the group of gauge transformations through this Riemannian Schrödinger operator. Hence, $\mathcal{A}$ and $\widetilde{\mathcal{A}}$ are gauge equivalent.

**3.1.7.** The difference between Problem 3 and Problem 4 clarifies the effect of *a priori* data in the invariant approach to inverse problems. Problem 3 and Theorem 3.3 deal with the general operators and give an answer in terms of the classes of equivalence of operators. In Problem 4 we have additional *a priori* information that specifies the structure of the operator and makes the inverse problem uniquely solvable. Clearly, there are intermediate cases, when we know some *a priori* information that does not make the problem uniquely solvable but decreases the admissible group of transformations. This is clarified by the following examples.

**Example 3.4** Assume that we know the boundary spectral data

$$\{\partial\mathcal{M},\ \lambda_j,\ \partial_\nu\varphi_j|_{\partial\mathcal{M}},\ j = 1, 2, \dots\},$$

rather than the class of the gauge equivalent boundary spectral data. It then follows from formula (2.55) and Theorem 3.3 that these data determine $\mathcal{M}$ and the class of operators $\sigma_0(\mathcal{A})$, where

$$\sigma_0(\mathcal{A}) = \{S_\kappa\mathcal{A} :\ \kappa > 0,\ \kappa|_{\partial\mathcal{M}} = 1\},$$

rather than the wider class $\sigma(\mathcal{A})$. Actually, $\sigma_0(\mathcal{A})$ is an orbit of the subgroup of gauge transformations that are normalized on the boundary.

**Example 3.5** Consider a more traditional inverse boundary spectral problem for a differential operator $\mathcal{A} = -c(\mathbf{x})^2\Delta + q(\mathbf{x})$ in a given bounded domain $\Omega \subset \mathbf{R}^2$. This operator is a Schrödinger operator with respect to the metric $g^{ij}(\mathbf{x}) = c^2(\mathbf{x})\delta^{ij}$. Therefore, by solving Problem 4, we can construct a Riemannian manifold $(\mathcal{M}, g)$ and a Schrödinger operator on it. This Riemannian manifold and the Schrödinger operator can be considered an abstract representation of the operator $\mathcal{A}$. In addition, we know that $\mathcal{A}$ is an operator of the form $-c(\mathbf{x})^2\Delta + q(\mathbf{x})$ in a given domain $\Omega \subset \mathbf{R}^2$, so that the corresponding metric is conformally Euclidean. Using this information we can uniquely reconstruct $c(\mathbf{x})$ and $q(\mathbf{x})$ in $\Omega$ from its abstract representation. This fact and some other examples are considered in detail in section 4.5.

## 3.2. Fourier coefficients of waves

In this and the next sections, we give a solution of the inverse boundary spectral problem for a Schrödinger operator. We start with the formulae for the Fourier coefficients of waves that are generated by boundary sources.

**3.2.1.** Consider the initial-boundary value problem (2.6)–(2.9) with $F = 0$, $\psi_0 = \psi_1 = 0$,

$$\partial_t^2 u - \Delta_g u + q(\mathbf{x})u = 0, \quad \text{in} \quad Q^T = \mathcal{M} \times [0, T], \tag{3.4}$$

$$u|_{\Sigma^T} = f(t), \quad \Sigma^T = \partial\mathcal{M} \times [0, T], \tag{3.5}$$

$$u|_{t=0} = \partial_t u|_{t=0} = 0. \tag{3.6}$$

By Theorem 2.30 and Lemma 2.42, we know that

$$u \in C([0, T]; H^1(\mathcal{M})) \cap C^1([0, T]; L^2(\mathcal{M})),$$

when $f \in H^1(\Sigma^T)$, $f|_{t=0} = 0$, and $u \in C([0, T]; L^2(\mathcal{M}))$ when $f \in L^2(\Sigma^T)$. Here and later we denote by $u^f(t)$ the solution of problem (3.4)–(3.6) corresponding to a boundary source $f$. The Fourier expansion of the wave $u^f(t)$ is given by the formula

$$u^f(t) = \sum_{k=1}^{\infty} u_k^f(t)\varphi_k, \tag{3.7}$$

where the Fourier coefficients $u_k^f(t) \in C^1([0,T])$ when $f \in H^1(\Sigma^T)$, $f|_{t=0} = 0$, and $u_k^f \in C([0,T])$ when $f \in L^2(\Sigma^T)$. Moreover, we have the following representation.

**3.2.2.**

**Lemma 3.6** *Let $u^f(t)$ be the solution of (3.4)–(3.6), where $f \in L^2(\Sigma^T)$. Then for $k = 1, 2, \ldots,$*

$$u_k^f(t) = \int_0^t \int_{\partial\mathcal{M}} f(\mathbf{z}, t') s_k(t - t') \partial_\nu \varphi_k(\mathbf{z})\, dS_g(\mathbf{z}) dt'. \tag{3.8}$$

*We remind the reader that $dS_g$ is the volume element of the boundary $\partial\mathcal{M}$, which is induced by the metric $g$, and*

$$s_k(t) = \begin{cases} \frac{\sin\sqrt{\lambda_k}t}{\sqrt{\lambda_k}}, & \lambda_k > 0, \\ t, & \lambda_k = 0, \\ \frac{\sinh\sqrt{|\lambda_k|}t}{\sqrt{|\lambda_k|}}, & \lambda_k < 0. \end{cases} \tag{3.9}$$

**Proof.** Let $f \in C^\infty(\Sigma^T)$ and $\partial_t^p f|_{t=0} = 0$ for all $p = 0, 1, \ldots$. Then by Theorem 2.45 $u^f \in C^\infty(Q^T)$ and $u_k^f \in C^\infty([0,T])$. Moreover,

$$\partial_t^p u^f(\mathbf{x}, t)|_{t=0} = 0, \ p = 0, 1, \ldots,$$

and

$$\partial_t^p u_k^f(t)|_{t=0} = 0, \ p = 0, 1, \ldots$$

Therefore,

$$\frac{d^2}{dt^2} u_k^f(t) = \int_{\mathcal{M}} \partial_t^2 u^f(\mathbf{x}, t) \varphi_k(\mathbf{x}) dV_g$$

$$= \int_{\mathcal{M}} (\Delta_g u^f(\mathbf{x}, t) - q(\mathbf{x}) u^f(\mathbf{x}, t)) \varphi_k(\mathbf{x}) dV_g$$

$$= -\int_{\partial\mathcal{M}} (\partial_\nu u^f(\mathbf{x}, t) \varphi_k(\mathbf{x}) - u^f(\mathbf{x}, t) \partial_\nu \varphi_k(\mathbf{x})) dS_g -$$

$$-\lambda_k \int_{\mathcal{M}} u^f(\mathbf{x}, t) \varphi_k(\mathbf{x}) dV_g$$

$$= \int_{\partial \mathcal{M}} f(\mathbf{x}, t) \partial_\nu \varphi_k(\mathbf{x}) dS_g - \lambda_k u_k^f(t). \tag{3.10}$$

Equation (3.10) is an ordinary differential equation for $u_k^f(t)$. Solving this equation together with the initial conditions $u_k^f(0) = \partial_t u_k^f(0) = 0$, we obtain representation (3.8) for smooth $f$'s. Lemma 2.42 implies the continuous dependence of $u^f(t)$ in $C([0,T]; L^2(\mathcal{M}))$ on the boundary source $f \in L^2(\Sigma^T)$. As even $C_0^\infty(\Sigma^T)$ is dense $L^2(\Sigma^T)$, representation (3.8) is valid for any $f \in L^2(\Sigma^T)$. □

Lemma 3.6 means that, given the boundary spectral data of a Schrödinger operator and the volume element $dS_g$, which is induced on $\partial\mathcal{M}$ by the metric $g$, it is possible to find the Fourier coefficients of any wave $u^f(t)$. This observation is crucial for the following fundamental result.

**3.2.3.**

**Theorem 3.7** *Let $u^f(t)$ and $u^h(s)$ be the solutions of system (3.4)–(3.6), which are generated by boundary sources $f, h \in L^2(\Sigma^T)$. Then, for any $0 \le t, s \le T$, the inner products of the waves $u^f(t)$ and $u^h(s)$ may be found by the formula,*

$$\langle u^f(t), u^h(s)\rangle = \sum_{k=1}^{\infty} u_k^f(t)\overline{u_k^h(s)}, \tag{3.11}$$

*where the Fourier coefficients $u_k^f(t)$ are given by formula (3.8) and the Fourier coefficients $u_k^h(s)$ by the same formula with $f$ replaced by $h$.*

**3.2.4.** In the inverse boundary spectral Problem 5, we do not assume the knowledge of the volume element $dS_g$. Namely, in the boundary spectral data, the boundary $\partial\mathcal{M}$ is considered just as a differentiable manifold. However, we can always choose a smooth positive volume element $d\mu$ on $\partial\mathcal{M}$. Thus, by the Radon-Nikodym theorem, there is a function $\eta \in C^\infty(\partial\mathcal{M})$, $\eta > 0$ such that

$$d\mu = \eta\, dS_g. \tag{3.12}$$

Moreover, we know the functions $\partial_\nu\varphi_k|_{\partial\mathcal{M}}$ only up to some unknown positive factor $\kappa_0$, $\kappa_0 \in C^\infty(\partial\mathcal{M})$. Hence, instead of the Fourier coefficients of the wave $u^f(t)$,

$$u_k^f(t) = \int_0^t \int_{\partial\mathcal{M}} f(\mathbf{z},t')s_k(t-t')\partial_\nu\varphi_k(\mathbf{z})\,dS_g dt', \tag{3.13}$$

we can evaluate the Fourier coefficients

$$u_k^{\kappa\eta f}(t) = \int_0^t \int_{\partial\mathcal{M}} \kappa_0(\mathbf{z})\eta(\mathbf{z})f(\mathbf{z},t')s_k(t-t')\partial_\nu\varphi_k(\mathbf{z})\,dS_g dt' \tag{3.14}$$

$$= \int_0^t \int_{\partial\mathcal{M}} f(\mathbf{z},t')s_k(t-t')(\kappa_0(\mathbf{z})\partial_\nu\varphi_k(\mathbf{z}))\,d\mu dt'.$$

These are the Fourier coefficients of the wave $u^{\kappa\eta f}(t)$ with unknown $\kappa$ and $\eta$. Introducing the function $\varrho \in C^\infty(\partial\mathcal{M})$,

$$\varrho(\mathbf{z}) = \kappa_0(\mathbf{z})\eta(\mathbf{z}), \quad \mathbf{z} \in \partial\mathcal{M}, \tag{3.15}$$

we see that, instead of the inner product $\langle u^f(t), u^h(s)\rangle$, we evaluate the inner product

$$\langle u^{\varrho f}(t), u^{\varrho h}(s)\rangle = \sum_{k=1}^{\infty} u_k^{\varrho f}(t)\overline{u_k^{\varrho h}(s)}, \tag{3.16}$$

where the Fourier coefficients in the right-hand side of this equation are given by formula (3.14).

## 3.3. Domains of influence

**3.3.1.** Let $\Gamma \subset \partial\mathcal{M}$ be a non-empty open set. We denote by $L^2(\Gamma \times [0,T])$ the subspace of $L^2(\Sigma^T)$ that consists of the functions $f$ with supp $(f) \subset \overline{\Gamma} \times [0,T]$.

**Definition 3.8** *The subset $\mathcal{M}(\Gamma,\tau) \subset \mathcal{M}$, $\tau > 0$,*

$$\mathcal{M}(\Gamma,\tau) = \{\mathbf{x} \in \mathcal{M} : \ d(\mathbf{x},\Gamma) \le \tau\}$$

*is called the domain of influence of $\Gamma$ at time $\tau$.*

Here, as usual, we denote by $d(\mathbf{x},\mathbf{y})$ the distance between $\mathbf{x}$ and $\mathbf{y}$ in $(\mathcal{M},g)$.

Figure 3.1: Domain of influence

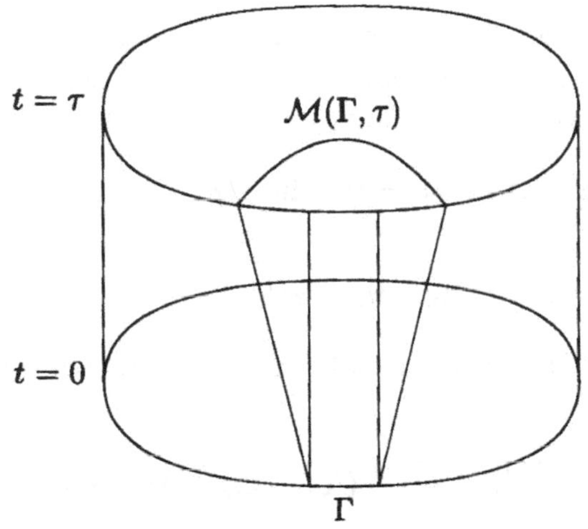


**Lemma 3.9** *Let $u^f(t)$ be the solution of initial boundary value problem (3.4)–(3.6) with $f \in L^2(\Gamma \times [0, T])$. Then*

$$supp\ (u^f(\tau)) \subset \mathcal{M}(\Gamma, \tau).$$

**Proof.** The result follows immediately from Theorem 2.47. □

Putting together the previous lemma and Lemma 2.42, we see that, for $f \in L^2(\Gamma \times [0, T])$,

$$u^f(\tau) \in L^2(\mathcal{M}(\Gamma, \tau)). \tag{3.17}$$

Here and below we denote by $L^2(\Omega)$, $\Omega \subset \mathcal{M}$, the subspace of $L^2(\mathcal{M})$, which consists of all functions $f \in L^2(\mathcal{M})$ that are equal to zero in $\mathcal{M} \setminus \Omega$.

**3.3.2.** Consider the space of the waves $u^f(t)$, $f \in L^2(\Gamma \times [0, T])$. Clearly, it is a linear subspace (not necessarily closed) of the space $L^2(\mathcal{M}(\Gamma, \tau))$ (see Fig. 3.1).

**Theorem 3.10** *Let $\tau > 0$. The linear subspace,*

$$\{u^f(\tau) \in L^2(\mathcal{M}(\Gamma, \tau)) : \ f \in L^2(\Gamma \times [0, \tau])\},$$

*is dense in* $L^2(\mathcal{M}(\Gamma,\tau))$.

**Proof.** Let $\psi \in L^2(\mathcal{M}(\Gamma,\tau))$ be such that

$$\langle u^f(\cdot\,,\tau),\psi\rangle = 0 \tag{3.18}$$

for all $f \in C_0^\infty(\Gamma \times [0,\tau])$. To prove the assertion, it is sufficient to show that $\psi = 0$. To this end, we consider the following initial boundary value problem for the wave equation,

$$(\partial_t^2 - \Delta_g + q)e = 0, \quad \text{in} \quad Q^\tau, \tag{3.19}$$

$$e|_{\Sigma^\tau} = 0, \quad e|_{t=\tau} = 0, \quad \partial_t e|_{t=\tau} = \psi. \tag{3.20}$$

Integrating by parts and using equations (3.5)–(3.6) and (3.18), we obtain that

$$0 = \int_{Q^\tau} [u^f\overline{(\partial_t^2 - \Delta_g + q)e} - ((\partial_t^2 - \Delta_g + q)u^f)\overline{e}]\, dV_g\, dt$$

$$= \int_{\mathcal{M}} u^f(\tau)\,\overline{\psi}\, dV_g + \int_{\Sigma^\tau} f\,\overline{\partial_\nu e}\, dS_g\, dt$$

$$= \int_{\Sigma^\tau} f\,\overline{\partial_\nu e}\, dS_g\, dt,$$

for all $f \in C_0^\infty(\Gamma \times [0,\tau])$. In deriving this formula, we have used Lemma 2.34, which guarantees that $\partial_\nu e \in L^2(\Sigma^\tau)$. Because

$$\int_{\Sigma^\tau} f\,\overline{\partial_\nu e}\, dS_g\, dt = 0$$

for any $f \in C_0^\infty(\Gamma \times [0,\tau])$, we see that

$$\partial_\nu e|_{\Gamma\times[0,\tau]} = 0.$$

Together with boundary condition (3.20), this equation yields that the Cauchy data of $e$ vanish on $\Gamma \times [0,\tau]$. Because $e(\tau) = 0$, we can continue $e$ onto the interval $[\tau, 2\tau]$ as

$$E(\mathbf{x},t) = \begin{cases} e(\mathbf{x},t), & \text{for } t \le \tau, \\ -e(\mathbf{x}, 2\tau - t), & \text{for } t > \tau. \end{cases}$$

Then $E \in C([0,2\tau];H^1(\mathcal{M})) \cap C^1([0,2\tau];L^2(\mathcal{M}))$ and

$$(\partial_t^2 - \Delta_g + q)E = 0 \text{ in } Q^{2\tau}.$$

Moreover, the Cauchy data of $E$ vanish on $\Gamma \times [0,2\tau]$,

$$E|_{\Gamma\times[0,2\tau]} = 0, \quad \partial_\nu E|_{\Gamma\times[0,2\tau]} = 0.$$

The fact that $\psi = 0$ will follow from the following theorem.

**Theorem 3.11** *Let $u \in C([0,2\tau];H^1(\mathcal{M})) \cap C^1([0,2\tau];L^2(\mathcal{M}))$ be a solution in $Q^{2\tau}$ of the wave equation (3.4), such that for an open set $\Gamma \subset \partial\mathcal{M}$,*

$$u|_{\Gamma\times[0,2\tau]} = 0, \quad \partial_\nu u|_{\Gamma\times(0,2\tau)} = 0. \tag{3.21}$$

*Then, at $t = \tau$, the function $u$ and its derivative $\partial_t u$ vanish in the domain of influence of $\Gamma$,*

$$u(\mathbf{x},\tau) = 0, \quad \partial_t u(\mathbf{x},\tau) = 0 \textit{ for } \mathbf{x} \in \mathcal{M}(\Gamma,\tau).$$

This theorem is the global Holmgren-John uniqueness theorem which will be proven later in section 3.4.

□

**3.3.3.** Clearly, the set

$$\{u^{\varrho f}(\tau) \in L^2(\mathcal{M}(\Gamma,\tau)) : \; f \in L^2(\Gamma\times[0,\tau])\}$$

is also dense in $L^2(\mathcal{M}(\Gamma,\tau))$. Thus, there are functions $f_j$, $j = 1,2,\ldots$, such that $\{u^{\varrho f_j}(\tau)\}_{j=1}^\infty$ form an orthonormal basis in the space $L^2(\mathcal{M}(\Gamma,\tau))$. In this section, we will construct such functions $f_j$ from the boundary spectral data. The corresponding basis $\{u^{\varrho f_j}(\tau)\}_{j=1}^\infty$ is then called the wave basis.

**Lemma 3.12** *Let $\tau > 0$. Given the gauge equivalent boundary spectral data it is possible to construct boundary sources $f_j \in L^2(\Gamma\times[0,\tau])$ such that*

$$v_j = u^{\varrho f_j}(\tau), \; j = 1,2,\ldots, \tag{3.22}$$

*form an orthonormal basis of $L^2(\mathcal{M}(\Gamma,\tau))$.*

**Proof.** The proof is analogous to the proof in the one-dimensional case, which is given in section 1.3.4. Indeed, let $\{h_j\}_{j=1}^{\infty}$ be a complete set in $L^2(\Gamma \times [0,\tau])$. Then $\{\varrho h_j\}_{j=1}^{\infty}$ is also complete in $L^2(\Gamma \times [0,\tau])$. By means of the procedure described in section 3.2.4, we find the inner products

$$c_{jk} = \langle u^{\varrho h_j}(\tau), u^{\varrho h_k}(\tau)\rangle.$$

Next we use the Gram-Schmidt orthogonalization procedure to construct $f_j$. More precisely, we define $f_j \in L^2(\Gamma \times [0,\tau])$ recursively by

$$g_j = h_j - \sum_{k=1}^{j-1} \langle u^{\varrho h_j}(\tau), u^{\varrho f_k}(\tau)\rangle f_k,$$

$$f_j = \frac{g_j}{\langle u^{\varrho g_j}(\tau), u^{\varrho g_j}(\tau)\rangle^{1/2}}.$$

When $g_j = 0$, we remove the corresponding $h_j$ from the original sequence and continue the procedure with the next $h_j$. The waves $\{u^{\varrho f_j}(\tau)\}_{j=1}^{\infty}$ obtained in this way form an orthonormal basis in $L^2(\mathcal{M}(\Gamma,\tau))$. □

We point out that, if we choose the original set $\{h_j\}$ to be in $C_0^{\infty}(\Gamma \times [0,\tau])$, then, obviously, $f_j \in C_0^{\infty}(\Gamma \times [0,\tau])$ and the corresponding wave basis consists of $C^{\infty}(\mathcal{M})$-functions.

**3.3.4.** Denote by $P_{\Gamma,\tau}$ the orthogonal projector in $L^2(\mathcal{M})$ onto the space $L^2(\mathcal{M}(\Gamma,\tau))$,

$$P_{\Gamma,\tau} : L^2(\mathcal{M}) \to L^2(\mathcal{M}(\Gamma,\tau)), \tag{3.23}$$

$$(P_{\Gamma,\tau}a)(\mathbf{x}) = \chi_{\mathcal{M}(\Gamma,\tau)}(\mathbf{x})a(\mathbf{x}),$$

where $\chi_{\mathcal{M}(\Gamma,\tau)}$ is the characteristic function of the domain of influence $\mathcal{M}(\Gamma,\tau)$,

$$\chi_{\mathcal{M}(\Gamma,\tau)}(\mathbf{x}) = \begin{cases} 1, & \text{for } \mathbf{x} \in \mathcal{M}(\Gamma,\tau), \\ 0, & \text{for } \mathbf{x} \notin \mathcal{M}(\Gamma,\tau). \end{cases}$$

**Lemma 3.13** *Let $f, h \in L^2(\Sigma^T)$ and $\Gamma \subset \partial\mathcal{M}$ be an open set. Then, given the gauge equivalent boundary spectral data, it is possible to find the inner product*

$$\langle P_{\Gamma,\tau} u^{\varrho f}(t), u^{\varrho h}(s)\rangle = \int_{\mathcal{M}(\Gamma,\tau)} u^{\varrho f}(\mathbf{x},t)\,\overline{u^{\varrho h}(\mathbf{x},s)}\,dV_g \tag{3.24}$$

*for any $0 \le t, s, \tau \le T$.*

**Proof.** By Lemma 3.12, we find functions $f_j \in C_0^\infty(\Gamma \times [0,\tau])$ such that the corresponding waves $v_j = u^{\varrho f_j}(\tau)$ form an orthonormal basis in $L^2(\mathcal{M}(\Gamma,\tau))$. Then, for any $a \in L^2(\mathcal{M}(\Gamma,\tau))$,

$$a = \sum_{j=1}^{\infty} \langle a, v_j\rangle\, v_j. \tag{3.25}$$

As

$$\langle P_{\Gamma,\tau} u^{\varrho f}(t), v_j\rangle = \langle u^{\varrho f}(t), v_j\rangle, \tag{3.26}$$

then

$$\langle P_{\Gamma,\tau} u^{\varrho f}(t), u^{\varrho h}(s)\rangle = \sum_{j=1}^{\infty} \langle u^{\varrho f}(t), v_j\rangle \overline{\langle u^{\varrho h}(s), v_j\rangle}. \tag{3.27}$$

However, $v_j = u^{\varrho f_j}(\tau)$ so that $\langle u^{\varrho f}(t), v_j\rangle$ and $\langle u^{\varrho h}(s), v_j\rangle$ can be computed by means of formula (3.16). □

Denote by $\mathcal{M}(\mathbf{y},\tau)$ the domain of influence of a point $\mathbf{y} \in \partial\mathcal{M}$,

$$\mathcal{M}(\mathbf{y},\tau) = \{\mathbf{x} \in \mathcal{M} :\ d(\mathbf{x},\mathbf{y}) \le \tau\}, \tag{3.28}$$

and by $P_{\mathbf{y},\tau}$ the orthoprojection onto $L^2(\mathcal{M}(\mathbf{y},\tau))$.

**Remark.** Domains $\mathcal{M}(\mathbf{y},\tau)$ are direct multidimensional analogs of domains $\mathcal{M}(0;0,\tau)$ which play crucial role in solving the one-dimensional inverse problem.

**Corollary 3.14** *Let $f, h \in L^2(\Sigma^T)$ and $\mathbf{y} \in \partial\mathcal{M}$ be given. Then the gauge equivalent boundary spectral data determine the inner product*

$$\langle P_{\mathbf{y},\tau} u^{\varrho f}(t), u^{\varrho h}(s)\rangle = \int_{\mathcal{M}(\mathbf{y},\tau)} u^{\varrho f}(\mathbf{x},t)\,\overline{u^{\varrho h}(\mathbf{x},s)}\,dV_g \tag{3.29}$$

*for any $0 \le t, s, \tau \le T$.*

**Proof.** Let $\Gamma_l$, $l = 1, 2, \ldots$, be a decreasing sequence of open sets which converge to the point $\mathbf{y}$,

$$\Gamma_{l+1} \subset \Gamma_l, \quad \bigcap_{l=1}^{\infty} \Gamma_l = \{\mathbf{y}\}.$$

Then,

$$\lim_{l\to\infty} \chi_{\mathcal{M}(\Gamma_l,\tau)}(\mathbf{x}) = \chi_{\mathcal{M}(\mathbf{y},\tau)}(\mathbf{x})$$

pointwise, and, by the Lebesgue dominated convergence theorem,

$$\lim_{l\to\infty} \langle P_{\Gamma_l,\tau} u^{ef}(t), u^{eh}(s)\rangle = \langle P_{\mathbf{y},\tau} u^{ef}(t), u^{eh}(s)\rangle.$$

□

The previous considerations yield also the following corollary.

**Corollary 3.15** *Let $f \in L^2(\Sigma^T)$ and $\mathbf{y} \in \partial\mathcal{M}$. Then the gauge equivalent boundary spectral data determine uniquely the inner product*

$$\langle P_{\mathbf{y},\tau}\varphi_k, u^{ef}(t)\rangle = \sum_{j=1}^{\infty} \langle \varphi_k, u^{ef_j}(\tau)\rangle \overline{\langle u^{ef}(\tau), u^{ef_j}(t)\rangle}, \qquad (3.30)$$

*where $\{u^{ef_j}(\tau)\}_{j=1}^{\infty}$ form an orthonormal basis in $L^2(\mathcal{M}(\mathbf{y},\tau))$.*

## 3.4. Global unique continuation from boundary

In this section, we will prove Theorem 3.11. Actually, we will prove a more general theorem which is called the global Holmgren-John uniqueness theorem. to this end, we will use the local unique continuation result given by Theorem 2.66.

**3.4.1.** Let $u \in H^1(\mathcal{M} \times [-T, T])$, $u = u(\mathbf{x}, t)$ be a weak solution of the hyperbolic equation (3.4),

$$P(\mathbf{x}, D)u = \partial_t^2 u - \Delta_g u + qu = 0, \quad \text{in} \quad \mathcal{M} \times [-T, T]. \qquad (3.31)$$

Figure 3.2: Double-cone of influence

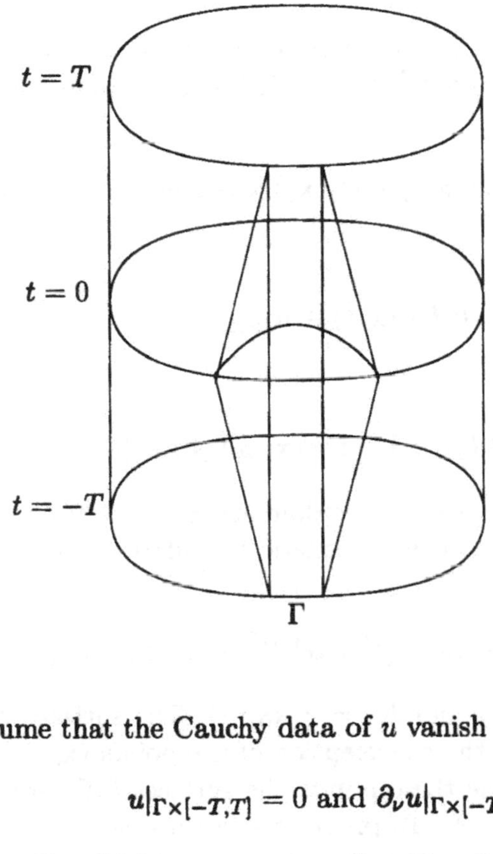


Assume that the Cauchy data of $u$ vanish on $\Gamma \times [-T, T]$,

$$u|_{\Gamma\times[-T,T]} = 0 \text{ and } \partial_\nu u|_{\Gamma\times[-T,T]} = 0, \tag{3.32}$$

where $\Gamma \subset \partial\mathcal{M}$ is an open set. Let $K \subset \mathcal{M} \times [-T, T]$ be the double-cone of influence of $\Gamma \times [-T, T]$,

$$K = K_{\Gamma,T} = \{(\mathbf{x}, t) \in \mathcal{M} \times [-T, T] : \ d(\mathbf{x}, \Gamma) \le T - |t|\} \tag{3.33}$$

(see Fig. 3.2). The main result of this section is

**Theorem 3.16** *Let $u$ be a solution of wave equation (3.31) that has zero Cauchy data on $\Gamma \times [-T, T]$, i.e., let (3.32) be satisfied. Then $u = 0$ in the double cone $K_{\Gamma,T}$.*

The proof of Theorem 3.16 relies on the following result.

### 3.4.2.

**Lemma 3.17** *Let* $\mathbf{x}_0 \in \mathcal{M}^{int}$ *and* $\rho, \delta \geq 0$ *be so small that*

$$B_{\rho+\delta}(\mathbf{x}_0) \cap \partial\mathcal{M} = \emptyset$$

*and that* $B_{\rho+\delta}(\mathbf{x}_0)$ *lies in the coordinate chart of the normal coordinates centered at* $\mathbf{x}_0$. *Let* $u \in H^1(\mathcal{M} \times [-T, T])$ *be a solution of wave equation (3.31) and*

$$u(\mathbf{x}, t) = 0 \text{ for } (\mathbf{x}, t) \in B_\rho(\mathbf{x}_0) \times [-\delta, \delta]. \tag{3.34}$$

*Then*

$$u(\mathbf{x}, t) = 0 \text{ for } (\mathbf{x}, t) \in K_{\mathbf{x}_0,\delta}, \tag{3.35}$$

*where*

$$K_{\mathbf{x}_0,\delta} = \{(\mathbf{x}, t) \in \mathcal{M} \times [-\delta, \delta] : \ d(\mathbf{x}, \mathbf{x}_0) \leq \delta - |t|\}. \tag{3.36}$$

**Proof.** *Step 1.* First we introduce the domains $K_\alpha$, $\alpha \in [0, \delta]$, which exhaust $K_{\mathbf{x}_0,\delta}$ and have non-characteristic boundaries. Denote by $K_\alpha$, $\alpha \in [0, \delta]$, the set

$$K_\alpha = \{(\mathbf{x}, t) \in \mathcal{M} \times [-\delta, \delta] : \ (t^2 + \alpha^2)^{1/2} + d(\mathbf{x}, \mathbf{x}_0) \leq \delta\}. \tag{3.37}$$

In particular, $K_0 = K_{\mathbf{x}_0,\delta}$ and $K_\delta = \{(\mathbf{x}_0, 0)\}$. The surfaces $\partial K_\alpha$, $\alpha \in [0, \delta)$, are smooth, with the exception of the points $(\mathbf{x}_0, \pm(\delta^2 - \alpha^2)^{1/2})$. Moreover, outside these points the surfaces $\partial K_\alpha$ are non-characteristic for any $\alpha < \delta$. To prove this, we use normal coordinates $\mathbf{y} = (y^1, \ldots, y^m)$ centered at $\mathbf{x}_0$. Let $\nu$ be the normal to the surface $\partial K_\alpha$ with respect to the metric $g \times$ can of $\mathcal{M} \times \mathbf{R}$ where can is the canonical metric $dt^2$ on $\mathbf{R}$. Then $\nu$ has the form

$$\nu(\mathbf{y}, t) = \sqrt{\frac{t^2 + \alpha^2}{2t^2 + \alpha^2}} \left( \frac{t}{(t^2 + \alpha^2)^{1/2}}, \frac{\mathbf{y}}{|\mathbf{y}|} \right), \tag{3.38}$$

where $(\mathbf{y}, t) \in \partial K_\alpha$, $\mathbf{y} \neq 0$. For any $\alpha < \delta$, $(J\nu, \nu) > 0$, where $J = \text{diag}\,(-1, 1, \ldots, 1)$. Therefore, the surfaces $\partial K_\alpha$ are time-like everywhere except $\mathbf{y} = 0$.

*Step 2.* Let $\beta$ be the infimum of all $\alpha > 0$ such that

$$u|_{K_\alpha} = 0. \tag{3.39}$$

Because $u(\mathbf{x},t) = 0$ for $(\mathbf{x},t) \in B_\rho(\mathbf{x}_0) \times [-\delta,\delta]$, we see that $\beta < \delta$. Let us show that $\beta = 0$. Assume, on the contrary, that $\beta > 0$. Then, by definition, $u = 0$ in $K_\beta^{int}$.

Due to the local unique continuation result given by Theorem 2.66, the fact that $\partial K_\beta$ is non-characteristic implies that $u(\mathbf{x},t) = 0$ in a neighborhood of any point $(\mathbf{y},t) \in \partial K_\beta \setminus (\{\mathbf{x}_0\} \times \mathbf{R})$.

However, if $(\mathbf{y},t) \in \partial K_\beta \cap \{\mathbf{x}_0\} \times \mathbf{R}$, $\beta > 0$, i.e. $\mathbf{y} = \mathbf{x}_0$, $t = \pm\sqrt{\delta^2 - \alpha^2}$, then $(\mathbf{y},t) \in B_\rho(\mathbf{x}_0) \times [-\delta,\delta]$. Therefore, $u(\mathbf{x},t) = 0$ also in a neighborhood of these points.

Thus, there is an open neighborhood $V$ of $K_\beta$ such that $u|_V = 0$. As $\partial K_\beta$ is compact, there is $\epsilon > 0$ such that $u = 0$ in $K_{\beta-\epsilon}$. Hence, $\beta = 0$. □

The previous result has the following corollary.

**Corollary 3.18** *Let* $\mathbf{x}_0 \in \mathcal{M}^{int}$ *and* $\rho, \delta \geq 0$ *be so small that*

$$B_{\rho+\delta}(\mathbf{x}_0) \cap \partial\mathcal{M} = \emptyset$$

*and that* $B_{\rho+\delta}(\mathbf{x}_0)$ *lies in the coordinate chart of the normal coordinates centered at* $\mathbf{x}_0$. *Let* $u \in H^1(\mathcal{M} \times [-T,T])$ *be a solution of wave equation (3.31) and*

$$u(\mathbf{x},t) = 0 \textit{ for } (\mathbf{x},t) \in B_\rho(\mathbf{x}_0) \times [-T,T].$$

*Then*

$$u(\mathbf{x},t) = 0 \textit{ for } (\mathbf{x},t) \in K_{\mathbf{x}_0,T} \cap (B_{\rho+\delta}(\mathbf{x}_0) \times [-T,T]). \quad (3.40)$$

**Proof.** The corollary follows by using Lemma 3.17 for point $(\mathbf{x},t)$ with $\mathbf{x} \in B_\rho(\mathbf{x}_0)$, $|t| < T - \delta$. □

**3.4.3.** In this section we will complete the proof of Theorem 3.16.

*Step 1.* Let $(\mathbf{x}_0, t_0) \in K_{\Gamma,T}^{int}$. Then $d(\mathbf{x}_0,\Gamma) < T - |t_0|$ and for any $\epsilon > 0$ there is a path $\mu(s)$ of length $l$, $l < d(\mathbf{x}_0,\Gamma) + \epsilon$ such that $\mu(0) = \mathbf{z} \in \Gamma$, $\mu(l) = \mathbf{x}_0$ and

$$\mu : (0,l] \to \mathcal{M}^{int}.$$

Moreover, $\mu$ is parametrized by the arclength so that $|\mu([a,b])| = b - a$. It follows from the definition of a manifold with boundary that there is a larger manifold $\widetilde{\mathcal{M}}$, $\mathcal{M} \subset \widetilde{\mathcal{M}}$, $\mathbf{z} \in \widetilde{\mathcal{M}}^{int}$ such that $\partial\mathcal{M} \setminus \Gamma \subset \partial\widetilde{\mathcal{M}}$.

Continue $u(\mathbf{x}, t)$ by zero onto $(\widetilde{\mathcal{M}} \setminus \mathcal{M}) \times [-T, T]$ to obtain a function $\widetilde{u}$ on $\widetilde{\mathcal{M}} \times [-T, T]$,

$$\widetilde{u}(\mathbf{x}, t) = \begin{cases} u(\mathbf{x}, t), & \text{for } (\mathbf{x}, t) \in \mathcal{M} \times [-T, T], \\ 0, & \text{for } (\mathbf{x}, t) \in \widetilde{\mathcal{M}} \setminus \mathcal{M} \times [-T, T]. \end{cases}$$

Due to boundary conditions (3.32), $\widetilde{u} \in H^1(\widetilde{\mathcal{M}} \times [-T, T])$ and satisfies there wave equation (3.31), where the metric tensor $g$ and the potential $q$ are continued smoothly onto $\widetilde{\mathcal{M}} \setminus \mathcal{M}$.

*Step 2.* Continue the path $\mu$ onto $\widetilde{\mathcal{M}} \setminus \mathcal{M}$. We obtain a path $\widetilde{\mu} : [-\epsilon, l] \to \widetilde{\mathcal{M}}^{int}$, $\widetilde{\mu}(-\epsilon) = \widetilde{\mathbf{z}} \in \widetilde{\mathcal{M}} \setminus \mathcal{M}$,

$$\widetilde{\mu}(s) = \mu(s) \text{ for } s \geq 0.$$

Due to the compactness of $\widetilde{\mu}([-\epsilon, l])$, there is a positive $\delta$ such that, for any $s \in [-\epsilon, l]$, we have $d(\widetilde{\mu}(s), \partial\widetilde{\mathcal{M}}) > 2\delta$, $d(\widetilde{\mu}(-\epsilon), \partial\mathcal{M}) > 2\delta$, and $B_{2\delta}(\widetilde{\mu}(s))$ is a domain of the normal coordinates. Let $N$ be a positive integer such that $N\delta > l+\epsilon$. Choose $s_j \in [0, l]$, $j = 0, \ldots, N$, such that $0 < s_{j+1} - s_j < \delta$, $s_0 = -\epsilon$, $s_N = l$ and denote $\mathbf{y}_j = \widetilde{\mu}(s_j)$.

We can apply Corollary 3.18 to complete the proof of Theorem 3.16. Indeed, $\widetilde{u} = 0$ in $B_\rho(\mathbf{y}_0) \times [-T, T]$, where $0 < \rho < \delta$ is, otherwise, arbitrary. Hence, by Corollary 3.18, $\widetilde{u} = 0$ in $K_{\mathbf{y}_0, T} \cap B_{\rho+\delta}(\mathbf{y}_0) \times [-T, T]$. In particular,

$$\begin{aligned} \widetilde{u} = 0 \;\text{ in }\; & B_\rho(\mathbf{y}_1) \times [-T + d(\mathbf{y}_1, \mathbf{y}_0) + \rho, T - d(\mathbf{y}_1, \mathbf{y}_0) - \rho] \\ & \subset B_\rho(\mathbf{y}_1) \times [-T + (s_1 + \epsilon) + \rho, T - (s_1 + \epsilon) - \rho]. \end{aligned}$$

Continuing this procedure, we obtain that

$$\widetilde{u} = 0 \quad \text{in} \quad B_\rho(\mathbf{y}_j) \times [-T + (s_j + \epsilon) + j\rho, T - (s_j + \epsilon) - j\rho].$$

Since $\epsilon$ and $\delta$ are arbitrary small and $d(\mathbf{x}, \Gamma) < T - |t_0|$, we see that $u(\mathbf{x}, t) = \widetilde{u}(\mathbf{x}, t) = 0$ in a neighborhood of $(\mathbf{x}_0, t_0)$. This proves Theorem 3.16. □

Theorem 3.11 follows immediately from Theorem 3.16.

## 3.5. Gaussian beams from the boundary

**3.5.1.** In section 3.3 we have obtained formula (3.30), which makes it possible to find the inner products of projections of waves generated by boundary sources. Later we will use this formula to reconstruct the geometry of the manifold. To do that, we will use the techniques of Gaussian beams. The general theory of Gaussian beams on manifolds $\mathcal{N}$ without boundary is given in section 2.4. In the current section, we will adapt this theory to construct the Gaussian beams generated by boundary sources.

Let us consider a manifold $(\mathcal{M}, g)$ with boundary that is isometrically embedded into a manifold $(\mathcal{N}, g)$ without boundary, i.e., $\mathcal{M} \subset \mathcal{N}$. We understand a formal Gaussian beam on the manifold $\mathcal{M}$ to be the restriction onto $\mathcal{M}$ of a formal Gaussian beam defined on $\mathcal{N}$. Analyzing the restrictions of the formal Gaussian beams on the boundary cylinder $\partial\mathcal{M} \times \mathbf{R}$, we determine the proper boundary conditions to obtain a Gaussian beam in $\mathcal{M}$ that is generated by a boundary source. For our purposes, we need only those Gaussian beams that correspond to the normal geodesics. Summarizing, in this section we will construct the boundary sources such that the corresponding solutions of the initial-boundary value problem are Gaussian beams.

**3.5.2.** Let $\mathbf{z}_0 \in \partial\mathcal{M}$, $t_0 > 0$, and let $\mathbf{z} = (z^1, \dots z^{m-1})$ be a local system of coordinates on $\partial\mathcal{M}$ near $\mathbf{z}_0$. Consider a class of functions $f_\epsilon = f_{\epsilon,\mathbf{z}_0,t_0}(\mathbf{z}, t)$ on the boundary cylinder $\partial\mathcal{M} \times \mathbf{R}$, where

$$f_\epsilon(\mathbf{z}, t) = (\pi\epsilon)^{-m/4}\chi(\mathbf{z}, t) \exp\{i\epsilon^{-1}\Theta(\mathbf{z}, t)\}V(\mathbf{z}). \tag{3.41}$$

Here $\chi$ is a smooth cut-off function near $(\mathbf{z}_0, t_0)$ and

$$\Theta(\mathbf{z}, t) = -(t - t_0) + \frac{1}{2}(H_0(\mathbf{z} - \mathbf{z}_0), (\mathbf{z} - \mathbf{z}_0)) + \frac{i}{2}(t - t_0)^2, \tag{3.42}$$

where $(\cdot\,,\cdot)$ is the real Euclidean inner product and $H_0$ is a symmetric matrix with a positive definite imaginary part, i.e. $H_0 = H_0^t$, $\Im H_0 > 0$. At last, let $V$ be a smooth function on $\partial\mathcal{M}$. Its Taylor expansion near $\mathbf{z}_0 \in \partial\mathcal{M}$ is

$$V \asymp \sum_{l=1}^{\infty} V_l \asymp \sum_{l=1}^{\infty} \sum_{|\alpha|=l} V_\alpha(\mathbf{z} - \mathbf{z}_0)^\alpha,$$

where $V_l$ are homogeneous polynomials of order $l$ with complex coefficients $V_\alpha \in \mathbf{C}$.

**3.5.3.** Consider the initial-boundary value problem

$$\partial_t^2 u - \Delta_g u + qu = 0,$$

$$u|_{t=0} = \partial_t u|_{t=0} = 0, \tag{3.43}$$

$$u|_{\partial\mathcal{M}} = f_\epsilon(\mathbf{z}, t).$$

In this section, we will prove that the solution $u = u_\epsilon(\mathbf{x}, t)$ of this initial-boundary value problem is a Gaussian beam. The corresponding geodesic starts at $\mathbf{z}_0$ and is normal to the boundary. To prove this result, we first construct a formal Gaussian beam that asymptotically has the right boundary value. Then we prove that the solution of problem (3.43) is close to the constructed formal Gaussian beam for $t < t_0 + l(\mathbf{z}_0)$. By $l(\mathbf{z}_0)$ we denote the maximal arclength of the normal geodesic, which starts at the point $\mathbf{z}_0 \in \partial\mathcal{M}$, until it hits the boundary. Clearly, $l(\mathbf{z}_0) > \tau_{\partial\mathcal{M}}(\mathbf{z}_0)$, where the function $\tau_{\partial\mathcal{M}}$ is defined in section 2.1.16.

**3.5.4.** In our constructions we use the boundary normal coordinates $(\mathbf{z}, n) = (z^1, \ldots, z^{m-1}, n)$ on $\mathcal{M}$, which are described in sections 2.1.16–2.1.18. We remind the reader that this system of coordinates is smooth until the cut locus, and that the length element in these coordinates has the form

$$ds^2 = g_{\alpha\beta}(\mathbf{z}, n) dz^\alpha dz^\beta + dn^2, \tag{3.44}$$

$\alpha, \beta = 1, \ldots, m-1$. Here the metric tensor $g_{\alpha\beta}(\mathbf{z}, 0)$ is the metric tensor on the boundary $\partial\mathcal{M}$. If we allow $n$ to be negative, these coordinates define also a coordinate system on $\mathcal{N}$ near $\partial\mathcal{M}$. Next, we show that there is a formal Gaussian beam on $\mathcal{N} \times \mathbf{R}$ that is equal to $f_\epsilon$ in a neighborhood of $(\mathbf{z}_0, t_0)$ on $\partial\mathcal{M} \times \mathbf{R}$. For $|t - t_0|$ be sufficiently small, we seek for the representation of $U_\epsilon^N(\mathbf{x}, t) = U_\epsilon^N(\mathbf{z}, n, t)$ in the boundary normal coordinates,

$$U_\epsilon^N(\mathbf{x}, t) = (\pi\epsilon)^{-m/4} \exp\{-(i\epsilon)^{-1}\theta(\mathbf{z}, n, t)\} \sum_{j=0}^{N} u_j(\mathbf{z}, n, t)(i\epsilon)^j. \tag{3.45}$$

**Theorem 3.19** *Let $\mathbf{z}_0 \in \partial\mathcal{M}$, $t_0 > 0$ and assume that the function $V \in C^\infty(\partial\mathcal{M})$ and the matrix $H_0$ are given. Then there is a formal Gaussian beam $U_\epsilon^N(\mathbf{x}, t)$ on $\mathcal{N} \times \mathbf{R}$ which is concentrated near $\mathbf{z}_0$ at time $t_0$ and satisfies the following conditions:*
*i) The formal Gaussian beam $U_\epsilon^N(\mathbf{x}, t)$ is propagating into $\mathcal{M}$, i.e., the geodesic $\gamma(t)$ corresponding to $U_\epsilon^N(\mathbf{x}, t)$ satisfies*

$$(\mathbf{v}, \nu)_g \geq 0,$$

*where $\mathbf{v} = \partial_s \gamma(t_0)$.*
*ii) In representation (3.45) the phase and the amplitude functions satisfy the boundary conditions*

$$\theta(\mathbf{z}, 0, t) \asymp -(t - t_0) + \frac{1}{2}(H_0(\mathbf{z} - \mathbf{z}_0), (\mathbf{z} - \mathbf{z}_0)) + \frac{i}{2}(t - t_0)^2, \quad (3.46)$$

$$u_0(\mathbf{z}, 0, t) \asymp V(\mathbf{z}), \quad u_k(\mathbf{z}, 0, t) \asymp 0, \quad k = 1, \ldots, N.$$

*iii) $U_\epsilon^N(\mathbf{x}, t)$ satisfies inequality (2.217) near $(\mathbf{z}, t_0) \in \mathcal{N} \times \mathbf{R}$.*
*Moreover, these conditions imply that $\gamma$ is the normal geodesic, which starts at the point $\mathbf{z}_0$ at time $t_0$. In the boundary normal coordinates $(\mathbf{z}, n)$, $\gamma(t)$ has the form*

$$\mathbf{z}(t) = \mathbf{z}_0, \quad n(t) = t - t_0, \quad (3.47)$$

*for $t - t_0 < \tau_{\partial\mathcal{M}}(\mathbf{z}_0)$.*

**Remark.** We note that conditions *i-iii* determine the formal Gaussian beam uniquely up to a term $\mathcal{O}(\epsilon^p)$, $p = N - m$.

**3.5.5. Proof.** Let us consider any Gaussian beam on $\mathcal{N}$. First we show that, if the Gaussian beam satisfies the right boundary conditions near $(\mathbf{z}_0, 0, t_0)$, then it propagates along the normal geodesic.

**Lemma 3.20** *Let $U_\epsilon^N(\mathbf{z}, n, t)$ be a formal Gaussian beam that satisfies boundary conditions (3.46). Let $\gamma$ be the geodesic corresponding to it. Then $\gamma$ is the normal geodesic $\gamma_{\mathbf{z}_0,\nu}$.*

**Proof.** *Step 1.* Let us consider a formal Gaussian beam in the boundary normal coordinates $(\mathbf{z}, n, t)$. As it was shown in sections 2.4.5, 2.4.6, the phase function $\theta$ of the Gaussian beam has the form

$$\theta(\mathbf{z}, n, t) = \sum_{l \geq 1} \theta_l(t),$$

where

$$\theta_1(t) = p_\alpha(t)(z^\alpha - z^\alpha(t)) + p_m(t)(n - n(t))$$

and

$$\theta_2(t) = \frac{1}{2}[H_{\alpha\beta}(t)(z^\alpha - z^\alpha(t))(z^\beta - z^\beta(t)) +$$

$$+2H_{\alpha m}(t)(z^\alpha - z^\alpha(t))(n - n(t)) + H_{mm}(t)(n - n(t))^2], \quad (3.48)$$

$\alpha$, $\beta = 1, \ldots, m-1$. We remind the reader that $\mathbf{p}(t)$ is the canonical transformation of the unit velocity vector along the geodesic $\gamma$ and $\theta_l(t)$ are homogeneous polynomials of order $l$ with respect to $(\mathbf{z} - \mathbf{z}(t), n - n(t))$. In particular, on the boundary $n = 0$,

$$\theta(\mathbf{z}, 0, t) \asymp p_\alpha(t)(z^\alpha - z^\alpha(t)) + p_m(t)(-n(t)) +$$

$$+\frac{1}{2}[H_{\alpha\beta}(t)(z^\alpha - z^\alpha(t))(z^\beta - z^\beta(t)) + 2H_{\alpha m}(t)(z^\alpha - z^\alpha(t))(-n(t)) +$$

$$+H_{mm}(t)(n(t))^2] + \sum_{l\geq 3} \theta_l(t)|_{n=0}\,. \quad (3.49)$$

*Step 2.* We represent the geodesic $\gamma$ in the boundary normal coordinates $(\mathbf{z}, n)$ as $\gamma(t) = (\mathbf{z}(t), n(t))$. Assume that $(\mathbf{z}(t_0), n(t_0)) = (\mathbf{z}_0, 0)$, that is, the geodesic starts at the point $\mathbf{z}_0$. Next, we find the initial data,

$$\mathbf{p}|_{t=0} = \mathbf{p}_0 = (p_{10}, \ldots, p_{m0}).$$

Formula (3.49) implies that the Taylor expansion of $\theta(\mathbf{z}, 0, t)$ near $(\mathbf{z}_0, t_0)$ has the form

$$\theta(\mathbf{z}, 0, t) \asymp p_{\alpha 0}(z^\alpha - z_0^\alpha) - p_{\alpha 0}\frac{dz^\alpha(t_0)}{dt}(t - t_0) -$$

$$-p_{m0}\frac{dn(t_0)}{dt}(t - t_0) + \sum_{l\geq 2} \widetilde{\theta}_l, \quad (3.50)$$

where $\widetilde{\theta}_l$, $l = 2, \ldots,$ are homogeneous polynomials of order $l$ with respect to $(\mathbf{z} - \mathbf{z}_0, t - t_0)$. Indeed, formula (3.49) gives a polynomial expansion of $\theta(\mathbf{z}, 0, t)$with respect to $\mathbf{z} - \mathbf{z}(t)$. The coefficients in this expansion are some functions of $t$. Re-expanding $\mathbf{z}(t)$ and these coefficients into the Taylor series near $t = t_0$, we obtain the polynomials of $(\mathbf{z} - \mathbf{z}_0, t - t_0)$. The obtained polynomials on the right-hand side of formula (3.50) must coincide with those in the boundary condition (3.46). Moreover, it follows from (3.49) that $\widetilde{\theta}_l$ depend only on $\theta_l(t_0)|_{n=0}$ and the derivatives of $\theta_j(t_0)|_{n=0}$, $j = 1, 2, \ldots l-1$.

Comparing formula (3.50) with boundary condition (3.46), we see that

$$p_{\alpha 0} = 0, \quad \alpha = 1, \ldots, m-1, \quad p_{m0}\frac{dn(t_0)}{dt} = 1. \tag{3.51}$$

We have shown in section 2.4.9 that $|\mathbf{p}(t)| = 1$ for all $t$. In particular,

$$|\mathbf{p}_0| = h(\mathbf{z}_0, 0, \mathbf{p}_0) = \sqrt{g^{\alpha\beta}(\mathbf{z}_0, 0)p_{\alpha 0}p_{\beta 0} + p_{m0}^2} = 1.$$

Due to formulae (3.51), this implies that $|p_{m0}| = 1$. Thus, the geodesic, $\gamma(t)$ is a normal geodesic. As $dn(t_0)/dt$ is positive, we see that the velocity vector of $\gamma(t)$ at $\mathbf{z}_0$ is the interior normal vector.

**Exercise 3.21** *Using the Hamilton system (2.136) and the derived initial conditions,*

$$\mathbf{z}|_{t=t_0} = \mathbf{z}_0, \quad n|_{t=t_0} = 0, \quad \mathbf{p}|_{t=t_0} = (0, \ldots, 0, 1),$$

*show that*

$$\mathbf{z}(t) = \mathbf{z}_0, \quad n(t) = t - t_0, \quad \mathbf{p}(t) = (0, \ldots, 0, 1). \tag{3.52}$$

To complete the proof, we point out that it is actually necessary that $\gamma(t_0) = \mathbf{z}_0$. Indeed, if $\gamma(t_0) = \widetilde{\mathbf{z}}_0 \neq \mathbf{z}_0$, then the quadratic terms in the Taylor expansion (3.50) could not match the quadratic terms in the Taylor expansion (3.46). □

**3.5.6.** In this and the next sections we will complete the proof of Theorem 3.19. Since the geodesic $\gamma(t)$ is normal to the boundary, $\theta_l(t)$ are homogeneous polynomials of order $l$ with respect to $(\mathbf{z} - \mathbf{z}_0, n - (t - t_0))$.

Consider the quadratic terms in $\theta_2(\mathbf{z}, 0, t)$ in (3.50) and compare them with the quadratic terms in expansion (3.46). Then, the same considerations as in Lemma 3.20, show that

$$H(t_0) = \widehat{H}_0,$$

where

$$(\widehat{H}_0)_{\alpha\beta} = (H_0)_{\alpha\beta}, \quad (\widehat{H}_0)_{\alpha m} = 0, \quad (\widehat{H}_0)_{mm} = i, \tag{3.53}$$

$\alpha, \beta = 1, \ldots, m-1$. Having found the initial data for the quadratic form $H(t)$, we can use the procedure described in sections 2.4.11–2.4.13 to find $H(t)$. Then,

$$H(t) = Z(t)Y^{-1}(t), \tag{3.54}$$

and, in the boundary normal coordinates,

$$Z(t) = \widehat{H}_0, \quad Y(t) = I + \int_{t_0}^{t} C(t')dt' \cdot \widehat{H}_0, \tag{3.55}$$

where

$$C^{ik}(t) = g^{ik}(\mathbf{z}_0, t - t_0) - \delta^{im}\delta^{km}. \tag{3.56}$$

**Exercise 3.22** *Consider the matrix Riccati equation (2.162) in the boundary normal coordinates. Show that the matrices $B$ and $D$ are equal to zero and the matrix $C$ has form (3.56). Using the system of equations (2.172), prove formulae (3.55), (3.56) .*

Let us now return to expansion (3.50). If we assume that we already know $\theta_1(t), \ldots, \theta_{l-1}(t)$, then by comparing expansion (3.50) with expansion (3.46) we obtain the initial data for $\theta_l(t_0)$. Solving initial value problem (2.204), (2.205), we find $\theta_l(t)$. Thus all $\theta_l(t)$ can be found inductively.

Similarly, using boundary values (3.46) and equations (2.208), we find the Taylor expansions of the amplitude functions $u_l(\mathbf{z}, n, t)|_{t=t_0}$, $l = 0, 1, \ldots$, where

$$u_l(\mathbf{z}, n, t) \asymp \sum_{k \geq 0} u_{l,k}(t).$$

Here $u_{lk}(t)$ are homogeneous polynomials of order $k$ with respect to $(\mathbf{z}-\mathbf{z}_0, n-(t-t_0))$. In particular,

$$u_{0,0}(t_0) = V(\mathbf{z}_0)$$

and

$$u_{0,0}(t) = (\det Y(t))^{-1/2} \left[\frac{g(\mathbf{z}_0, 0)}{g(\mathbf{z}_0, t-t_0)}\right]^{1/4} V(\mathbf{z}_0). \qquad (3.57)$$

□

**3.5.7.** In the previous sections, we have constructed the phase and amplitude functions of a formal Gaussian beam $U_\epsilon^N(\mathbf{x}, t)$ that is defined in a neighborhood of $(\mathbf{z}_0, t_0)$ in $\mathcal{N} \times \mathbf{R}$ and satisfies boundary conditions (3.46). Moreover, using the considerations of section 2.4.22, we can find $U_\epsilon^N(\mathbf{x}, t)$ everywhere on $\mathcal{N} \times \mathbf{R}$. As the normal geodesic $\gamma(t)$ is on $\mathcal{M}$, for $t < t_0 + l(\mathbf{z}_0)$, we have constructed the desired formal Gaussian beam particularly on $\mathcal{M} \times [t_0, t_0 + l(\mathbf{z}_0)]$. This formal Gaussian beam $U_\epsilon^N(\mathbf{x}, t)$ satisfies boundary conditions (3.46). This proves Theorem 3.19. □

**3.5.8.** In this section, we will construct the Gaussian beam that corresponds to the formal Gaussian beam constructed in the previous sections.

Let $u_\epsilon(\mathbf{x}, t)$ be the solution of initial-boundary value problem (3.43). Then the following result holds.

**Theorem 3.23** *Let $T < l(\mathbf{z}_0)$. Then, for any $j \geq 0$ and multi-index $\alpha$,*

$$|\partial_t^j \partial_{\mathbf{x}}^\alpha (u_\epsilon(\mathbf{x}, t) - \chi(\mathbf{x}, t) U_\epsilon^N(\mathbf{x}, t))| \leq C_{j,\alpha} \epsilon^{N-(j+|\alpha|+m/4)}, \qquad (3.58)$$

$$\mathbf{x} \in \mathcal{M}, \ 0 < t < T.$$

*Here $U_\epsilon^N(\mathbf{x}, t)$ is the formal Gaussian beam of Theorem 3.19 and $\chi(\mathbf{x}, t)$ is an arbitrary cut-off function that is equal to 1 near the trajectory $(\mathbf{x}(t), t)$ of the Gaussian beam.*

**Proof.** Because the formal Gaussian beam $U_\epsilon^N(\mathbf{z}, n, t)$ satisfies inequality (2.217) and the boundary conditions (3.46), the assertion follows from Theorems 2.45 and 2.64. □

## 3.6. Domains of influence and Gaussian beams

**3.6.1.** The technique of Gaussian beams is instrumental in finding the distance between any point on a normal geodesic and any point on the boundary, i.e., in finding $d(\mathbf{z}_1, \gamma_{\mathbf{z}_0,\nu}(n))$, $\mathbf{z}_0, \mathbf{z}_1 \in \partial\mathcal{M}$. To this end, we use the fact that, at any time $t$, a Gaussian beam is concentrated near the point $\mathbf{x}(t)$ on its trajectory. Therefore, for any open set $\Omega \subset \mathcal{M}$ such that $\mathbf{x}(t) \notin \overline{\Omega}$, we have

$$\|\chi_\Omega u_\epsilon(\cdot\,,t)\| \le c_j\epsilon^j, \quad \|(1-\chi_\Omega)u_\epsilon(\cdot\,,t)\| \ge c_1 > 0, \tag{3.59}$$

for any $j$. Moreover, when $\Omega = \mathcal{M}(\Gamma, \tau_1)$ (see Fig 3.3), we can use the boundary spectral data to compute the above norms.

**Lemma 3.24** *For any* $\Gamma \subset \partial\mathcal{M}$, $t_0 < t < t_0 + l(\mathbf{z}_0)$ *and* $\tau_1 > 0$,

$$\lim_{\epsilon\to 0}\langle P_{\Gamma,\tau_1} u_\epsilon(\cdot,t), u_\epsilon(\cdot,t)\rangle = \begin{cases} \alpha(\mathbf{z}_0), & \mathbf{x}(t) \in \mathcal{M}^{int}(\Gamma,\tau_1), \\ 0, & \mathbf{x}(t) \in \mathcal{M}\setminus\mathcal{M}(\Gamma,\tau_1), \end{cases} \tag{3.60}$$

*where* $f = f_{\epsilon,\mathbf{z}_0,t_0}$ *is given by formula (3.41) and*

$$\begin{aligned} \alpha(\mathbf{z}_0) &= \frac{|V(\mathbf{z}_0)|^2[g(\mathbf{z}_0,0)]^{1/2}}{\sqrt{det(\Im H(t))}\,|det(Y(t))|} \\ &= \frac{|V(\mathbf{z}_0)|^2[g(\mathbf{z}_0,0)]^{1/2}}{\sqrt{det(\Im H(0))}\,|det(Y(0))|} > 0. \end{aligned} \tag{3.61}$$

**Remark.** Lemma 3.24 is the multidimensional analog of Lemma 1.24.

**Proof.** The proof is based on Lemma 2.58. We start by observing that

$$\langle P_{\Gamma,\tau_1} u_\epsilon(t), u_\epsilon(t)\rangle = \begin{cases} \|u_\epsilon(t)\|^2 + O(\epsilon^K), & \mathbf{x}(t) \in \mathcal{M}^{int}(\Gamma,\tau_1), \\ O(\epsilon^K), & \mathbf{x}(t) \in \mathcal{M}\setminus\mathcal{M}(\Gamma,\tau_1), \end{cases}$$

where $K > 0$ is arbitrary.

To evaluate $\|u_\epsilon(\cdot,t)\|^2$, we can restrict the integral, which represents the norm, to a coordinate neighborhood of the point $\mathbf{x}(t)$. Indeed, due to (3.59), the resulting error will be of the order $\epsilon^K$ for any $K > 0$. Moreover, we can replace $u_\epsilon(\mathbf{x},t)$ by the formal Gaussian beam $U_\epsilon^N(\mathbf{x},t)$, because this will also cause an error of the order $\epsilon^K$, when $N$ is sufficiently large. Thus,

$$\|u_\epsilon(\cdot,t)\|^2 = \int_{\mathbf{R}^m} |U_\epsilon^N(\mathbf{x},t)|^2 \chi(d(\mathbf{x},\mathbf{x}(t)))\sqrt{g(\mathbf{x})}d\mathbf{x} + O(\epsilon^K), \tag{3.62}$$

Figure 3.3: Gaussian beams and domains of influence

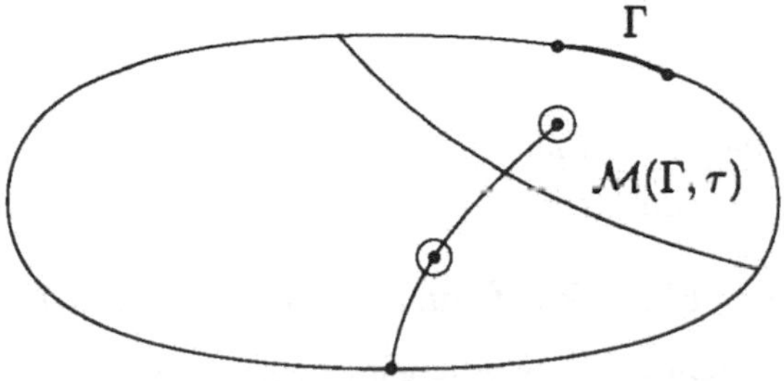


where $\chi(s)$ is, as usual, a cut-off function satisfying $\chi(s) = 1$ in a vicinity of $s = 0$. This integral contains the rapidly decreasing function $\exp\{-2\epsilon^{-1}\Im\theta(\mathbf{x},t)\}$, where

$$\Im\theta(\mathbf{x},t) = \frac{1}{2}(\Im H(t)\mathbf{y},\mathbf{y}) + \Im\Theta(\mathbf{x},t),$$

with

$$\Theta(\mathbf{x},t) = \sum_{l=3}^{N_\theta} \theta_l(t),$$

and $\theta_l(t)$, $l = 3, \dots, N_\theta$, being some homogeneous polynomials of order $l$ with respect to $\mathbf{y} = \mathbf{x} - \mathbf{x}(t)$. Using the fact that

$$|\epsilon^{-1}\Theta(\mathbf{x},t)| \le c\epsilon^{3\sigma},$$

when $|\mathbf{x} - \mathbf{x}(t)| \le \epsilon^{1/3+\sigma}$ with some small $\sigma > 0$, we can decompose $\exp\{-2\epsilon^{-1}\Theta(\mathbf{x},t)\}$ into the Taylor series in the ball $B_{\epsilon^{1/3+\sigma}}(\mathbf{x}(t))$. Outside this ball, $|\exp\{-2\epsilon^{-1}\Im\theta(\mathbf{x},t)\}| \le C_K\epsilon^K$, for any $K > 0$. Therefore, to within an error of order $\epsilon$, we can replace $\theta$ and $u_0(x,t)$ by the leading terms of their Taylor expansions. This leads to the following equation,

$$\|u_\epsilon(\cdot,t)\|^2 =$$

$$= (\pi\epsilon)^{-\frac{m}{2}} \int_{\mathbf{R}^m} e^{-\epsilon^{-1}(\Im H(t)\mathbf{y},\mathbf{y})} |u_{0,0}(t)|^2 \sqrt{g(\mathbf{x}(t))} dy^1 \cdots dy^m + O(\epsilon)$$

$$= \pi^{-m/2} |u_{0,0}(t)|^2 \sqrt{g(\mathbf{x}(t))} \int_{\mathbf{R}^m} e^{-(\Im H(t)\mathbf{z},\mathbf{z})} dz^1 \cdots dz^m + O(\epsilon)$$

$$= \frac{|u_{0,0}(t)|^2 \sqrt{g(\mathbf{x}(t))}}{\sqrt{\det(\Im H(t))}} + O(\epsilon). \tag{3.63}$$

Formulae (3.60) and (3.61) follow from representation (3.57). □

Let $\Gamma_l \to \mathbf{z}_1$ be a decreasing sequence of open sets, $\bigcap \Gamma_l = \mathbf{z}_1 \in \partial\mathcal{M}$. Then Lemma 3.24 implies the following result.

**Corollary 3.25** *For any $\mathbf{z}_1 \in \partial\mathcal{M}$, $t_0 < t < t_0 + l(\mathbf{z}_0)$ and $\tau_1 > 0$,*

$$\lim_{\epsilon\to 0} \langle P_{\mathbf{z}_1,\tau_1} u_\epsilon(\cdot,t), u_\epsilon(\cdot,t)\rangle = \begin{cases} \alpha(t), & \mathbf{x}(t) \in \mathcal{M}^{int}(\mathbf{z}_1,\tau_1), \\ 0, & \mathbf{x}(t) \in \mathcal{M} \setminus \mathcal{M}(\mathbf{z}_1,\tau_1), \end{cases} \tag{3.64}$$

*where $f$ and $\alpha(t)$ are the same as in Lemma 3.24.*

**3.6.2.** In the future, we will also need a more detailed information about $P_{\Gamma,\tau} u_\epsilon(t)$, when $t = t_0 + \tau$. Then the center of the Gaussian beam lies exactly on $\partial\mathcal{M}(\Gamma,\tau)$ (see Fig. 3.4).

**Lemma 3.26** *Let $\mathbf{z}_0 \in \partial\mathcal{M}$ and let $\Gamma \subset \partial\mathcal{M}$ be an open neighborhood of $\mathbf{z}_0$. Let $u_\epsilon(x,t)$ be the Gaussian beam that starts from the point $\mathbf{z}_0$ at time $t_0$ and propagates in the normal direction to $\partial\mathcal{M}$. Then, for any $\tau < \tau_{\partial\mathcal{M}}(\mathbf{z}_0)$ and any eigenfunction $\varphi_j$, $j = 1, 2, \ldots$,*

$$\langle P_{\Gamma,\tau}\varphi_j, u_\epsilon(t)\rangle =$$

$$= -i\epsilon^{\frac{m+2}{4}} 2^{\frac{m-1}{2}} \pi^{\frac{m-2}{4}} [det(-iH(t))]^{-\frac{1}{2}} \overline{u_{0,0}(t)} \varphi_j(\mathbf{z}_0,\tau) [g(\mathbf{z}_0,\tau)]^{\frac{1}{2}}$$

$$+ O(\epsilon^{(m+6)/4}), \tag{3.65}$$

*where $t = \tau + t_0$. Here $u_{0,0}(t)$ is given by formula (3.57) and $H(t)$ is given by formulae (3.53)–(3.55).*

Figure 3.4: Gaussian beam at the boundary of the domain of influence

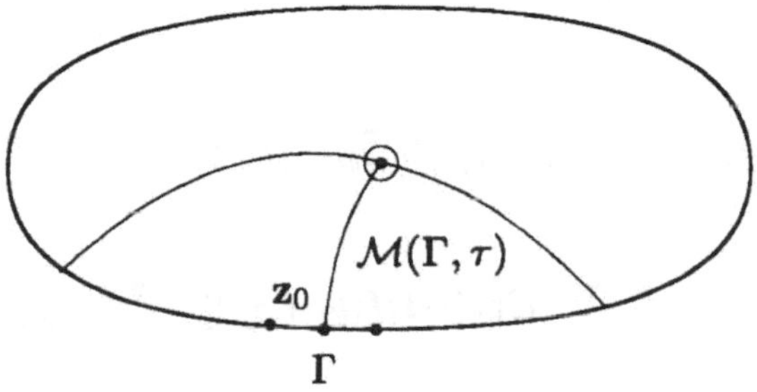


**Remark.** This lemma is an analog of the last equation in formula (1.77) of Lemma 1.24. As in Chapter 1, Lemma 3.26 will be used to find the potential $q$.

**Proof.** When deriving this formula, we can replace the Gaussian beam $u_\epsilon$ by the corresponding formal Gaussian beam $U_\epsilon^N$ and use the cut-off function that is equal to 1 in the $\epsilon^{1/2-\sigma}$-neighborhood of point $(\mathbf{z}_0, \tau)$, $0 < \sigma < 1/6$.

Thus, in the boundary normal coordinates $(\mathbf{z}, n)$, we have

$$\langle P_{\Gamma,\tau}\varphi_j, u_\epsilon^f(t_0+\tau)\rangle = \tag{3.66}$$

$$(\pi\epsilon)^{-m/4}\int_{\mathbf{R}^{m-1}}\int_{-\infty}^{\tau}\exp\left(i\epsilon^{-1}\theta(\mathbf{z},n,t_0+\tau)\right)\cdot$$

$$\cdot\left(\sum_{k=0}^{N}u_k(\mathbf{z},n,t_0+\tau)(i\epsilon)^k\right)\varphi_j(\mathbf{z},n)\cdot$$

$$\cdot\sqrt{g(\mathbf{z},n)}\chi((|\mathbf{z}-\mathbf{z}_0|^2+(n-\tau)^2)\epsilon^{-1+2\sigma})dz^1\cdots dz^{m-1}dn+O(\epsilon^K),$$

where $K>0$ can be taken arbitrarily large, when $N$ is sufficiently large, and $\chi(s)\in C_0^\infty(\mathbf{R})$ is equal to 1 near $s=0$. In the calculations below, we will replace $\sum_{k=0}^{N} u_k(i\epsilon)^k$ by the main term $u_{00}(t)$ and $\theta$ by $n-\tau+\frac{1}{2}\left(H(t)(\mathbf{z}-\mathbf{z}_0),(\mathbf{z}-\mathbf{z}_0)\right)$. We leave as an exercise to prove that the resulting error is of the order $\epsilon^{(m+6)/4}$. The main term of integral (3.66) is then given by

$$(\pi\epsilon)^{-\frac{m}{4}} \int_{\mathbf{R}^{m-1}} \int_{-\infty}^{T} \exp(i\epsilon^{-1}[n-\tau+\frac{1}{2}(H(t_0+\tau)(\mathbf{z}-\mathbf{z}_0),(\mathbf{z}-\mathbf{z}_0))])\cdot$$

$$\cdot\, u_{0,0}(t_0+\tau)\varphi_j(\mathbf{z}_0,n)\sqrt{g(\mathbf{z}_0,n)}d\mathbf{z}dn =$$

$$= -i\epsilon^{(m+2)/4}2^{(m-1)/2}\pi^{(m-2)/4}[\det(-iH(t_0+\tau))]^{-1/2}u_{0,0}(t_0+\tau)\cdot$$

$$\cdot\, \varphi_j(\mathbf{z}_0,\tau)\sqrt{g(\mathbf{z}_0,\tau)},$$

where we have used the formula

$$\int_{\mathbf{R}^{m-1}} e^{i(H\mathbf{z},\mathbf{z})/2}d\mathbf{z} = (2\pi)^{(m-1)/2}[\det(-iH)]^{-1/2},$$

when $\Im H > 0$. □

**Exercise 3.27** *Show that the remainder terms in formula (3.66) can be estimated by $C\epsilon^{(m+6)/4}$. This estimate can be obtained by integration by parts with respect to $n$ and further analysis of the $(m-1)$-dimensional integral over $\mathbf{z}$, taking into account that*

$$\int_{\mathbf{R}^{m-1}} e^{i(H\mathbf{z},\mathbf{z})/2}\mathbf{z}^{\alpha}d\mathbf{z} = 0$$

*for odd $|\alpha|$.*

## 3.7. Boundary distance functions

In section 3.6, we noted that Gaussian beams can be used to find the distance between any point on a normal geodesic and any boundary point. In this section, we will show how to do that to construct the set of the boundary distance functions.

**3.7.1.**

**Lemma 3.28** *For any* $\mathbf{z}_0, \mathbf{y} \in \partial\mathcal{M}$ *and* $s \in [0, l(\mathbf{z}_0))$, *the gauge equivalent boundary spectral data uniquely determines* $d(\gamma_{\mathbf{z}_0,\nu}(s), \mathbf{y})$.

**Proof.** By Corollary 3.14, the boundary spectral data determine uniquely $\|P_{\mathbf{y},\tau} u^{\varrho f}(t)\|$. Let $f = f_\epsilon$ be of form (3.41)–(3.42) with $V = 1$, i.e.

$$f_\epsilon(\mathbf{z}, t) = (\pi\epsilon)^{-m/4}\chi(\mathbf{z}, t)\exp\left(i\epsilon^{-1}\Theta(\mathbf{z}, t)\right), \ \epsilon > 0.$$

We point out that, if $u^f(t)$, $f = f_\epsilon$ is the Gaussian beam propagating along the normal geodesic, so is the wave $u_{\epsilon,\varrho}(t) = u^{\varrho f}(t)$, $f = f_\epsilon$. As $V(\mathbf{z}) = \rho(\mathbf{z})$ for this Gaussian beam, Lemma 3.24 implies that, for $s < l(\mathbf{z}_0)$,

$$\lim_{\epsilon \to 0} \|P_{\mathbf{y},\tau} u_{\epsilon,\varrho}(s + t_0)\| \tag{3.67}$$

$$= \begin{cases} |\varrho(\mathbf{z}_0)|^2 (g(\mathbf{z}_0, 0))^{1/2} h_Y(t), & d(\gamma_{\mathbf{z}_0,\nu}(s), \mathbf{y}) < \tau, \\ 0, & d(\gamma_{\mathbf{z}_0,\nu}(s), \mathbf{y}) > \tau, \end{cases}$$

where

$$h_Y(t) = [\det(\Im H(t))]^{-1/2} |\det(Y(t)|^{-1}.$$

As $|\varrho(\mathbf{z}_0)|^2 (g(\mathbf{z}_0, 0))^{1/2} [\det(\Im H(t))]^{-1/2} |\det(Y(t)|^{-1}$ is strictly positive, we can find $d(\gamma_{\mathbf{z}_0,\nu}(s), \mathbf{y})$. □

In particular, the minimum of the function $\mathbf{y} \mapsto d(\gamma_{\mathbf{z}_0,\nu}(s), \mathbf{y})$ gives the distance, $d(\gamma_{\mathbf{z}_0,\nu}(s), \partial\mathcal{M})$, from $\gamma_{\mathbf{z}_0,\nu}(s)$ to the boundary. Thus, by increasing $s$, we can find the arclength $l(\mathbf{z}_0)$, when the geodesic $\gamma_{\mathbf{z}_0,\nu}(s)$ hits the boundary for the first time. Similarly, we can find the function, $\tau_{\partial\mathcal{M}}(\mathbf{z}_0)$ defined in section 2.1.16,

$$\tau_{\partial\mathcal{M}}(\mathbf{z}_0) = \left\{ \sup_{s \geq 0} s : d(\gamma_{\mathbf{z}_0,\nu}(s), \mathbf{z}_0) = d(\gamma_{\mathbf{z}_0,\nu}(s), \partial\mathcal{M}) \right\}.$$

**3.7.2.** Let $\mathbf{x} \in \mathcal{M}$. Then the boundary distance function $r_\mathbf{x}$, which corresponds to the point $\mathbf{x}$ is given by the formula

$$r_\mathbf{x}(\mathbf{y}) = d(\mathbf{x}, \mathbf{y}), \quad \mathbf{y} \in \partial\mathcal{M},$$

Figure 3.5: Boundary distance function

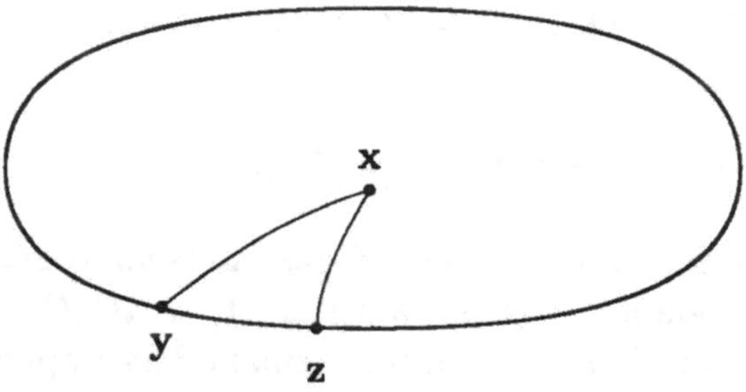


(see Fig. 3.5). The functions $r_{\mathbf{x}} : \partial\mathcal{M} \to \overline{\mathbf{R}_+}$ are continuous functions. We use the mapping

$$R : \mathcal{M} \mapsto L^\infty(\partial\mathcal{M}), \ \mathbf{x} \mapsto r_{\mathbf{x}},$$

which assigns to any point $\mathbf{x}$ the corresponding boundary distance function. Here $L^\infty(\partial\mathcal{M})$ is the metric space with the norm

$$\|r\|_{L^\infty(\partial\mathcal{M})} = \sup_{\mathbf{z}\in\partial\mathcal{M}} |r(\mathbf{z})|.$$

The image $R(\mathcal{M})$ of the mapping $R$,

$$R(\mathcal{M}) = \{r_{\mathbf{x}} \in L^\infty(\partial\mathcal{M}) : \ \mathbf{x} \in \mathcal{M}\}$$

and is called the set of the boundary distance functions.

### 3.7.3.

**Theorem 3.29** *The gauge equivalent boundary spectral data*

$$\{\partial\mathcal{M}, \ \lambda_j, \ \kappa_0\partial_\nu\varphi_j|_{\partial\mathcal{M}}, \ j = 1, 2, \dots\}$$

*determine the set* $R(\mathcal{M})$.

**Proof.** By Lemma 2.10, for any $\mathbf{x} \in \mathcal{M}$ there is at least one nearest boundary point $\mathbf{z} = \mathbf{z}(\mathbf{x})$. The point $\mathbf{x}$ lies on the normal geodesic from $\mathbf{z}$ at the arclength $s = s(\mathbf{x}) = d(\mathbf{x}, \partial\mathcal{M})$,

$$\mathbf{x} = \gamma_{\mathbf{z},\nu}(s)$$

(see Fig. 3.5). In particular, $s \le \tau_{\partial\mathcal{M}}(\mathbf{z})$. Thus,

$$\mathcal{M} = \bigcup_{\mathbf{z}\in\partial\mathcal{M}} \{\gamma_{\mathbf{z},\nu}(s) : \; s \in [0, \tau_{\partial\mathcal{M}}(\mathbf{z})]\}.$$

Note that this is not a disjoint union, because there may be several shortest geodesics that meet at the cut locus (see Theorem 2.12). By means of Lemma 3.28, for any $\mathbf{z} \in \partial\mathcal{M}$ we can find $\tau_{\partial\mathcal{M}}(\mathbf{z})$, and further for any $s \in [0, \tau_{\partial\mathcal{M}}(\mathbf{z})]$ we can construct the function

$$r_{\mathbf{z},s} = d(\gamma_{\mathbf{z},\nu}(s), \mathbf{y}), \quad \mathbf{y} \in \partial\mathcal{M}.$$

Clearly, $r_{\mathbf{z},s} = r_{\mathbf{x}}$ for $\mathbf{x} = \gamma_{\mathbf{z},\nu}(s)$. Changing $\mathbf{z}$ and $s$, we construct the set

$$R(\mathcal{M}) = \{r_{\mathbf{z},s} \in L^\infty(\partial\mathcal{M}) \; : \; \mathbf{z} \in \partial\mathcal{M}, \; s \in [0, \tau_{\partial\mathcal{M}}(\mathbf{z})]\}.$$

□

## 3.8. Reconstruction of the Riemannian manifold

In this section, we will construct a differential and Riemannian structures on the set $R(\mathcal{M})$ of the boundary distance functions. We will show that the resulting Riemannian manifold $(R(\mathcal{M}), \widetilde{g})$ is isometric to $(\mathcal{M}, g)$. Since all isometric manifolds are considered to be identical, constructing of $(R(\mathcal{M}), \widetilde{g})$ is equivalent to the reconstruction of $(\mathcal{M}, g)$. We emphasize that just the knowledge of $R(\mathcal{M})$ as a subset of $L^\infty(\partial\mathcal{M})$ is sufficient to find the differential and Riemannian structures on $\mathcal{M}$. [1]

**3.8.1.** As $R(\mathcal{M}) \subset C(\partial\mathcal{M}) \subset L^\infty(\partial\mathcal{M})$, it is a topological space with topology inherited from $L^\infty(\partial\mathcal{M})$.

**Lemma 3.30** *The mapping $R : \mathcal{M} \to R(\mathcal{M})$ is a homeomorphism.*

---

[1] It is well known that any manifold $\mathcal{M}$ can be embedded into Euclidean space $\mathbf{R}^d$ of a sufficiently large dimension, for instance, $d = 2m + 1$. In our case, we go even further and embed $\mathcal{M}$ into an infinite dimensional space $L^\infty(\partial\mathcal{M})$. For this end, we use the embedding $R : \mathcal{M} \to R(\mathcal{M})$.

**Proof.** Let $\mathbf{x}, \mathbf{y} \in \mathcal{M}$. By triangular inequality,

$$\|r_{\mathbf{x}} - r_{\mathbf{y}}\|_{L^\infty(\partial\mathcal{M})} \leq d(\mathbf{x}, \mathbf{y}), \ \mathbf{x}, \mathbf{y} \in \mathcal{M}. \tag{3.68}$$

Thus, $R$ is continuous. Next we show that $R$ is injective.

Assume that $r_{\mathbf{x}} = r_{\mathbf{y}}$. Let

$$s = \min_{\mathbf{z}' \in \partial\mathcal{M}} r_{\mathbf{x}}(\mathbf{z}'),$$

and $\mathbf{z} \in \partial\mathcal{M}$ be a point, where $r_{\mathbf{x}}(\mathbf{z}) = s$. Then, by Lemma 2.10, the point $\mathbf{x}$ lies on the normal geodesic, which starts from the point $\mathbf{z}$, $\mathbf{x} = \gamma_{\mathbf{z},\nu}(s)$. Since the same is valid for $\mathbf{y}$, we see that $\mathbf{x} = \mathbf{y}$.

Since, by definition, the mapping $R$ maps $\mathcal{M}$ onto $R(\mathcal{M})$, then $R$ is a bijective, continuous mapping defined on a compact set $\mathcal{M}$. These properties imply that $R^{-1}$ is also continuous and, therefore, $R$ is a homeomorphism. Although this is an elementary result of general topology, we recall its proof for the convenience of the reader.

The continuity of $R^{-1}$ is equivalent to the fact that, for any closed $K \subset \mathcal{M}$, its inverse image with respect to the mapping $R^{-1}$ is closed. In other words, this means that the set $R(K)$ is closed. However, $K$ is compact, so that its image by a continuous mapping is compact and, therefore, closed.

Hence, $R^{-1} : R(\mathcal{M}) \to \mathcal{M}$ is continuous and $R$ is a homeomorphism. □

**3.8.2. Remark** Let $(\mathcal{M}, g)$ be a geodesically regular manifold, i.e.,

i) For any $\mathbf{x}, \mathbf{y} \in \mathcal{M}$ there is a unique geodesic $\gamma$ joining these points.

ii) Any geodesic $\gamma([a, b])$ can be continued to a geodesic $\gamma([a', b'])$ whose end-points lie on the boundary.

Consider $R(\mathcal{M})$ as a metric space $(R(\mathcal{M}), d_\infty)$ with the distance inherited from $L^\infty(\partial\mathcal{M})$,

$$d_\infty(r_{\mathbf{x}}, r_{\mathbf{y}}) = \|r_{\mathbf{x}} - r_{\mathbf{y}}\|_{L^\infty(\partial\mathcal{M})}.$$

Then the mapping $R$ is an isometry, i.e.

$$d_\infty(r_{\mathbf{x}}, r_{\mathbf{y}}) = d(\mathbf{x}, \mathbf{y}).$$

Indeed, let $\gamma([a,b])$ be the shortest geodesic from $\mathbf{y}$ to $\mathbf{x}$. Continue this geodesic to the shortest geodesic $\gamma([a',b])$, where $\mathbf{z} = \gamma(a') \in \partial\mathcal{M}$. Then

$$r_{\mathbf{x}}(\mathbf{z}) - r_{\mathbf{y}}(\mathbf{z}) = |\gamma([a',b])| - |\gamma([a',a])| = b - a = d(\mathbf{x},\mathbf{y}).$$

Hence,

$$d(\mathbf{x},\mathbf{y}) \leq d_\infty(r_{\mathbf{x}}, r_{\mathbf{y}}).$$

Together with inequality (3.68) this yields that $d(x,y) = d_\infty(r_x, r_y)$. This means that, if we know *a priori* that $(\mathcal{M}, g)$ is geodesically regular, then $(R(\mathcal{M}), d_\infty))$ is isometric to $(\mathcal{M}, g)$ and the problem of the reconstruction of $(\mathcal{M}, g)$ is solved.

However, in the general case,

$$R : (\mathcal{M}, g) \to (R(\mathcal{M}), d_\infty)$$

is not an isometry. Think, for example, about a two-dimensional sphere with a small circular segment removed.

Nevertheless, as we will show later in this section, we can always equip $R(\mathcal{M})$ with a distance function that makes $R$ an isometry.

**3.8.3.** As $R : \mathcal{M} \to R(\mathcal{M})$ is a homeomorphism and $\mathcal{M}$ is a differentiable manifold, also $R(\mathcal{M})$ can be made a differentiable manifold. However, there may be several differentiable structures on the topological manifold $R(\mathcal{M})$. In this section, we will provide $R(\mathcal{M})$ with the structure of a differentiable manifold so that $R$ becomes a diffeomorphism from $\mathcal{M}$ onto $R(\mathcal{M})$.

Let $r \in R(\mathcal{M})$. Let $S(r) \in \mathbf{R}_+$ and $Z(r) \in \partial\mathcal{M}$ be defined as

$$S(r) = \min_{\mathbf{z}\in\partial\mathcal{M}} r(\mathbf{z})$$

and $Z(r) \in \partial\mathcal{M}$ as a point, where $r$ attains its minimum, i.e.

$$r(Z(r)) = S(r).$$

We point out that this point $Z(r)$ may not be unique. In this case, we choose $Z(r)$ to be any of these points. At last, let

$$\Gamma_{\mathbf{z}} = \{r \in R(\mathcal{M}) : \; r(\mathbf{z}) = S(r)\}, \; \mathbf{z} \in \partial\mathcal{M},$$

and

$$\mathcal{T}_{\partial\mathcal{M}}(\mathbf{z}) = \max_{r\in\Gamma_{\mathbf{z}}} r(\mathbf{z}).$$

Let us explain the geometrical meaning of the functions $S, Z, \mathcal{T}_{\partial\mathcal{M}}$ and the set $\Gamma_{\mathbf{z}}$. Let $\mathbf{x} = R^{-1}(r)$, i.e. $r = r_{\mathbf{x}}$. Then $S(r) = d(\mathbf{x}, \partial\mathcal{M})$ and $Z(r)$ is a nearest point to x on $\partial\mathcal{M}$ so that, in particular,

$$\mathbf{x} = \gamma_{Z(r),\nu}(S(r)).$$

Moreover, $\mathcal{T}_{\partial\mathcal{M}}(\mathbf{z}) = \tau_{\partial\mathcal{M}}(\mathbf{z})$ is the critical value defined in section 2.1.16 and

$$\Gamma_{\mathbf{z}} = R(\gamma_{\mathbf{z},\nu}([0, \tau_{\partial\mathcal{M}}(\mathbf{z})])$$

is a part of the normal geodesic before the cut locus, $\omega$.

It is clear from these considerations that we can find the image of $\omega$ in $R(\mathcal{M})$. Indeed,

$$R(\omega) = \{r \in R(\mathcal{M}) : \; S(r) = \mathcal{T}_{\partial\mathcal{M}}(Z(r))\}.$$

Next, we consider an analog of the boundary normal coordinates on $R(\mathcal{M}\setminus\omega) = R(\mathcal{M})\setminus R(\omega)$. First note that, since $R$ is a homeomorphism and the function $s(\mathbf{x})$, defined in section 2.1.16, is continuous, $s \in C(\mathcal{M})$, the function $S = s \circ R^{-1}$ is continuous on $R(\mathcal{M})$. Analogously, the function $\mathbf{z}(\mathbf{x})$, also defined in section 2.1.16, is continuous on $\mathcal{M}\setminus\omega$, so that the function $Z = z \circ R^{-1}$ is continuous on $R(\mathcal{M}\setminus\omega)$. Let $r_0 \in R(\mathcal{M}\setminus\omega)$ and $\mathbf{z}_0 = Z(r_0)$. Let $V \subset \partial\mathcal{M}$ be a coordinate neighborhood of $\mathbf{z}_0$ having coordinates $(z^1, \dots, z^{m-1})$. Then $(Z(r), S(r)) = (z^1, \dots, z^{m-1}, s)$ define coordinates on $\{r \in R(\mathcal{M}\setminus\omega) : \; Z(r) \in V\}$. Since $R : \mathcal{M}\setminus\omega \to R(\mathcal{M}\setminus\omega)$ is the identity mapping from the $(\mathbf{z}, s)$-coordinates to the $(Z, S)$-coordinates, we obtain the following lemma.

**Lemma 3.31** *The set of the boundary distance functions $R(\mathcal{M})$ determine uniquely the sets $R(\omega)$ and $R(\mathcal{M}\setminus\omega)$ and the function $\tau_{\partial\mathcal{M}}$. The pair $(Z(r), S(r))$ determines a system of coordinates on $R(\mathcal{M}\setminus\omega)$, which makes the mapping $R : \mathcal{M}\setminus\omega \to R(\mathcal{M}\setminus\omega)$,*

$$S(R(\mathbf{x})) = s(\mathbf{x}), \;\; Z(R(\mathbf{x})) = \mathbf{z}(\mathbf{x}),$$

*a diffeomorphism.*

**3.8.4.** As we know, the boundary normal coordinates are not appropriate near the cut locus, $\omega$. However, near $\omega$ we can use the boundary distance coordinates introduced in section 2.1.21.

In this section, we define on $R(\mathcal{M})$ an analog of the boundary distance coordinates. To this end, we use the evaluation functions

$$E_{\mathbf{z}} : R(\mathcal{M}) \to \mathbf{R}_+,$$

$$E_{\mathbf{z}}(r) = r(\mathbf{z}).$$

**Lemma 3.32** *For any $r_0 \in R(\mathcal{M}^{int})$ there are points $\mathbf{z}_1, \dots, \mathbf{z}_m \in \partial\mathcal{M}$ such that the functions $E^j = E_{\mathbf{z}_j}$, $j = 1, \dots, m$, form a local system of coordinates in a neighborhood of $r_0$. These coordinates, together with the boundary normal coordinates, form a differential structure of $R(\mathcal{M})$, which makes the mapping $R : \mathcal{M} \to R(\mathcal{M})$ a diffeomorphism.*

**Proof.** By Lemma 2.14, for any $\mathbf{x} \in \mathcal{M}^{int}$, e.g. $\mathbf{x} = R^{-1}(r_0)$, there are points $\mathbf{z}_1, \dots, \mathbf{z}_m \in \partial\mathcal{M}$ such that $\rho^j(\mathbf{x}) = d(\mathbf{x}, \mathbf{z}_j)$, $j = 1, \dots, m$, define local coordinates near $R^{-1}(r_0)$. Clearly,

$$(E_{\mathbf{z}_1}(r), \dots, E_{\mathbf{z}_m}(r)) = (d(R^{-1}(r), \mathbf{z}_1), \dots, d(R^{-1}(r), \mathbf{z}_m))$$

$$= (\rho^1(R^{-1}(r)), \dots, \rho^m(R^{-1}(r)))$$

can be used as as local coordinates near $r_0$ on $R(\mathcal{M})$ and the functions $(\rho^1, \dots, \rho^m)$ as coordinates near $R^{-1}(r_0)$. We see that $R : \mathbf{x} \mapsto r_{\mathbf{x}}$ is the identity mapping in these coordinates. Thus, the claim follows. □

**3.8.5.** Now we are in the position to complete the reconstruction of the Riemannian manifold. In this section, we use the evaluation functions and the boundary distance coordinates to equip the differentiable manifold $R(\mathcal{M})$ with a metric $\widetilde{g}$ such that $R : (\mathcal{M}, g) \to (R(\mathcal{M}), \widetilde{g})$ becomes an isometry.

Let us first show that such metric exists. Indeed, since $R : \mathcal{M} \to R(\mathcal{M})$ is a diffeomorphism, its differential $dR|_{\mathbf{x}} : T_{\mathbf{x}}\mathcal{M} \to T_{r_{\mathbf{x}}}R(\mathcal{M})$ is an isomorphism (see Section 2.1.3). Hence, the formula

$$(dR|_{\mathbf{x}}v, dR|_{\mathbf{x}}w)_{\widetilde{g}} = (v, w)_g, \quad v, w \in T_{\mathbf{x}}\mathcal{M},$$

Figure 3.6: Boundary distance coordinates

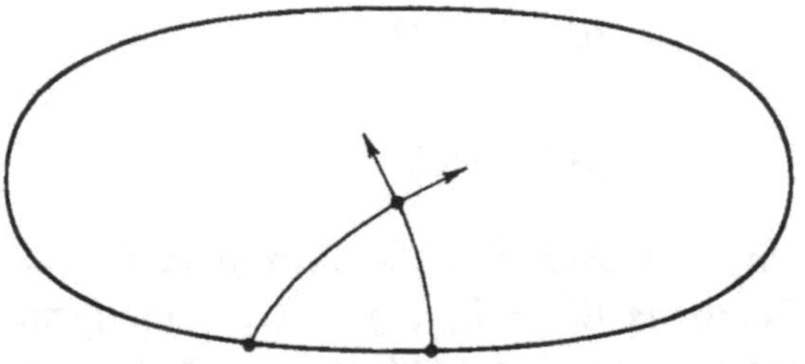


defines an inner product $(\cdot,\cdot)_{\widetilde{g}}$ on $T_{r_{\mathbf{x}}}(R(\mathcal{M}))$ and a metric tensor $\widetilde{g}$ on $R(\mathcal{M})$. This metric tensor $\widetilde{g}$ is the push forward of the metric tensor $g$ by the mapping $R$. It is then clear that the mapping $R : (\mathcal{M}, g) \to (R(\mathcal{M}), \widetilde{g})$ is an isometry. However, since we do not know $(\mathcal{M}, g)$, it is our goal to construct the metric tensor $\widetilde{g}$ from $R(\mathcal{M})$.

Let $r_0 \in R(\mathcal{M})^{int}$ and let $\mathbf{z}_1, \ldots, \mathbf{z}_m$ be such points of $\partial\mathcal{M}$ that

$$(\rho^1, \ldots, \rho^m) = (E_{\mathbf{z}_1}(r), \ldots, E_{\mathbf{z}_m}(r))$$

are local coordinates near $r_0$ (see Fig. 3.6). Consider the evaluation functions $E_{\mathbf{z}}(r)$, where $\mathbf{z}$ lies in a neighborhood $V$ of $\mathbf{z}_0 = Z(r_0)$ and $r$ lies in a neighborhood of $U$ of $\mathbf{x}_0 = R^{-1}(r_0)$. As we know from section 3.8.4,

$$E_{\mathbf{z}}(r) = d(\mathbf{x}, \mathbf{z}),$$

where $\mathbf{x} = R^{-1}(r)$. By Lemma 2.15, the function $d(\mathbf{x}, \mathbf{z})$ is a smooth function of $\mathbf{x} \in U$ and $\mathbf{z} \in V$, when $U$ and $V$ are sufficiently small, and the vectors

$$\{\,\mathrm{Grad}_{\mathbf{x}} d(\mathbf{x}, \mathbf{z})|_{\mathbf{x}_0} \in S_{\mathbf{x}_0}\mathcal{M} :\ \mathbf{z} \in V\}$$

form an open set $W$ of the unit ball $S_{\mathbf{x}_0}\mathcal{M} \subset T_{\mathbf{x}_0}\mathcal{M}$.

Hence, the set $\mathcal{W}$,

$$\mathcal{W} = \{\,\mathrm{Grad}_r E_{\mathbf{z}}|_{r_0} \in T_{r_0}R(\mathcal{M}) :\ \mathbf{z} \in V\},$$

is an $(m-1)$-dimensional submanifold of $T_{r0}R(\mathcal{M})$. With respect to the yet unknown metric $\widetilde{g}$, the submanifold $\mathcal{W}$ lies on $S_{r_0}(R(\mathcal{M}))$, i.e.,

$$|\operatorname{Grad}_r E_{\mathbf{z}}|_{r0}|_{\widetilde{g}} = |\operatorname{Grad}_{\mathbf{x}} d(\mathbf{x},\mathbf{z})|_{\mathbf{x}0}|_g = 1, \ \mathbf{z} \in V.$$

We do not know $\widetilde{g}$ and, therefore, the canonical transformation $I_{\widetilde{g}}$ from covectors to vectors on $(R(\mathcal{M}), \widetilde{g})$, defined in section 2.1.4. Hence, although we can find $d_r E_{\mathbf{z}}$, we cannot find

$$\operatorname{Grad}_r E_{\mathbf{z}} = I_{\widetilde{g}}\, d_r E_{\mathbf{z}}.$$

Let us, however, work with the known set $\mathcal{W}^*$,

$$\mathcal{W}^* = \{d_r E_{\mathbf{z}}|_{r0} \in T^*_{r0} R(\mathcal{M}) : \ \mathbf{z} \in V\}.$$

Previous observations imply that $\mathcal{W}^* \subset S^*_{r0} R(\mathcal{M})$ is open, where $S^*_{r0} R(\mathcal{M}) \subset T^*_{r0} R(\mathcal{M})$ is the unit sphere with respect to the metric tensor $\widetilde{g}$. We use this observation to actually construct the metric tensor $\widetilde{g}$.

**Lemma 3.33** *The submanifold $\mathcal{W}^* \subset T^*_{r0}(R(\mathcal{M}))$ determines the metric tensor $\widetilde{g}^{jk}(r_0)$.*

**Proof.** We remind the reader that we work in the local coordinates $\rho^1, \ldots, \rho^m$ in $R(\mathcal{M})$ so that $E_{\mathbf{z}} = E_{\mathbf{z}}(\rho^1, \ldots, \rho^m)$. The 1-forms $d\rho^1, \ldots, d\rho^m$ form a basis in $T^*_{r0} R(\mathcal{M})$. As $\mathcal{W}^* \subset S^*_{r0}(R(\mathcal{M}))$ is open, the set

$$\mathbf{R}_+ \mathcal{W}^* = \left\{\alpha \left(\partial_{\rho^1} E_{\mathbf{z}}, \ldots, \partial_{\rho^m} E_{\mathbf{z}}\right) : \alpha \in \mathbf{R}_+, \, \mathbf{z} \in V\right\},$$

is an open cone in the cotangent space $T^*_{r0} R(\mathcal{M})$. Therefore, for any $\mathbf{p} = \alpha\left(\partial_{\rho^1} E_{\mathbf{z}}, \ldots, \partial_{\rho^m} E_{\mathbf{z}}\right) \in \mathbf{R}_+ \mathcal{W}^*$,

$$F(\mathbf{p}) = (\mathbf{p}, \mathbf{p})_{\widetilde{g}} = \widetilde{g}^{jk}(r_0) p_j p_k = \alpha^2. \tag{3.69}$$

This equation can be used to find the metric tensor $\widetilde{g}$. Indeed, since $F(\mathbf{p})$ is known in an open set $\mathbf{R}_+ \mathcal{W}^*$, we can find the differentials of $F$. In view of (3.69),

$$\partial_{p_j} \partial_{p_k} F(\mathbf{p}) = \widetilde{g}^{jk}(r_0).$$

This equation, which is valid for any $\mathbf{p} \in \mathbf{R}_+ \mathcal{W}^*$, determines $\widetilde{g}$ in the boundary distance coordinates $\rho^1, \ldots, \rho^m$ at the point $r_0$. □

**Exercise 3.34** *Prove that the diagonal elements* $\widetilde{g}^{jj}$ *of the metric tensor* $\widetilde{g}$ *are equal to 1.*

Since the point $r_0 \in R(\mathcal{M})^{int}$ is arbitrary, we have constructed $\widetilde{g}$ in $R(\mathcal{M})^{int}$. Rewriting the metric tensor $\widetilde{g}$ in the boundary normal coordinates and using its smoothness on $R(\mathcal{M})$, we can find $\widetilde{g}$ on whole $R(\mathcal{M})$.

**3.8.6.** Summarizing the previous considerations we obtain the following result.

**Lemma 3.35** *Given the gauge equivalent boundary spectral data*

$$\{\partial\mathcal{M},\ \lambda_j,\ \kappa_0\partial_\nu\varphi_j|_{\partial\mathcal{M}},\ j=1,2,\dots\}$$

*of a Riemannian Schrödinger operator* $-\Delta_g + q$, *it is possible to reconstruct the Riemannian manifold* $(\mathcal{M}, g)$.

## 3.9. Reconstruction of the potential

**3.9.1.** Since we have already constructed the metric $\widetilde{g}$, we can, in particular, find the boundary volume element $dS_g$ and, therefore, the factor $\eta$ in the equation $d\mu = \eta\, dS_g$.

**Lemma 3.36** *The gauge equivalent boundary spectral data of a Schrödinger operator,*

$$\{\partial\mathcal{M},\ \lambda_j,\ \kappa_0\partial_\nu\varphi_j|_{\partial\mathcal{M}},\ j=1,2,\dots\},$$

*determines the function* $\kappa_0$.

**Proof.** Consider the Gaussian beam $u_{\epsilon,\rho}(\mathbf{x},t)$ that is generated by the boundary source $f_\epsilon$ given by formula (3.41) with $V = 1$ and some matrix $H_0$. Using Corollary 3.14, we can find $||u_{\epsilon,\rho}(\cdot\,,t)||$ and evaluate

$$\lim_{\epsilon\to 0} ||u_{\epsilon,\rho}(t)||^2 = \frac{|\varrho(\mathbf{z}_0)|^2\sqrt{g(\mathbf{z}_0,0))}}{\sqrt{\det(\Im H(t))}\,|\det(Y(t))|}.$$

Since the metric tensor $g$ is already known and $H_0$ is in our disposal, we can find $\Im H(t)$ and $Y(t)$. Using the above equation, this gives us $\varrho(\mathbf{z}_0) > 0$. Since $\varrho = \eta\kappa_0$, we find $\kappa_0$. □

**3.9.2.** The previous result reduces Problem 5 to Problem 4, where the boundary measure $dS_g$ is known. Since the metric $g$ is already reconstructed, it remains to show how to find the potential $q$.

By Corollary 3.15, we can find the inner product

$$\langle \varphi_k, P_{\mathbf{y},\tau} u^f(t)\rangle = \langle P_{\mathbf{y},\tau}\varphi_k, u^f(t)\rangle$$

for any $f$, in particular, for $f = f_\epsilon$ of form (3.41)–(3.42) with $V(\mathbf{z}) = 1$. We remind the reader that, in this case, $u^f(t)$ is a Gaussian beam that we have denoted by $u_\epsilon(t)$. Due to Lemma 3.26,

$$\lim_{\epsilon\to 0} \epsilon^{-(m+2)/4} \langle P_{\mathbf{y},\tau}\varphi_k, u_\epsilon(t)\rangle = \tag{3.70}$$

$$= -i\pi^{(m-2)/2} 2^{(m-1)/2} \frac{\overline{u_{0,0}(t)\sqrt{g(\mathbf{z}_0,\tau)}}}{\sqrt{\det(-iH(t))}} \varphi_k(\mathbf{z}_0,\tau),$$

where $t = \tau + t_0$ and $\tau < \tau_{\partial\mathcal{M}}(\mathbf{z}_0)$ and the right-hand side of (3.70) is written in the boundary normal coordinates. As $u_{0,0}(t)$ and $H(t)$ depend only on the metric tensor $g$, the boundary spectral data determine $\varphi_k(\mathbf{x})$ on $\mathcal{M} \setminus \omega$, and, due to the continuity of $\varphi_k(\mathbf{x})$, everywhere on $\mathcal{M}$. As by Theorem 2.21, $\varphi_1(\mathbf{x}) \neq 0$, the potential $q$ is determined by the formula

$$q(x) = \frac{\Delta_g\varphi_1(\mathbf{x}) + \lambda_1\varphi_1(\mathbf{x})}{\varphi_1(\mathbf{x})}, \quad \mathbf{x} \in \mathcal{M}^{int}.$$

**3.9.3.** Now we can formulate the final result for Problem 5 of section 3.1.

**Theorem 3.37** *Given the gauge equivalent boundary spectral data*

$$\{\partial\mathcal{M},\ \lambda_j,\ \kappa_0\partial_\nu\varphi_j|_{\partial\mathcal{M}},\ j = 1, 2, \dots\}$$

*of a Schrödinger operator $-\Delta_g + q$ on a manifold $\mathcal{M}$, it is possible to reconstruct the Riemannian manifold $(\mathcal{M}, g)$ and potential $q$.*

In particular, this result implies Theorem 3.3 as explained in section 3.1.6.

**Notes.** The interest of the international mathematical community toward the multidimensional inverse boundary spectral problems was attracted by Gel'fand [Ge] in 1954. These problems were first considered by Berezanskii. He studied the inverse problem for the Schrödinger operator in a domain of $\mathbf{R}^m$ in the case of a small potential [Br1]–[Br3]. A breakthrough in the study of these problems was achieved in mid-eighties. At that time, there appeared two different new approaches to solve the inverse boundary spectral problems.

The first approach is based on the use of exponentially growing solutions. It was originally developed to study the fixed frequency inverse problems, rather the inverse boundary spectral ones. We will provide a more extensive description of the history of this method in Notes to Chapter 4. Using this method, Nachman, Sylvester, Uhlmann [NaSyU] and Novikov [Nv1] studied inverse boundary spectral problems for the Schrödinger and some other types of isotropic operators in $\mathbf{R}^m$ and proved global uniqueness results.

The other approach to inverse problems, which we will mainly follow in this book, is based on the boundary control method. This method was developed by Belishev [Be1] to study the inverse boundary spectral problem for the acoustic operator $-c^2(\mathbf{x})\Delta$ in a domain in $\mathbf{R}^m$. It heavily utilizes control theory for the wave equation. The method is based on the study of domains of influence and inner products of waves and goes back to the one-dimensional method of Krein-Blagovestchenskii [Kr1]–[Kr4], [Bl1]–[Bl3]. The boundary control method was also used to study some other types of inverse boundary problems (see, e.g., [Bl2] and Notes to Chapter 4).

The coordinate invariance of the approach was first understood in the study of the inverse boundary spectral problem for the Laplace-Beltrami operator on a Riemannian manifold [BeKu3]. The gauge invariance of the approach was observed later in the study of the inverse boundary spectral problem for a general elliptic operator on a manifold with boundary [Ku1]–[Ku4]. The next logical step in the development of the approach was related to Gaussian beams. They were originally used for the dynamical inverse problem (see [BeKa1] and also Notes to Chapter 4). In the spectral case, Gaussian beams were introduced to study the inverse boundary spectral problem with incomplete data [KaKu1], [KaKu2]. The boundary distance function representation of a Riemannian manifold was introduced in [Ku5]. Our purpose in this chapter has been to combine these ideas to solve the inverse boundary spectral problem for a general operator.

Both these approaches have various extensions and generalizations. The approach based on the exponentially growing solutions has been used to study inverse boundary spectral problems with incomplete data [Iz], [Rm3], non-selfadjoint inverse boundary spectral problems [La1], stability of inverse boundary spectral problems [Al], and inverse problems for the Maxwell equations [La2].

The approach based on the boundary control method has been used also to study inverse boundary spectral problems with incomplete data [KaKu1], [KaKu2], non-selfadjoint inverse boundary spectral problems and inverse problems for operator pencils [KuLa1], [KuLa2], stability of the inverse boundary spectral problems [KKuLa] and to construct numerical solutions of inverse boundary spectral problems [BeKa2], [BeRFi]. The inverse boundary spectral problems for the acoustic equation was also analyzed by the moments method in [KuSr], [KuPe].

# Chapter 4

# Inverse problems for the wave and other types of equations

This chapter is devoted to various generalizations and extensions of the approach developed in Chapter 3. We will pay special attention to the inverse problems for the wave equation, due to both the importance of such problems and the hyperbolic nature of our approach.

In section 4.1, we will consider a number of inverse boundary-value problems for the wave and heat equations and also for the original elliptic operator. We will reduce them to the inverse boundary spectral problem. In particular, we will show the solvability of the inverse boundary-value problem with data given by the energy flux.

Section 4.2 deals with the inverse boundary-value problem for the wave equation with data given on a finite time interval. It is shown that, in this case, the reconstruction is possible in a collar neighborhood of the boundary.

In section 4.3, we will show how to continue the boundary data for the wave equation, which are given on a sufficiently large time interval, onto the infinite time interval. For the heat equation, the construction is possible for any finite time interval.

In section 4.4, we will develop a variant of the approach described in Chapter 3, when the boundary spectral data are given on a part of the boundary. In the last section, we will apply the results of Chapter

3 to study some practically important inverse boundary problems in domains of $\mathbf{R}^m$. In particular, we will analyze, how additional information about the operator can be used to reduce the admissible group of transformations and, in many cases, prove the uniqueness.

## 4.1. Inverse problems with different types of data

**4.1.1.** In this section, we will consider inverse boundary value problems, that are equivalent to the inverse boundary spectral problem. In particular, we will consider inverse boundary-value problems for the wave equation,

$$\partial_t^2 u + a(\mathbf{x}, D)u = 0, \quad \text{in} \quad Q^T = \mathcal{M} \times [0, T], \tag{4.1}$$

(see section 2.3.1), where $a(\mathbf{x}, D)$ is an elliptic differential expression given by formula (2.30),

$$a(\mathbf{x}, D)u = -m^{-1} g^{-1/2} \partial_i m g^{1/2} g^{ij} \partial_j u + qu. \tag{4.2}$$

We will also consider an inverse boundary value-problem for the heat equation,

$$\partial_t w + a(x, D)w = 0, \quad \text{in} \quad Q^T.$$

Moreover, we will consider different kinds of physically meaningful boundary data for these equations, as well as for the corresponding elliptic operator introduced in sections 2.2.3-2.2.4,

$$\mathcal{A}u = a(\mathbf{x}, D)u, \quad \mathcal{D}(A) = H^2(\mathcal{M}) \cap H_0^1(\mathcal{M}).$$

Our aim is to explain how various inverse problems for all these operators can be transformed to the inverse boundary spectral problem. Before that, in sections 4.1.2–4.1.5, we will consider different sets of boundary data used in inverse problems.

**4.1.2.** One of the most important concepts of physics is that of energy. For wave equation (4.1), an appropriate energy form $E(u, t)$ is given by the formula

$$E(u, t) = E_{int}(u, t) + E_b(u, t). \tag{4.3}$$

Here

$$E_{int}(u,t) = \frac{1}{2}\int_{\mathcal{M}} (|\partial_t u(\mathbf{x},t)|^2 + |\operatorname{Grad} u(\mathbf{x},t)|_g^2 +$$

$$+q(\mathbf{x})|u(\mathbf{x},t)|^2)m(\mathbf{x})dV_g(\mathbf{x}), \tag{4.4}$$

$$E_b(u,t) = \frac{1}{2}\int_{\partial\mathcal{M}} \sigma(\mathbf{x})|u(\mathbf{x},t)|^2 m(\mathbf{x})dS_g(\mathbf{x})$$

are the interior and the boundary energy forms. The real valued function $\sigma$, $\sigma \in C^\infty(\partial\mathcal{M})$ in the boundary energy form $E_b$ is called the Robin coefficient.

Simple computations, based on integration by parts, show that for a wave $u(\mathbf{x},t)$,

$$E(u,T) - E(u,0) = \Re\int_0^T \int_{\partial\mathcal{M}} (-\partial_\nu u + \sigma u)\overline{\partial_t u}\, m dS_g dt. \tag{4.5}$$

Note that the sign minus in formula (4.5) appears because $\nu$ is the interior normal vector. Formula (4.5) makes natural the following definition of the energy flux $\Pi^T$ through the boundary,

$$\Pi^T(u) = \Re\int_0^T \int_{\partial\mathcal{M}} (-\partial_\nu u + \sigma u)\overline{\partial_t u}\, m dS_g dt. \tag{4.6}$$

Physically, the energy flux $\Pi^T$ tells how much energy comes through the boundary during the time interval from 0 to $T$.

**4.1.3.** Inverse boundary-value problems for wave equation (4.1), which we study in this section, are related to the Dirichlet initial-boundary value problem,

$$\partial_t^2 u + a(x,D)u = 0, \quad \text{in} \quad Q^T, \tag{4.7}$$

$$u|_{\Sigma^T} = f, \quad \Sigma^T = \partial\mathcal{M} \times [0,T], \tag{4.8}$$

$$u|_{t=0} = 0, \quad \partial_t u|_{t=0} = 0. \tag{4.9}$$

As in Chapter 3, we denote the solution of this problem by $u = u^f$ to indicate the boundary source $f$. For this initial boundary-value problem we define the response operator $\Lambda^T$, which is also called the hyperbolic Dirichlet-Robin map,

$$\Lambda^T f = (-\partial_\nu u^f + \sigma u^f)|_{\Sigma^T}. \tag{4.10}$$

This map corresponds to the energy form (4.3)–(4.4).

We consider $\Lambda^T$ as an operator in the space $\overset{\circ}{C}{}^\infty(\Sigma^T)$,

$$\overset{\circ}{C}{}^\infty(\Sigma^T) = \{f \in C^\infty(\Sigma^T) : \text{ supp } (f) \subset \partial\mathcal{M} \times (0, T]\}. \tag{4.11}$$

The graph of the operator $\Lambda^T$ is the set of all admissible Cauchy data,

$$(u|_{\Sigma^T},\ (-\partial_\nu u + \sigma u)|_{\Sigma^T}), \tag{4.12}$$

of the solutions of initial-boundary value problem (4.7)–(4.9).

Obviously, the Dirichlet-Robin map $\Lambda^T$ determines the energy flux $\Pi^T(u^f)$, for any given $f$, if we know $mdS_g$. Another important object, determined by $\Lambda^T$, is the boundary form $\mathcal{B}^T$, which is given by

$$\mathcal{B}^T[f,h] \; = \; \int_0^T \int_{\partial\mathcal{M}} (\partial_\nu u^f \overline{u^h} - u^f\, \overline{\partial_\nu u^h})\, mdS_g dt \tag{4.13}$$

$$= - \int_0^T \int_{\partial\mathcal{M}} (\Lambda^T f\, \overline{h} - f\, \overline{\Lambda^T h})\, mdS_g dt.$$

The boundary form $\mathcal{B}^T$ has a semi-symplectic structure in the following sense. Consider the pairs $(f, h)$, $f, h \in C^\infty(\Sigma^T)$ with the natural symplectic form,

$$\omega((f_1, h_1), (f_2, h_2)) = \int_0^T \int_{\mathcal{M}} (f_1 \overline{h_2} - h_1 \overline{f_2})\, mdS_g dt.$$

Then, $\mathcal{B}^T$ is the restriction of the symplectic form $\omega$ to the graph of $\Lambda^T$. We note also that the form $\mathcal{B}^T$ does not depend on the Robin coefficient $\sigma$.

**4.1.4.** The inverse boundary-value problem for the heat equation is related to the initial boundary-value problem for $w = w^f(\mathbf{x}, t)$, satisfying

$$\partial_t w + a(\mathbf{x}, D)w = 0, \quad \text{in} \quad Q^T, \tag{4.14}$$

$$w|_{\Sigma^T} = f, \quad w|_{t=0} = 0.$$

Analogously to the hyperbolic case, we define the parabolic Dirichlet-Robin map, i.e., the response operator for the heat equation, by the formula,

$$L^T f = (-\partial_\nu w^f + \sigma w^f)|_{\Sigma^T}.$$

Clearly,

$$L^T : \overset{\circ}{C}{}^\infty(\Sigma^T) \to \overset{\circ}{C}{}^\infty(\Sigma^T).$$

**4.1.5.** Other types of inverse boundary-value problems that we will discuss in this section are related to the Dirichlet problem for the Helmholtz equation,

$$(-a(\mathbf{x}, D) + \lambda)v = 0, \tag{4.15}$$

$$v|_{\partial\mathcal{M}} = h,$$

where $\lambda \in \mathbf{C}$ is a given parameter and $h \in C^\infty(\partial\mathcal{M})$. We denote the solution of this problem by $v = v^h(\mathbf{x}, \lambda)$. The Dirichlet problem (4.15) is closely related to the operator $\mathcal{A}$, introduced in sections 2.2.3–2.2.4,

$$\mathcal{A}u = a(\mathbf{x}, D)u, \quad \mathcal{D}(\mathcal{A}) = H^2(\mathcal{M}) \cap H_0^1(\mathcal{M}). \tag{4.16}$$

In particular, problem (4.15) has a unique solution, when $\lambda \notin \{\lambda_j\} = \{\lambda_j : j = 1, 2, \dots\}$. We remind the reader that $\lambda_j$ are the eigenvalues of the operator $\mathcal{A}$, while $\varphi_j$ are the corresponding orthonormal eigenfunctions. The elliptic Dirichlet-Robin map $\Lambda_\lambda$ is given by

$$\Lambda_\lambda h = -\partial_\nu v^h + \sigma v^h|_{\partial\mathcal{M}}, \ \lambda \notin \{\lambda_j\}.$$

This map is continuous in $C^\infty(\partial\mathcal{M})$. Later we will show that $\Lambda_\lambda$ depends analytically on $\lambda \in \mathbf{C} \setminus \{\lambda_j\}$. Hence, its Schwartz kernel is a distribution $\Lambda_\lambda(\mathbf{x}, \mathbf{y}) \in \mathcal{D}'(\partial\mathcal{M} \times \partial\mathcal{M})$, which also depends analytically on $\lambda \in \mathbf{C} \setminus \{\lambda_j\}$.

To give a representation of this kernel, we use Green's function that corresponds to the operator $\mathcal{A}$,

$$(-a(\mathbf{x}, D) + \lambda)G_\lambda(\mathbf{x}, \mathbf{y}) \quad = \quad -\delta_\mathbf{y}(\mathbf{x}), \tag{4.17}$$

$$G_\lambda(\mathbf{x}, \mathbf{y})|_{\mathbf{x}\in\partial\mathcal{M}} = 0,$$

where $\delta_\mathbf{y}$ is the Dirac delta-distribution, corresponding to the volume element $mdV_g$ on $\mathcal{M}$. Distribution $G_\lambda(\mathbf{x}, \mathbf{y})$ with $\mathbf{x}$, $\mathbf{y} \in \mathcal{M}$ is, in fact, the Schwartz kernel of the resolvent $R_\lambda$, defined in section 2.2.5.

Green's function $G_\lambda(\mathbf{x}, \mathbf{y})$ is a smooth function of $(\mathbf{x}, \mathbf{y})$ outside the diagonal $D = \{(\mathbf{x}, \mathbf{x}) : \ \mathbf{x} \in \mathcal{M}\}$. Therefore, we can define the Robin derivative,

$$(-\frac{\partial}{\partial s} + \sigma(\mathbf{x}))(-\frac{\partial}{\partial t} + \sigma(\mathbf{y}))\, G_\lambda(\gamma_{\mathbf{x},\nu}(s), \gamma_{\mathbf{y},\nu}(t)),$$

when $(\mathbf{x}, s) \neq (\mathbf{y}, t)$, where $\mathbf{x}, \mathbf{y} \in \partial\mathcal{M}$ and $t, s > 0$ are small enough. We remind the reader that $\gamma_{\mathbf{x},\nu}$ is the normal geodesic from the boundary (see section 2.1.14).

**Lemma 4.1** *The Dirichlet-Robin map $\Lambda_\lambda$ has the representation,*

$$\Lambda_\lambda h(\mathbf{y}) = \tag{4.18}$$

$$= -\int_{\partial\mathcal{M}} (-\partial_{\nu(\mathbf{x})} + \sigma(\mathbf{x}))(-\partial_{\nu(\mathbf{y})} + \sigma(\mathbf{y}))G_\lambda(\mathbf{x}, \mathbf{y})h(\mathbf{x})\, m(\mathbf{x})dS(\mathbf{x}),$$

*The kernel $-(\partial_{\nu(\mathbf{x})} - \sigma(\mathbf{x}))(\frac{\partial}{\partial\nu(\mathbf{y})} - \sigma)G_\lambda(\mathbf{x}, \mathbf{y})$ of the operator $\Lambda_\lambda$ is understood as the limit,*

$$-(\partial_{\nu(\mathbf{x})} - \sigma(\mathbf{x}))(\partial_{\nu(\mathbf{y})} - \sigma(\mathbf{y}))G_\lambda(\mathbf{x}, \mathbf{y}) =$$

$$= \lim_{t\to 0}\lim_{s\to 0} -(\frac{\partial}{\partial s} - \sigma(\mathbf{x}))(\frac{\partial}{\partial t} - \sigma(\mathbf{y}))G_\lambda(\gamma_{\mathbf{x},\nu}(s), \gamma_{\mathbf{y},\nu}(t)), \tag{4.19}$$

*which is defined in $\mathcal{D}'(\partial\mathcal{M} \times \partial\mathcal{M})$.*

**Proof.** Let $h \in C^\infty(\partial\mathcal{M})$ and let $v = v^h(\mathbf{y})$ be the solution of problem (4.15). Integrating by parts and using definition (4.17) of Green's function, we see that

$$v(\mathbf{y}) = \int_{\partial\mathcal{M}} \partial_{\nu(\mathbf{x})} G_\lambda(\mathbf{x},\mathbf{y})\, h(\mathbf{x})\, m(\mathbf{x}) dS_g(\mathbf{x}).$$

Since $G_\lambda(\mathbf{x},\mathbf{y})$ satisfies boundary condition (4.17) and is symmetric with respect to $\mathbf{x}$ and $\mathbf{y}$,

$$v(\mathbf{y}) = \int_{\partial\mathcal{M}} (\partial_{\nu(\mathbf{x})} - \sigma(\mathbf{x})) G_\lambda(\mathbf{x},\mathbf{y}) h(\mathbf{x})\, m(\mathbf{x}) dS_g(\mathbf{x}).$$

Thus,

$$\frac{\partial}{\partial t} v(\gamma_{\mathbf{y},\nu}(t)) =$$

$$= \int_{\partial\mathcal{M}} \frac{\partial}{\partial t}\left(\frac{\partial}{\partial s} - \sigma(\mathbf{x})\right) G_\lambda(\gamma_{\mathbf{x},\nu}(s), \gamma_{\mathbf{y},\nu}(t)) h(\mathbf{x})\, m(\mathbf{x}) dS_g(\mathbf{x})\Big|_{s=0}.$$

By letting $t \to 0$, the claim follows. □

Lemma 4.1 implies that the Schwartz kernel of the operator $\Lambda_\lambda$ is given by limit (4.19). When $\lambda$ runs over $\mathbf{C} \setminus \{\lambda_j\}$, these data, i.e., the distribution,

$$-(\partial_{\nu(\mathbf{x})} - \sigma(\mathbf{x}))(\partial_{\nu(\mathbf{y})} - \sigma(\mathbf{y}))G_\lambda(\mathbf{x},\mathbf{y})|_{\mathbf{x},\mathbf{y}\in\partial\mathcal{M}}, \quad \lambda \in \mathbf{C}\setminus\{\lambda_j\},$$

are called the Gel'fand data. Hence, the Dirichlet-Robin map $\Lambda_\lambda$, where $\lambda \in \mathbf{C} \setminus \{\lambda_j\}$, determines the Gel'fand data and vice versa.

**4.1.6.** We are now in the position to formulate the inverse problems to be considered in this section.

Let $\mathcal{M}$ be an unknown compact manifold with boundary and $\mathcal{A} = a(\mathbf{x}, D)$ be an unknown differential operator on $\mathcal{M}$ of form (4.2), (4.16). Let $\sigma$ be an unknown smooth real function on $\partial\mathcal{M}$. We consider the case when the boundary measurements are given on the set $\Sigma = \Sigma^\infty$,

$$\Sigma = \partial\mathcal{M} \times (0, \infty),$$

i.e., for $T = \infty$. In this case, we denote by $\Pi = \Pi^\infty$, $\Lambda = \Lambda^\infty$, and $L = L^\infty$ the corresponding energy flux and the hyperbolic and parabolic Dirichlet-Robin maps and by $\mathcal{B} = \mathcal{B}^\infty$ the corresponding boundary form.

**Problem 6** *Is it possible to determine the manifold $\mathcal{M}$ and the orbit $\sigma(\mathcal{A})$ of the operator $\mathcal{A}$ with respect to the group $\mathcal{G}$ of gauge transformations, that is, the set*

$$\{a_\kappa(\mathbf{x}, D): \ \kappa \in C^\infty(\mathcal{M}), \kappa > 0 \text{ in } \mathcal{M}\},$$

*when we are given $\partial\mathcal{M}$ and, in addition, one of the following boundary data:*

**I. Hyperbolic data.**

*i.* **Energy flux through boundary.** *In this case, we are given the energy flux $\Pi(u^f)$,*

$$\Pi(u^f) = \Re \int_0^\infty \int_{\partial\mathcal{M}} (-\partial_\nu u^f + \sigma u^f)\overline{\partial_t u^f}\, m dS_g dt,$$

*where $u^f$ is the solution of the initial-boundary value problem (4.7)–(4.9) for an arbitrary $f \in C_0^\infty(\Sigma)$.*

*ii.* **Hyperbolic Dirichlet-Robin form.** *In this case, we are given the form $\Lambda$ corresponding to the operator $\Lambda$,*

$$\Lambda[f, h] = \int_0^\infty \int_{\partial\mathcal{M}} (-\partial_\nu u^f + \sigma u^f)\overline{u^h}\, m dS_g dt,$$

*where $f, h \in C_0^\infty(\Sigma)$.*

*iii.* **Hyperbolic boundary form.** *In this case, we are given the boundary form $\mathcal{B}$,*

$$\mathcal{B}[f, h] = \int_0^\infty \int_{\partial\mathcal{M}} (\partial_\nu u^f \overline{u^h} - u^f \overline{\partial_\nu u^h})\, m dS_g dt,$$

*where $f, h \in C_0^\infty(\Sigma)$.*

**II. Parabolic data.**

*iv. In this case, we are given the form $L$ corresponding to the operator $L$,*

$$L[f, h] = \int_0^\infty \int_{\partial\mathcal{M}} (-\partial_\nu w^f + \sigma w^f)\overline{w^h}\, m dS_g dt,$$

*where $w^f$ and $w^h$ are solutions of initial-boundary value problem (4.14) corresponding to $f, h \in C_0^\infty(\Sigma)$.*

**III. Elliptic data.**

*v.* *Elliptic Dirichlet-Robin form or Gel'fand data. In this case, we are given the form $\Lambda_\lambda$ for $\lambda \in \mathbf{C} \setminus \{\lambda_j\}$,*

$$\Lambda_\lambda[h, f] = \int_{\partial\mathcal{M}} (-\partial_\nu v^h + \sigma v^h)\overline{v^f}\, m dS_g,$$

*where $h, f \in C^\infty(\partial\mathcal{M})$, or, alternatively, the Robin derivative of Green's function,*

$$(-\partial_{\nu(\mathbf{x})} + \sigma(\mathbf{x}))\,(-\partial_{\nu(\mathbf{y})} + \sigma(\mathbf{y}))\, G_\lambda(\mathbf{x}, \mathbf{y})|_{(\mathbf{x},\mathbf{y}) \in \partial\mathcal{M} \times \partial\mathcal{M}}, \quad (4.20)$$

*for $\lambda \in \mathbf{C} \setminus \{\lambda_j\}$.*

*vi.* **Boundary spectral data.** *In this case, we are given the eigenvalues $\lambda_j$ and the normal derivatives, $\partial_\nu \varphi_j|_{\partial\mathcal{M}}$, of the normalized eigenfunctions of $\mathcal{A}$,*

$$\{\lambda_j,\ \partial_\nu \varphi_j|_{\partial\mathcal{M}} : \quad j = 1, 2, \ldots\}.$$

To clarify some details of these problems, we make the following remarks.

**Remark.** We note that, in data *v.*, the Dirichlet-Robin form depends analytically on $\lambda \in \mathbf{C} \setminus \{\lambda_j\}$. Actually, we will show that it is a meromorphic function of $\lambda \in \mathbf{C}$ with values in the space of boundary forms. This implies that $\Lambda_\lambda$, $\lambda \in \mathbf{C}$, is uniquely determined by its values on any set $\mathcal{X} \subset \mathbf{C}$, which has an accumulation point. We note also that the form $\Lambda_\lambda$ and the Robin derivative (4.20) of Green's function are equivalent, when the boundary measure $m dS_g$ is given.

**Remark.** Often in practice, the given information is the Dirichlet-Robin operator, rather than the corresponding form. This means that we know *a priori* the boundary measure $m dS_g$. As we are interested in gauge-invariant solutions, we prefer to work with forms.

**Remark.** Note that the Robin coefficient $\sigma$ is, in general, unknown. In cases *iii.* and *vi.*, the data is independent of $\sigma$, so that this coefficient cannot be found from these data. However, in the other cases, $\sigma$ can be found and its determination is a part of solution of the corresponding inverse problem.

Finally, we note that the inverse problems with data, analogous to ***i.-vi.***, can be posed for the Robin initial-boundary value problem,

$$\partial_t^2 u + a(\mathbf{x}, D)u = 0, \quad \text{in} \quad Q^T,$$

$$\partial_\nu u + \sigma u|_{\Sigma^T} = f,$$

$$u|_{t=0} = 0, \quad \partial_t u|_{t=0} = 0.$$

This case can be considered similar to the Dirichlet one.

**4.1.7.** As in Chapter 3, we would like to reconstruct $(\mathcal{M}, \mathcal{A})$ in a gauge-invariant way. Therefore, we need also to consider what happens to the data ***i.-vi.*** in gauge transformations. We remind the reader that a gauge transformation is given by the formula,

$$S_\kappa u(\mathbf{x}, t) = \kappa(\mathbf{x}) u(\mathbf{x}, t),$$

where $\kappa \in C^\infty(\mathcal{M})$,and $\kappa > 0$ on $\mathcal{M}$.

First we will consider the hyperbolic data. Let $u(\mathbf{x}, t)$ be a solution of wave equation (4.1) Since $\kappa(\mathbf{x})$ is independent of $t$, then $\widetilde{u}(\mathbf{x}, t) = S_\kappa u(\mathbf{x}, t)$ is a solution of the following wave equation,

$$\partial_t^2 \widetilde{u} + a_\kappa(\mathbf{x}, D)\widetilde{u} = 0, \quad \text{in} \quad Q^T, \tag{4.21}$$

where $a_\kappa(\mathbf{x}, D)$ is the gauge transformation of the operator $a(\mathbf{x}, D)$ given by (2.49). Then, the Cauchy data of the solution $\widetilde{u}$ takes the form,

$$\widetilde{u}|_{\Sigma^T} = \kappa|_{\partial\mathcal{M}}\, u|_{\Sigma^T}, \tag{4.22}$$

$$(-\partial_\nu \widetilde{u} + \sigma_\kappa \widetilde{u})|_{\Sigma^T} = \kappa|_{\partial\mathcal{M}}(-\partial_\nu u + \sigma u)|_{\Sigma^T},$$

where the Robin coefficient $\sigma_\kappa$ for wave equation (4.21) is given by the formula,

$$\sigma_\kappa = \sigma + \kappa^{-1}\, \partial_\nu \kappa|_{\partial\mathcal{M}}. \tag{4.23}$$

This formula defines the transformation rule for the Dirichlet-Robin map $\Lambda^T$, $T > 0$, due to a gauge transformation. Namely, by formula (4.22),

$$\widetilde{\Lambda}^T = S_{\kappa_0} \circ \Lambda^T \circ S_{\kappa_0}^{-1}, \quad \kappa_0 = \kappa|_{\partial\mathcal{M}}, \tag{4.24}$$

where $\widetilde{\Lambda}^T$ is the Dirichlet-Robin map for wave equation (4.21) with the Robin parameter $\sigma_\kappa$. The transformation rule for the Dirichlet-Robin map implies the transformation rule for the Dirichlet-Robin form and the hyperbolic boundary form. Namely,

$$\widetilde{\Lambda}^T[\kappa_0 f, \kappa_0 h] = \Lambda^T[f, h], \tag{4.25}$$

where $\widetilde{\Lambda}^T$ is the Dirichlet-Robin form for the transformed problem, and

$$\widetilde{\mathcal{B}}^T[\kappa_0 f, \kappa_0 h] = \mathcal{B}^T[f, h], \tag{4.26}$$

where $\widetilde{\mathcal{B}}^T$ is the hyperbolic boundary form for the transformed problem,

$$\widetilde{\mathcal{B}}^T[f, h] = -\int_0^T \int_{\partial\mathcal{M}} (\widetilde{\Lambda}^T f\, \overline{h} - f\, \overline{\widetilde{\Lambda}^T h})\, m_\kappa dS_g dt. \tag{4.27}$$

Two forms $F$ and $\widetilde{F}$ are gauge-equivalent with respect to a gauge transformation $S_{\kappa_0}$ on the boundary, if

$$\widetilde{F}[\cdot, \cdot] = F[S_{\kappa_0}^{-1} \cdot, S_{\kappa_0}^{-1} \cdot]. \tag{4.28}$$

In this case, we say that $\widetilde{F} = S_{\kappa_0} F$. In particular, $\widetilde{\Lambda}^T = S_{\kappa_0} \Lambda^T$ and $\widetilde{\mathcal{B}}^T = S_{\kappa_0} \mathcal{B}^T$.

It follows from definition (4.10) of the Dirichlet-Robin map, that

$$\Pi^T(u^f) = \Re \int_{\Sigma^T} \Lambda^T f\, \overline{\partial_t f}\, m dS_g dt.$$

Thus, the transformation rule for $\Lambda^T$, given by formula (4.24), implies that

$$\widetilde{\Pi}^T(\widetilde{u}^{\widetilde{f}}) = \Pi^T(u^f), \quad f = S_{\kappa_0}^{-1} \widetilde{f}.$$

Here $\widetilde{u}^{\widetilde{f}}$ is the solution of wave equation (4.21) with the initial and boundary data,

$$\widetilde{u}^{\widetilde{f}}|_{\Sigma^T} = \widetilde{f}, \quad \widetilde{u}^{\widetilde{f}}|_{t=0} = 0, \quad \partial_t \widetilde{u}^{\widetilde{f}}|_{t=0} = 0,$$

and $\widetilde{\Pi}^T$ is the energy flux for wave equation (4.21).

**4.1.8.** The transformation rules for $\Lambda^T$, $\mathcal{B}^T$, and $\Pi^T$ were obtained in the previous section in a rather formal way. However, these transformation rules have a deep physical foundation. Indeed, the notions of the energy flux and, therefore, the Dirichlet-Robin and hyperbolic boundary forms, are based on the notion of energy (see sections 4.1.2–4.1.3). It is the gauge-invariance of energy, which underlies the transformation rules of $\Lambda^T$, $\mathcal{B}^T$, and $\Pi^T$. As the energy is a basic physical concept, the gauge-invariance of energy means that all gauge-equivalent differential operators have the same physical meaning. Hence, the boundary data, which corresponds to the gauge-equivalent operators have to be gauge-equivalent.

The invariance of the energy with respect to gauge transformations is described in the following lemma.

**Lemma 4.2** *Let $u(\mathbf{x},t)$ and $\widetilde{u}(\mathbf{x},t) = S_\kappa u(x,t)$ be solutions of wave equations (4.1) and (4.21), correspondingly. Then,*

$$\widetilde{E}(\widetilde{u},T) = E(u,T),$$

*where*

$$\widetilde{E}(\widetilde{u},T) =$$

$$= \frac{1}{2}\int_{\mathcal{M}} \left(|\partial_t \widetilde{u}(\mathbf{x},t)|^2 + |\,Grad\,\widetilde{u}(\mathbf{x},t)|_g^2 + q_\kappa(\mathbf{x})|\widetilde{u}(\mathbf{x},t)|^2\right) m_\kappa(\mathbf{x}) dV_g(\mathbf{x})$$

$$+\frac{1}{2}\int_{\partial\mathcal{M}} \sigma_\kappa(\mathbf{x})|\widetilde{u}(\mathbf{x},t)|^2 m_\kappa(\mathbf{x}) dS_g(\mathbf{x})$$

*and $q_\kappa$, $m_\kappa$, and $\sigma_\kappa$ are given by (2.52) and (4.23).*

**Proof.** Integrating by parts the term in the right-hand side of this equation, which corresponds to $|\,\mathrm{Grad}\,\widetilde{u}(\mathbf{x},t)|_g^2$, and using wave equation (4.21) yields the claim. □

**4.1.9.** To consider the behaviour of the parabolic and spectral data with respect to gauge transformations, we first notice that the Cauchy data of the solutions $w^f$ of heat equation (4.14) and $v^h$ of elliptic boundary-value problem (4.15) are also transformed according to transformation rule (4.22). This immediately implies the transformation formulae for the Dirichlet-Robin maps,

$$\widetilde{L}^T = S_{\kappa_0} \circ L^T \circ S_{\kappa_0}^{-1}, \qquad (4.29)$$

$$\widetilde{\Lambda}_\lambda = S_{\kappa_0} \circ \Lambda_\lambda \circ S_{\kappa_0}^{-1},$$

where $\widetilde{L}^T$ and $\widetilde{\Lambda}_\lambda$ are, correspondingly, the parabolic and elliptic Dirichlet-Robin maps related to the operator $\mathcal{A}_\kappa$, $\kappa_0 = \kappa|_{\partial\mathcal{M}}$. Then, for the corresponding boundary forms, $\widetilde{L}^T$ and $\widetilde{\Lambda}_\lambda$, we have that $\widetilde{L}^T = S_{\kappa_0} L^T$ and $\widetilde{\Lambda}_\lambda = S_{\kappa_0}\Lambda_\lambda$. Because the Gel'fand data is the Schwartz kernel of the operator $\Lambda_\lambda$ with respect to the volume measure $mdS_g$, formula (4.29) implies that

$$-(-\partial_{\nu(\mathbf{x})} + \sigma(\mathbf{x}))\,(-\partial_{\nu(\mathbf{y})} + \sigma_\kappa(\mathbf{y}))\,\widetilde{G}_\lambda(\mathbf{x},\mathbf{y})|_{\partial\mathcal{M}\times\partial\mathcal{M}} =$$

$$= -\kappa_0(\mathbf{x})\kappa_0(\mathbf{y})(-\partial_{\nu(\mathbf{x})} + \sigma(\mathbf{x}))(-\partial_{\nu(\mathbf{y})} + \sigma(\mathbf{y}))G_\lambda(\mathbf{x},\mathbf{y})|_{\partial\mathcal{M}\times\partial\mathcal{M}},$$

where $\widetilde{G}_\lambda(\mathbf{x},\mathbf{y})$ is Green's function of the operator $\mathcal{A}_\kappa$.

**4.1.10.** The rest of this section will be devoted to the transformation of boundary data *i.-v.* to the boundary spectral data.

We start with the hyperbolic data. The bilinear energy flux form, which we continue to denote by $\Pi$, is given by the formula

$$\Pi[f,h] = \frac{1}{2}\Re\int_0^\infty \int_{\partial\mathcal{M}} (\Lambda f \overline{\partial_t h} + \partial_t f \overline{\Lambda h}) mdS_g dt.$$

Clearly, the energy flux $\Pi(u^f)$, given for any $f \in C_0^\infty(\Sigma)$, determines $\Pi[f,h]$ for any real valued $f, h \in C_0^\infty(\Sigma;\mathbf{R})$. Since the coefficients of $a(\mathbf{x}, D)$ are real, $\Lambda f$ and $\Lambda h$ are also real, so that

$$\Pi[f,h] = \frac{1}{2}\int_0^\infty \int_{\partial\mathcal{M}} (\Lambda f\, \partial_t h + \partial_t f\, \Lambda h) mdS_g dt,$$

where $f, h \in C_0^\infty(\Sigma;\mathbf{R})$. Therefore, it is possible to compexify the form $\Pi$ to obtain a complex-bilinear form $\Pi_\mathbf{C}$,

$$\Pi_\mathbf{C}[f,h] = \frac{1}{2}\int_0^\infty \int_{\partial\mathcal{M}} (\Lambda f\, \partial_t h + \partial_t f\, \Lambda h) mdS_g dt, \qquad (4.30)$$

where $f, h \in C_0^\infty(\Sigma;\mathbf{C})$.

**Lemma 4.3** *Given any of the forms $\Lambda$, $\mathcal{B}$ or $\Pi$, it is possible to uniquely determine the bilinear form $\Pi_{\mathbf{C}}$.*

**Proof.** We have already proven the claim for the energy flux $\Pi$, and, therefore, for the form $\Lambda$.

Let $f, h \in C_0^\infty(\Sigma; \mathbf{C})$. Integrating by parts the term in the right-hand side of (4.30), which corresponds to $\Lambda f\, \partial_t h$, and taking into account that $\partial_t(\Lambda f) = \Lambda(\partial_t f)$, shows that

$$\mathcal{B}[\partial_t f, \overline{h}] = 2\Pi_{\mathbf{C}}[f, h].$$

□

**4.1.11.** To construct the boundary spectral data from the form $\Pi_{\mathbf{C}}$, we need to consider the relations between the hyperbolic and elliptic Dirichlet-Robin forms. Corollary 2.46 implies that, for any $f \in C_0^\infty(\partial\mathcal{M} \times \mathbf{R})$ with supp $(f) \subset \Sigma$,

$$\|u(t)\|_2 + \|\partial_t^2 u(t)\|_0 \le C_1 e^{C_2 t} \|f\|_{(2,\partial\Sigma^t)}, \tag{4.31}$$

$$\|\partial_\nu u(t)\|_{(1,\partial\Sigma^t)} \le C_1 e^{C_2 t} \|f\|_{(2,\partial\Sigma^t)}.$$

We remind the reader that we denote by $\|v\|_{(s,\Omega)}$ the norm of a function $v$ in $H^s(\Omega)$ and by $\|v\|_s$ the norm of $v$ in $H^s(\mathcal{M})$ (see the end of section 2.2.6).

Hence, the Fourier transform $\widehat{u}(\mathbf{x}, k)$ of $u = u^f(\mathbf{x}, t)$ with respect to time,

$$\widehat{u}(\mathbf{x}, k) = \mathcal{F}(u(\mathbf{x}, \cdot))(k) = \int_{\mathbf{R}} e^{-ikt} u(\mathbf{x}, t)dt$$

is an analytic $H^2(\mathcal{M})$-valued function of $k$ for $\Im k < -C_2$. Analogously, $\mathcal{F}(\partial_\nu u|_{\partial\mathcal{M}\times\mathbf{R}})(k)$ is an analytic $H^{1/2}(\partial\mathcal{M})$-valued function of $k$ in the same half plane.

Moreover, for $f, h \in C_0^\infty(\Sigma)$, the Plancherel theorem implies that

$$\int_{\partial\mathcal{M}} \int_0^\infty \Lambda f(t) h(t) dt m dS_g = \int_{\partial\mathcal{M}} \int_{-\infty}^\infty e^{-\mu t} \Lambda f(t) e^{\mu t} h(t) dt m dS_g$$

$$= \frac{1}{2\pi} \int_{\partial\mathcal{M}} \int_{-\infty+i\mu}^{\infty+i\mu} \widehat{\Lambda f}(-k) \widehat{h}(k)\, dk\, m dS_g, \tag{4.32}$$

for any $\mu > C_2$. Taking the Fourier transform with respect to $t$ in the initial-boundary value problem (4.7)–(4.9) and using estimate (4.31), we see that

$$(a(\mathbf{x}, D) - k^2)\widehat{u} = 0, \tag{4.33}$$

$$\widehat{u}|_{\partial\mathcal{M}} = \widehat{f}(k),$$

for $\Im k < -C_2$. This yields that

$$\widehat{\Lambda f}(k) = \Lambda_{k^2}(\widehat{f}(k)). \tag{4.34}$$

Next we consider the energy form.

**Lemma 4.4** *Let $f, h \in C_0^\infty(\Sigma)$. Then, there is $\mu_0 \geq 0$, such that, for $\mu > \mu_0$,*

$$\Pi_{\mathrm{C}}[f, h] = \frac{-i}{4\pi}\int_{\partial\mathcal{M}}\int_{L_\mu} \Lambda_{k^2}\widehat{f}(-k)\,\widehat{h}(k)\,kdk\,mdS_g, \tag{4.35}$$

*where the contour $L_\mu$ is the boundary of the strip $\{k : |\Im k| < \mu\}$, which is oriented in the positive direction (see Fig. 4.1).*

**Proof.** *Step 1.* Formula (4.32) and definition (4.30) of $\Pi_{\mathrm{C}}$, show that

$$\Pi_{\mathrm{C}}[f, h] = \tag{4.36}$$

$$\frac{i}{4\pi}\int_{\partial\mathcal{M}}\int_{-\infty+i\mu}^{\infty+i\mu}\left(\Lambda_{k^2}\widehat{f}(-k)\,k\widehat{h}(k) + \Lambda_{k^2}\widehat{h}(-k)\,k\widehat{f}(k)\right)dk\,mdS_g$$

$$= \frac{i}{4\pi}\int_{\partial\mathcal{M}}\left(\int_{-\infty+i\mu}^{\infty+i\mu}\Lambda_{k^2}\widehat{f}(-k)\,k\widehat{h}(k)\,dk - \right.$$

$$\left. - \int_{-\infty-i\mu}^{\infty-i\mu}\Lambda_{k^2}\widehat{h}(k)\,k\widehat{f}(-k)\,dk\right)mdS_g.$$

*Step 2.* We observe that

$$\int_{\partial\mathcal{M}}\widehat{f}(-k)\,\Lambda_{k^2}\widehat{h}(k)mdS_g = \int_{\partial\mathcal{M}}\Lambda_{k^2}\widehat{f}(-k)\,\widehat{h}(k)mdS_g. \tag{4.37}$$

Indeed, $\widehat{u^f}(\mathbf{x}, k)$ and $\widehat{u^h}(\mathbf{x}, k)$ satisfy elliptic equation (4.33). By the second Green identity, this implies relation (4.37).

Combining (4.36) and (4.37), we get the claim. □

Figure 4.1: Contour $L_\mu$

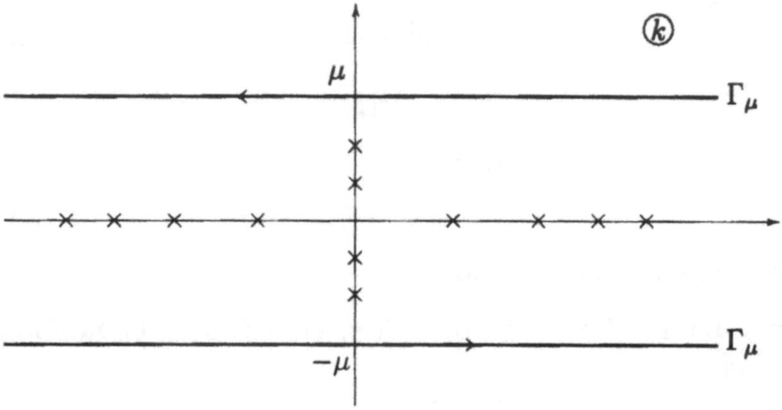


**4.1.12.** In this section, we will show that the elliptic Dirichlet-Robin map is a meromorphic operator-valued function with first-order poles at the eigenvalues $\lambda_j$ of $\mathcal{A}$. This will make it possible to later evaluate the integral in the right-hand side of (4.35) as a sum of residues.

Let us group together all equal eigenvalues, so that we have a sequence of sets $L_n \subset \mathbf{Z}_+$, $n = 1, 2, \ldots$, such that $\bigcup_{n=1}^{\infty} L_n = \mathbf{Z}_+$ with $l_1$ and $l_2$ belonging to the same set $L_n$, if and only if $\lambda_{l_1} = \lambda_{l_2}$. Sometimes we will use notations $\lambda_{(n)}$ for the eigenvalue corresponding to $L_n$, and $L(\lambda_j)$ for the set $L_n$ with $\lambda_{(n)} = \lambda_j$. We remind the reader that, for functions $f, h \in L^2(\partial\mathcal{M}, m dS_g)$, we denote

$$\langle f, h \rangle = \int_{\partial\mathcal{M}} f\,\overline{h}\, m dS_g, \tag{4.38}$$

(see (2.27)).

**Lemma 4.5** *Let $f, h \in H^{3/2}(\partial\mathcal{M})$. Then $\Lambda_\lambda[f,h]$ is a meromorphic function of $\lambda$, which may have first-order poles only at the eigenvalues $\lambda_j$ of $\mathcal{A}$. Moreover, in a vicinity of any eigenvalue $\lambda_j$,*

$$\Lambda_\lambda[f,h] = \frac{1}{\lambda - \lambda_j} \sum_{l \in L(\lambda_j)} \langle f, \partial_\nu \varphi_l \rangle \, \langle \overline{h}, \partial_\nu \varphi_l \rangle + H_\lambda[f,h], \tag{4.39}$$

*where $H_\lambda[f,h]$ is an analytic function of $\lambda$ in the vicinity of $\lambda_j$.*

**Proof.** *Step 1.* Let $h \in H^{3/2}(\partial\mathcal{M})$. Denote by $E^h \in H^2(\mathcal{M})$ an extension of $h$ into $\mathcal{M}$, such that $E^h|_{\partial\mathcal{M}} = h$ and

$$\|E^h\|_2 \leq c\|h\|_{(3/2,\partial\mathcal{M})}. \tag{4.40}$$

The existence of $E^h$ is guaranteed by the trace theorem (see section 2.2.2). Let $v^h(\lambda)$ be the solution of Dirichlet problem (4.15). Then, $v^h - E^h \in \mathcal{D}(\mathcal{A})$ and

$$(\mathcal{A}-\lambda)(v^h - E^h) = -(a(\mathbf{x},D)-\lambda)E^h.$$

Using spectral representation (2.37) of $\mathcal{R}(\lambda) = (\mathcal{A}-\lambda)^{-1}$, we see that $\mathcal{R}(\lambda)$ is a meromorphic operator-valued function of $\lambda$. Thus,

$$v^h(\lambda) = E^h - \mathcal{R}(\lambda)(a(\mathbf{x},D)-\lambda)E^h, \tag{4.41}$$

is a meromorphic $H^2(\mathcal{M})$ -valued function of $\lambda$, which may have first-order poles only at $\lambda \in \{\lambda_j\}$.

*Step 2.* Compute the Fourier coefficients of $v^h(k)$ with respect to the eigenfunctions $\varphi_j$, which are assumed, without loss of generality, to be real-valued. Integrating by parts, we see that

$$0 = \langle (a(\mathbf{x},D)-\lambda)v^h(\lambda), \varphi_l\rangle =$$

$$= (\lambda_l - \lambda)\langle v^h(\lambda), \varphi_l\rangle - \int_{\partial\mathcal{M}} h\,\partial_\nu\varphi_l\, m dS_g.$$

Thus,

$$\langle v^h(\lambda), \varphi_l\rangle = \frac{1}{\lambda_l - \lambda}\int_{\partial\mathcal{M}} h\,\partial_\nu\varphi_l\, m dS_g.$$

Let

$$P_j : L^2(\mathcal{M}) \to L^2(\mathcal{M}), \quad P_j(v) = \sum_{l\in L(\lambda_j)} \langle v, \varphi_l\rangle\varphi_l$$

be the orthoprojection to the eigenspace corresponding to $\lambda_j$.

The formula

$$(1-P_j)\mathcal{R}(\lambda)F = \sum_{l\notin L(\lambda_j)} \frac{1}{\lambda_l - \lambda}\langle F, \varphi_l\rangle\varphi_l,$$

defines a bounded operator from $L^2(\mathcal{M})$ to $H^2(\mathcal{M})$, which depends analytically on $\lambda$ near $\lambda_j$. Hence, it follows from formula (4.41), that $\partial_\nu(1-P_j)v^h(\lambda)|_{\partial\mathcal{M}} \in H^{1/2}(\partial\mathcal{M})$ depends analytically on $\lambda$ in a vicinity of $\lambda_j$, and

$$\partial_\nu(P_j v^h(\lambda))|_{\partial\mathcal{M}} = \sum_{l\in L(\lambda_j)} \left(\frac{1}{\lambda_j-\lambda}\int_{\partial\mathcal{M}} h\,\partial_\nu\varphi_l\, m dS_g\right)\partial_\nu\varphi_l|_{\partial\mathcal{M}}.$$

Thus,

$$\operatorname*{res}_{\lambda=\lambda_j}\Lambda_\lambda h = \sum_{l\in L(\lambda_j)}\left(\int_{\partial\mathcal{M}} h\,\partial_\nu\varphi_l\, m dS_g\right)\partial_\nu\varphi_l|_{\partial\mathcal{M}} \in C^\infty(\partial\mathcal{M}).$$

Equation (4.39) immediately follows from this equation. □

**4.1.13.** Lemma 4.5 shows that integral representation (4.35) can be used to find the energy flux $\Pi_{\mathbf{C}}$ by means of the residue theorem. Applying this idea, we will prove an analog of the Trotter formula.

**Lemma 4.6** *Let $f, h \in C_0^\infty(\Sigma)$. Then, the complexified energy flux form, $\Pi_{\mathbf{C}}[f,h]$ can be represented as the following sum,*

$$\Pi_{\mathbf{C}}[f,h] = \frac{1}{2}\sum_{j=1}^{\infty}\Big(\left\langle\widehat{f}(\sqrt{\lambda_j}),\partial_\nu\varphi_j\right\rangle\left\langle\widehat{h}(-\sqrt{\lambda_j}),\partial_\nu\varphi_j\right\rangle$$
$$+\left\langle\widehat{f}(-\sqrt{\lambda_j}),\partial_\nu\varphi_j\right\rangle\left\langle\widehat{h}(\sqrt{\lambda_j}),\partial_\nu\varphi_j\right\rangle\Big). \quad (4.42)$$

*We remind the reader that $\varphi_j$ are taken to be real-valued. The branch of the square root is obtained by using the cut along the half-line $\arg(z) = -\pi/4$.*

**Proof.** *Step 1.* We start by proving the estimate

$$\|\Lambda_\lambda h\|_{(1/2,\partial\mathcal{M})} \le C\frac{|\lambda|^2+1}{\operatorname{dist}(\lambda,\{\lambda_j\})}\|h\|_{(3/2,\partial\mathcal{M})}. \quad (4.43)$$

Using spectral representation (2.37) of the resolvent $\mathcal{R}(\lambda)$, we observe that

$$\|\mathcal{R}(\lambda)\|_{L^2(\mathcal{M})\to H^2(\mathcal{M})} \le C\frac{|\lambda|+1}{\operatorname{dist}(\lambda,\{\lambda_j\})}.$$

Figure 4.2: Contour $\Gamma_n$

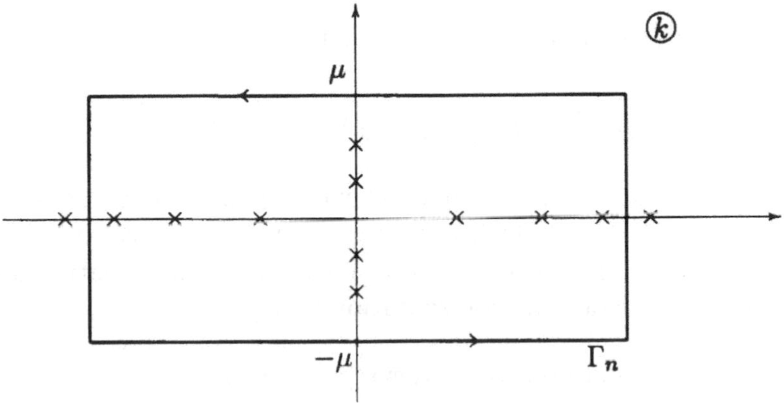


Estimate (4.43) follows from this inequality, if we take into account representation (4.41) of $v^h(\lambda)$ together with estimate (4.40).

*Step 2.* Since $f, h \in C_0^\infty(\Sigma)$, the Paley-Wiener theorem yields that

$$\|\widehat{f}(k)\|_{(3/2,\partial\mathcal{M})} \leq C_N(1+|k|)^{-N}, \tag{4.44}$$

$$\|\widehat{h}(k)\|_{(3/2,\partial\mathcal{M})} \leq C_N(1+|k|)^{-N},$$

when $k$ lies in the strip, $\{k : |\Im k| < \mu\}$, and $N > 0$ is arbitrary. Thus,

$$\left|\int_{\partial\mathcal{M}} \Lambda_{k^2}\widehat{f}(x,-k)\,\widehat{h}(x,k)\,m dS_g\right| \leq C_N' \frac{(1+|k|)^{-N}}{\operatorname{dist}(k^2,\{\lambda_j\})}. \tag{4.45}$$

*Step 3.* Due to Weyl's asymptotics (see Theorem 2.21), the number $n(r^2)$ of the eigenvalues of the operator $\mathcal{A}$, which are less than $r^2$, satisfy the estimate

$$n(r^2) \leq C(1+r^2)^{m/2}, \quad r > 0. \tag{4.46}$$

Therefore, there is an interval $(a_r, b_r) \subset [r^2, r^2+1]$, such that $b_r - a_r \geq c(1+r)^{-m}$ and this interval does not contain any eigenvalues of $\mathcal{A}$.

This makes it possible to choose a sequence $r_n \to \infty$, such that

$$d(r_n^2, \{\lambda_j\}) \geq c(1 + r_n)^{-m}.$$

Using the above estimate, we obtain that

$$\Pi_{\mathbf{C}}[f, h] = \frac{1}{4\pi i} \lim_{n \to \infty} \int_{\Gamma_n} \int_{\partial\mathcal{M}} \Lambda_{k^2} \widehat{f}(-k)\, \widehat{h}(k)\, m dS_g\, k dk,$$

where the contours $\Gamma_n$ are the boundary curves of the rectangles $\{k : |\Re k| < r_n, |\Im k| < \mu\}$, which are oriented in the positive direction (see Fig. 4.2). Applying the residue theorem to this equation and taking $n \to \infty$, we obtain representation (4.39). □

In future, we will need the following corollary.

**Corollary 4.7** *For any* $f \in C_0^\infty(\Sigma)$ *and* $N > 0$,

$$\left|\langle \widehat{f}(\sqrt{\lambda_j}), \partial_\nu \varphi_j \rangle\right| \leq C_{N,f} j^{-N}.$$

*Thus, the sum in the right-hand side of (4.42) converges absolutely.*

**Proof.** The Gårding inequality (see Theorem 2.22) implies that

$$\|\partial_\nu \varphi_j\|_{(0,\partial\mathcal{M})} \leq c\|\varphi_j\|_2 \leq c(1 + |\lambda_j|) \leq c j^{2/m},$$

where the last inequality follows from Weyl's asymptotics. Hence, the assertion follows from the Payley-Wiener theorem. □

**4.1.14.** In this section, we will show how to use formula (4.42) to find all eigenvalues $\lambda_j$ of $\mathcal{A}$ and the corresponding sums

$$\sum_{l \in L_j} \partial_\nu \varphi_l(\mathbf{x})\, \partial_\nu \varphi_l(\mathbf{y}), \quad \mathbf{x}, \mathbf{y} \in \partial\mathcal{M}. \tag{4.47}$$

We start with the following equation

$$\Pi_{\mathbf{C}}[Y_\tau f, Y_\tau h] = \Pi_{\mathbf{C}}[f, h], \quad \tau > 0, \tag{4.48}$$

where $Y_\tau$ is the time-delay operator,

$$Y_\tau f(\mathbf{x}, t) = f(\mathbf{x}, t - \tau), \quad \tau \in \mathbf{R}. \tag{4.49}$$

This equation is obvious for $f, h \in C_0^\infty(\Sigma)$ and can be used to define the form $\Pi_{\mathbf{C}}[f,h]$ for $f, h \in C_0^\infty(\partial\mathcal{M} \times \mathbf{R})$.

Consider $\Pi_{\mathbf{C}}[Y_\tau f, h]$. By Lemma 4.6,

$$\Pi_{\mathbf{C}}[Y_\tau f, h] = \frac{1}{2}\sum_{n=1}^{\infty}\Big(e^{-ik_n\tau}Q_n[\widehat{f}(k_n), \widehat{h}(-k_n)] + \\ +e^{ik_n\tau}Q_n[\widehat{f}(-k_n), \widehat{h}(k_n)]\Big), \tag{4.50}$$

where $k_n = \sqrt{\lambda_{(n)}} = \sqrt{\lambda_j}$ for $j \in L_n$, and

$$Q_n[h_1, h_2] = \sum_{j\in L_n}\langle h_1, \partial_\nu\varphi_j\rangle\langle h_2, \partial_\nu\varphi_j\rangle,$$

when $h_1, h_2 \in L^2(\partial\mathcal{M})$. Formula (4.50) can be used to determine $k_n$ and the forms $Q_n$. Indeed, take functions $f$ and $h$ of the form

$$f(\mathbf{x}, t) = F(\mathbf{x})\chi(t), \quad h(\mathbf{x}, t) = H(\mathbf{x})\chi(t),$$

where $\chi \in C_0^\infty(\mathbf{R})$ and $F, H \in C^\infty(\partial\mathcal{M})$. By direct computations, we obtain that

$$\Pi_{\mathbf{C}}[Y_\tau f, h] \; = \; \sum_{n=1}^{\infty}\cos(k_n\tau)\widehat{\chi}(-k_n)\widehat{\chi}(k_n)Q_n[F, H] = \pi_e(\tau) + \pi_b(\tau).$$

Here,

$$\pi_e(\tau) \; = \sum_{k_n^2<0}\cosh(ik_n\tau)\widehat{\chi}(-k_n)\widehat{\chi}(k_n)Q_n[F, H], \tag{4.51}$$

and

$$\pi_b(\tau) = \sum_{k_n^2\geq 0}\cos(k_n\tau)|\widehat{\chi}(k_n)|^2 Q_n[F, H]. \tag{4.52}$$

The sum $\pi_e(\tau)$ consists of a finite number of terms that grow exponentially when $\tau \to \infty$ or $\tau \to -\infty$. By Corollary 4.7, the sum $\pi_b(\tau)$ converges uniformly with respect to $\tau$ and is uniformly bounded when $\tau \in \mathbf{R}$. Analyzing the asymptotics of $\Pi_{\mathbf{C}}[Y_\tau\chi F, \chi H]$, when $\tau \to \infty$ or $\tau \to -\infty$, we find the exponentially growing terms, $\pi_e(\tau)$. Varying $\chi$, $F$, and $H$, we then find all eigenvalues $\lambda_{(n)} = k_n^2 < 0$ and the bilinear forms $Q_n[\cdot\,,\cdot]$ that correspond to these $\lambda_{(n)}$. As we know

$\Pi_C[Y_\tau \chi F, \chi H]$ and $\pi_e(\tau)$, we find $\pi_b(\tau)$. The uniform boundedness of $\pi_b(\tau)$ implies that its Fourier transform is a temperate distribution,

$$\widehat{\pi}_b(\zeta) = \sum_{k_n^2 \geq 0} \frac{1}{4\pi}(\delta(\zeta - k_n) + \delta(\zeta + k_n))|\widehat{\chi}(k_n)|^2 Q_n[F, H].$$

Varying $\chi$, $F$, and $H$, we find all eigenvalues $\lambda_{(n)} = k_n^2 \geq 0$ and the corresponding bilinear forms $Q_n[\cdot\,,\cdot]$.

Summarizing, we obtain the following result.

**Theorem 4.8** *The energy flux form $\Pi_C$ uniquely determines the eigenvalues $\lambda_j$ of $\mathcal{A}$ and the quadratic forms $Q_n$, i.e., their kernels*

$$\sum_{l \in L(\lambda_j)} \partial_\nu \varphi_l(\mathbf{x})\, \partial_\nu \varphi_l(\mathbf{y})\, m(\mathbf{x}) m(\mathbf{y}) dS_g(\mathbf{x}) dS_g(\mathbf{y}). \tag{4.53}$$

*We note that the multiplicity of an eigenvalue $\lambda_j$ is just the dimension of the maximal subspace of $L^2(\partial\mathcal{M})$, where $Q_n$, $j \in L_n$, is non-degenerate.*

**4.1.15.** The knowledge of kernels of quadratic forms $Q_n$ given by formula (4.53) does not allow us to find the functions $\partial_\nu \varphi_l|_{\partial\mathcal{M}}$ uniquely. There are two reasons for this. First, if the eigenvalues $\lambda_{(n)}$ corresponding to $L_n$ are not simple, that is, their multiplicity $N = N(n) > 1$, then the choice of the corresponding orthonormal eigenfunctions $\varphi_l$, $l \in L_n$, is not unique. Indeed, we can find $\varphi_l$ only up to an orthogonal transformation from the group $O(N)$. Second, if we do not know the boundary measure $mdS_g$, we can find $\partial_\nu \varphi_l|_{\partial\mathcal{M}}$ only up to a gauge transformation on the boundary. Next, we will show that this is the only non-uniqueness in the reconstruction of $\partial_\nu \varphi_l|_{\partial\mathcal{M}}$ from $Q_n$.

**Lemma 4.9** *Let the form $Q_n[\cdot\,,\cdot]$ be given on $C^\infty(\partial\mathcal{M}) \times C^\infty(\partial\mathcal{M})$. Then it is possible to determine functions $\zeta_k$, $l = 1, \ldots, N(n)$, such that*

$$\zeta_k(\mathbf{x}) = \eta(\mathbf{x}) \partial_\nu \psi_k(\mathbf{x}), \quad \mathbf{x} \in \partial\mathcal{M}, \tag{4.54}$$

*where $\eta \in C^\infty(\partial\mathcal{M})$ is positive and $\psi_k$, $k = 1, \ldots, N(n)$, form an orthonormal basis in the eigenspace of $\mathcal{A}$, corresponding to the eigenvalues $\lambda_{(n)}$.*

**Proof.** It is sufficient to show that $Q_n$ determine the functions $\zeta_k$, such that

$$\zeta_k(\mathbf{x}) = \eta(\mathbf{x}) \sum_{l=1}^{N} \alpha_{kl} \partial_\nu \varphi_{s+l}(\mathbf{x}), \quad \mathbf{x} \in \partial\mathcal{M}.$$

Here $\eta(\mathbf{x}) > 0$, the matrix $[\alpha_{kl}]_{k,l=1}^{N} \in \mathcal{O}(N)$, $\varphi_{s+l}$ are the eigenfunctions of $\mathcal{A}$ used in section 4.1.14, and $s$ is a integer such that $L_n = \{s+1, \ldots, s+N\}$.

First, let $d\mu$ be an arbitrary smooth positive measure on $\partial\mathcal{M}$. Then there is $\eta \in C^\infty(\partial\mathcal{M})$, $\eta > 0$, such that $mdS_g = \eta d\mu$. Therefore, the quadratic form $Q_n$ determines an integral operator $\mathcal{Q}$ in $L^2(\partial\mathcal{M}, d\mu)$, such that

$$Q_n[f,h] = \int_{\partial\mathcal{M}} (\mathcal{Q}f)(x)\, \overline{h}(x)\, d\mu(x).$$

The operator $\mathcal{Q}$ has a distribution kernel,

$$\sum_{l=1}^{N} \eta(\mathbf{x})\eta(\mathbf{y}) \partial_\nu \varphi_{s+l}(\mathbf{x})\, \partial_\nu \varphi_{s+l}(\mathbf{y}), \tag{4.55}$$

with respect to the measure $d\mu$. Thus, the knowledge of $Q_n$ determines the kernel (4.55). Since $\partial_\nu \varphi_{s+l}|_{\partial\mathcal{M}}$, $l = 1, \ldots, N$ are linearly independent, the dimension of the range of $\mathcal{Q}$ is $N$. Thus, we can choose points $\mathbf{x}_1, \ldots, \mathbf{x}_N$ and a basis of real-valued functions $e_1, \ldots, e_N \in \mathcal{Q}(L^2(\partial\mathcal{M})) \subset C^\infty(\partial\mathcal{M})$, such that

$$e_k(\mathbf{x}_j) = \delta_{kj}. \tag{4.56}$$

Therefore, there is a non-degenerate matrix $B = [\beta_{lk}]_{l,k=1}^{N}$, such that

$$\eta \partial_\nu \varphi_{s+l} = \sum_{k=1} \beta_{lk} e_k. \tag{4.57}$$

Considering kernel (4.55) at points $\mathbf{x}_1, \ldots, \mathbf{x}_N$ and taking into account equations (4.56) and (4.57), we obtain that

$$\sum_{l=1}^{N} \eta(x_p)\eta(x_q) \partial_\nu \varphi_{s+l}(x_p)\, \partial_\nu \varphi_{s+l}(x_q) =$$

$$= \sum_{k=1}^{N} \sum_{j=1}^{N} \left( \sum_{l=1}^{N} \beta_{lk}\beta_{lj} \right) e_k(x_p) e_j(x_q) = \sum_{l=1}^{N} \beta_{lp}\beta_{lq}.$$

Hence, we know the matrix $B^tB$. Next we use the polar decomposition, $B = UP$, of the matrix $B$. In this decomposition, $P = [P_{kj}] = (B^tB)^{1/2}$ is known and $U \in \mathcal{O}(N)$. Let

$$\zeta_k(\mathbf{x}) = \sum_{j=1}^{N} P_{kj} e_j(\mathbf{x}).$$

Then the functions $\zeta_k$ satisfy equation (4.54) with $[\alpha_{kl}] = U^{-1}$. □

**Remark.** Analyzing the procedure of constructing the manifold $\mathcal{M}$ and the operator $\mathcal{A}$ given in Chapter 3, we see that it is not the individual Fourier coefficients of the waves, but the inner products,

$$\langle u^f(t), u^h(s)\rangle = \sum_{j=1}^{\infty} u_l^f(t)\overline{u_l^h(s)}, \tag{4.58}$$

that are needed in this reconstruction procedure (see, e.g., Theorem 3.7 and thereafter). We observe that, by using just the quadratic forms $Q_n$, we are able to find the sum (4.58).

Summarizing the previous results, we have shown the following lemma.

**Lemma 4.10** *Assume that any hyperbolic data, i.e., either of the forms $\Lambda$, $\mathcal{B}$, or the energy flux $\Pi$ is given. Then it is possible to determine the boundary spectral data up to a gauge transformation.*

**4.1.16.** In this section, we will consider the parabolic Dirichlet-Robin form $L$. We use the Laplace transform,

$$\check{f}(\mathbf{x}, \lambda) = \mathcal{L}f(\lambda) = \int_0^\infty e^{-\lambda t} f(\mathbf{x}, t)\, dt.$$

Then, the same considerations as in section 4.1.14 show that

$$\mathcal{L}(Lf)(\lambda) = \int_0^\infty e^{-\lambda t}(Lf)(\mathbf{x}, t)\, dt = \Lambda_{-\lambda}\check{f}(\mathbf{x}, \lambda), \quad \Re\lambda > -\lambda_1.$$

We consider the inner product,

$$\langle Y_\tau Lf, h\rangle_{L^2(\Sigma)} = \int_0^\infty \int_{\partial\mathcal{M}} (Lf)(\mathbf{x}, t)\, h(\mathbf{x}, t-\tau)\, m dS_g dt,$$

where $\tau > 0$ and $f, h \in C_0^\infty(\Sigma)$. By the Plancherel's formula for the Laplace transform,

$$\int_0^\infty \int_{\partial\mathcal{M}} (Lf)(\mathbf{x}, t-\tau)\, h(\mathbf{x}, t)\, m dS_g dt = \tag{4.59}$$

$$\frac{1}{2\pi i} \int_{\partial\mathcal{M}} \int_{\sigma - i\infty}^{\sigma + i\infty} e^{\lambda\tau} \Lambda_{-\lambda} \check{f}(\mathbf{x}, \lambda)\, \check{h}(\mathbf{x}, -\lambda)\, d\lambda\, m dS_g,$$

when $\sigma > -\lambda_1$. Together with $\Lambda_{-\lambda}$, the integral in the right-hand side of (4.59) has poles at points $-\lambda_j$ with $-\lambda_j \to -\infty$. Moreover, when supp $(f) \subset \partial\mathcal{M} \times (0, a)$, then for arbitrary $N > 0$,

$$\|\check{f}(\lambda)\| \le C_N e^{-a\Re\lambda}(1 + |\lambda|)^{-N},$$

in the half plane $\Re\lambda < 0$. As for any $h \in C_0^\infty(\Sigma)$,

$$\|\check{h}(-\lambda)\|_{(3/2, \partial\mathcal{M})} \le C_N(1 + |\lambda|)^{-N},$$

in this half plane. Therefore, using considerations analogous to those in Steps 2 and 3 of the proof of Lemma 4.6,

$$\int_{\sigma - i\infty}^{\sigma + i\infty} e^{\lambda\tau} \Lambda_{-\lambda} \check{f}(\mathbf{x}, \lambda)\, \check{h}(\mathbf{x}, -\lambda)\, d\lambda =$$

$$= \lim_{n\to\infty} \int_{\Gamma_n} e^{\lambda\tau} \Lambda_{-\lambda} \check{f}(\mathbf{x}, \lambda)\, \check{h}(\mathbf{x}, -\lambda)\, d\lambda,$$

where $\Gamma_n$ is the boundary of the half disks $B(\sigma, r_n) \cap \{z : \Re z \le \sigma\}$ and $r_n \to \infty$ when $n \to \infty$.

Thus, as in Lemma 4.6, the integral in the right-hand side of (4.59) can be evaluated by means of the residue theorem. These considerations give rise to the following formula,

$$\langle Lf, Y_\tau h\rangle_{L^2(\Sigma)} = \sum_{n=1}^\infty e^{-\lambda_{(n)}\tau} Q_n[\check{f}(-\lambda_{(n)}), \check{h}(\lambda_{(n)})]. \tag{4.60}$$

Since $\lambda_{(n)} \to +\infty$ when $n \to \infty$, different terms in the right-hand side of (4.60) have different exponential behaviour when $\tau \to \infty$. Therefore, varying $f$ and $h$, we can find the eigenvalues $\lambda_{(n)}$ and the quadratic forms $Q_n$. Combining this result with Lemma 4.9, we obtain the following lemma.

**Lemma 4.11** *Assume that the parabolic Dirichlet-Robin form L is given. Then it is possible to determine the boundary spectral data up to a gauge transformation.*

**4.1.17.** At last, we will consider the elliptic Dirichlet-Robin form $\Lambda_\lambda[\cdot,\cdot]$ given for $\lambda \in \mathbf{R}$. It follows from Lemma 4.5 that, for given $f, h \in C^\infty(\partial\mathcal{M})$, the function $\Lambda_\lambda[f,h]$, $\lambda \in \mathbf{R}$, can have first-order singularities only at points $\lambda = \lambda_{(n)}$, $n = 1, 2, \ldots$. Varying $f$ and $h$ and using representation (4.39), we see that the singularities of the function $\Lambda_\lambda[f,h]$ determine the eigenvalues $\lambda_{(n)}$ and the quadratic forms $Q_n$.

When the Gel'fand data (4.20) is given, we can find

$$\widetilde{\Lambda}_\lambda[f,h] = \int_{\partial\mathcal{M}} \Lambda_\lambda f\,\overline{h}\,d\mu,$$

where $d\mu$ is an arbitrary smooth measure on $\partial\mathcal{M}$. This is a gauge transform of the form $\Lambda_\lambda$. Thus, we can find $\lambda_{(n)}$ and $Q_n$ as above. Hence, in both cases, using Lemma 4.5 again, we find the boundary spectral data to within a gauge transformation.

**4.1.18.** We summarize the previous considerations in the following theorem

**Theorem 4.12** *Let $\partial\mathcal{M}$ be given as a differentiable manifold. Assume that we know the gauge equivalence class of either of the forms $\Pi$, $\Lambda$, $\mathcal{B}$, $L$, or $\Lambda_\lambda$. Then it is possible to find the gauge equivalence class of the boundary spectral data and, henceforth, the orbit $\sigma(\mathcal{A})$ of the operator $\mathcal{A}$.*

Theorem 4.12 implies the following corollary, which is valid when the boundary forms, rather then their equivalence classes, are known.

**Corollary 4.13** *Let $\partial\mathcal{M}$ be given as a differentiable manifold. Assume that we know either of the forms $\Pi$, $\Lambda$, $\mathcal{B}$, $L$, or $\Lambda_\lambda$. Then it is possible to find the orbit of $\sigma(\mathcal{A})$ of $\mathcal{A}$ in the group of normalized gauge transformations, that is,*

$$\sigma_0(\mathcal{A}) = \{\mathcal{A}_\kappa : \ \kappa \in C^\infty(\mathcal{M}),\ \kappa > 0,\ \kappa|_{\partial\mathcal{M}} = 1\}.$$

**Proof.** The previous considerations show that we can find the Riemannian manifold $(\mathcal{M}, g)$ and the unique Schrödinger operator $\widetilde{\mathcal{A}} = -\Delta_g + q \in \sigma(\mathcal{A})$. Consider the lowest eigenvalue $\lambda_1$, which, due to Theorem 2.21, has multiplicity 1. Since the metric induced by $g$

on $\partial\mathcal{M}$ is already known, we can find the integral kernels of the form $Q_1$ and, therefore, the function $m\partial_\nu\varphi_1|_{\partial\mathcal{M}}$ for the operator $\mathcal{A}$. Let $\widetilde{\varphi}_1$ be the normalized eigenfunction of $\widetilde{\mathcal{A}}$ in $L^2(\mathcal{M}, dV_g)$. When $\kappa$ is such that $\mathcal{A} = \widetilde{\mathcal{A}}_\kappa$, then

$$(\kappa|_{\partial\mathcal{M}})^{-1}\partial_\nu\widetilde{\varphi}_1|_{\partial\mathcal{M}} = m\partial_\nu\varphi_1|_{\partial\mathcal{M}}.$$

Since $\partial_\nu\widetilde{\varphi}_1|_{\partial\mathcal{M}}$ and $m\partial_\nu\varphi_1|_{\partial\mathcal{M}}$ are known and $\partial_\nu\widetilde{\varphi}_1|_{\partial\mathcal{M}}$ is positive almost everywhere, we can find $\kappa|_{\partial\mathcal{M}}$. □

## 4.2. Dynamical inverse problem for the wave equation

**4.2.1.** In this section, we study inverse boundary problem *iii.* in Problem 6, when the boundary form $\mathcal{B}^{2T}[f,h]$,

$$\mathcal{B}^{2T}[f,h] = \int_{\Sigma^{2T}} \left(\partial_\nu u^f \overline{u^h} - u^f \partial_\nu \overline{u^h}\right) m dS_g dt, \quad f,h \in \mathring{C}^\infty(\Sigma^{2T}),$$

is given on a finite time interval $[0, 2T]$.

**4.2.2.**

**Problem 7** *Assume that two pairs $(\mathcal{M}, \mathcal{A})$ and $(\widetilde{\mathcal{M}}, \widetilde{\mathcal{A}})$, such that $\partial\mathcal{M} = \partial\widetilde{\mathcal{M}}$, have the gauge-equivalent boundary forms $\mathcal{B}^{2T}$ and $\widetilde{\mathcal{B}}^{2T}$. Do then*

$$\mathcal{M}^T = \widetilde{\mathcal{M}}^T \tag{4.61}$$

*and*

$$\widetilde{a}(\mathbf{x}, D) = a_\kappa(\mathbf{x}, D), \quad \text{in } \mathcal{M}^T = \widetilde{\mathcal{M}}^T \tag{4.62}$$

*for some positive $\kappa \in C^\infty(\mathcal{M}^T)$?*

We remind the reader that an operator $\mathcal{A}$ of form (3.1) determines a Riemannian metric on $\mathcal{M}$. Then the corresponding boundary layer $\mathcal{M}^T$ has the form

$$\mathcal{M}^T = \{\mathbf{x} \in \mathcal{M} : d(\mathbf{x}, \partial\mathcal{M}) < T\}$$

(see section 2.1.16). Equations $\partial\mathcal{M} = \partial\widetilde{\mathcal{M}}$ and $\mathcal{M}^T = \widetilde{\mathcal{M}}^T$ mean that these manifolds are diffeomorphic. Then relation (4.62) implies, in particular, that $\mathcal{M}^T$ and $\widetilde{\mathcal{M}}^T$ are isometric.

**Remark.** Let us explain the physical meanings of Problem 7. It corresponds to probing an unknown medium $\mathcal{M}$ by waves sent from the boundary. These waves are scattered by inhomogeneities of the medium and return to the boundary affecting the Cauchy data of the wave. Hence, to get information about a point $\mathbf{x} \in \mathcal{M}$, the observation time $2T$ should satisfy the inequality

$$2T \geq 2d(\mathbf{x}, \partial\mathcal{M}).$$

In other words, the waves should have enough time to go from the boundary to $\mathbf{x}$ and return back to the boundary. Thus, there is no way to obtain information about any $\mathbf{x} \notin \mathcal{M}^T$ from the knowledge of $\mathcal{B}^{2T}$.

**4.2.3.** According to the considerations in section 4.1.3, the boundary forms $\mathcal{B}^{2T}$, which correspond to differential expressions $a(\mathbf{x}, D)$ lying in the same orbit $\sigma(\mathcal{A})$ of the group of gauge transformations, are gauge-equivalent. Due to the existence of the unique Schrödinger differential expression, $-\Delta_g+q$, in $\sigma(\mathcal{A})$, Problem 7 can be reduced to the inverse boundary-value problem for the Schrödinger wave equation,

$$\partial_t^2 u - \Delta_g u + qu = 0, \text{ in } Q^{2T},$$

$$u|_{\Sigma^{2T}} = f, \tag{4.63}$$

$$u|_{t=0} = 0, \quad \partial_t u|_{t=0} = 0.$$

**Problem 8** *Let $\mathcal{B}^{2T}$ be the boundary form corresponding to a Schrödinger operator $-\Delta_g + q$. Does a form $S_{\kappa_0}\mathcal{B}^{2T}$ with some $\kappa_0 \in C^\infty(\partial\mathcal{M})$, $\kappa_0 > 0$, determine the boundary layer $\mathcal{M}^T$, and metric $g$ and potential $q$ on $\mathcal{M}^T$?*

**4.2.4.** The rest of this section is devoted to the solution of Problem 8. Our considerations follow the pattern given in Chapter 3. We will prove the following result:

Figure 4.3: Domain of integration in the Blagovestchenskii identity

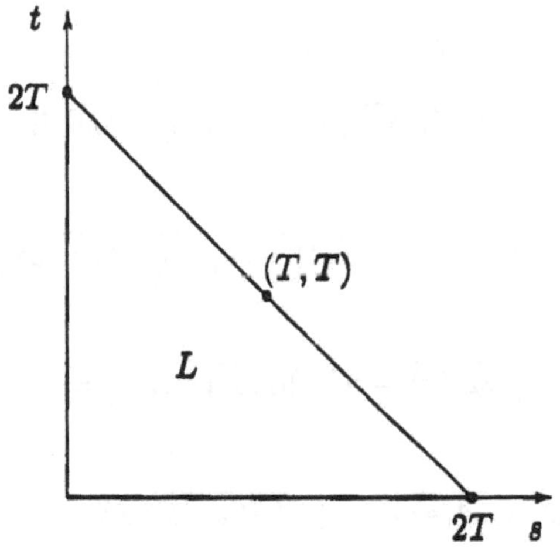


**Theorem 4.14** *Let $\widetilde{\mathcal{B}}^{2T}$ be gauge equivalent to the boundary form $\mathcal{B}^{2T}$ of a Schrödinger differential operator $-\Delta_g + q$. Then $\partial\mathcal{M}$ and $\widetilde{\mathcal{B}}^{2T}$ determine $\mathcal{M}^T$, metric $g$, and potential $q$ on $\mathcal{M}^T$.*

The proof of this theorem is given as a sequence of lemmas.

**4.2.5.** We start with the Blagovestchenskii identity (see Fig.4.3).

**Lemma 4.15** *Let $f, g \in \mathring{C}^\infty(\Sigma^{2T})$. Then*

$$\int_{\mathcal{M}} u^f(T)\overline{u^h(T)}\, dV_g = \tag{4.64}$$

$$= \frac{1}{4}\int_L sign(t-s)\int_{\partial\mathcal{M}} (f(t)\overline{(\Lambda^{2T}h)(s)} - (\Lambda^{2T}f)(t)\overline{h(s)})\, dS_g dt ds,$$

*where*

$$L = \{(s,t):\ 0 \le t+s \le 2T,\ t,s > 0\}.$$

**Proof.** Let

$$w(t,s) = \int_{\mathcal{M}} u^f(t)\overline{u^h(s)}\, dV_g.$$

Then, by integration by parts, we see that

$$(\partial_t^2 - \partial_s^2)w(t,s) = \int_{\mathcal{M}} [\partial_t^2 u^f(t)\overline{u^h(s)} - u^f(t)\overline{\partial_s^2 u^h(s)}]\, dV_g =$$

$$= -\int_{\mathcal{M}} [(-\Delta_g + q)u^f(t)\overline{u^h(s)} - u^f(t)\overline{(-\Delta_g + q)u^h(s)}]\, dV_g =$$

$$= -\int_{\partial\mathcal{M}} [\partial_\nu u^f(t)\overline{u^h(s)} - u^f(t)\overline{\partial_\nu u^h(s)}]\, dS_g =$$

$$= \int_{\partial\mathcal{M}} [\Lambda^{2T} u^f(t)\overline{u^h(s)} - u^f(t)\overline{\Lambda^{2T} u^h(s)}]\, dS_g.$$

Moreover,

$$w|_{t=0} = w|_{s=0} = 0,$$

$$\partial_t w|_{t=0} = \partial_s w|_{s=0} = 0.$$

Thus, $w$ is the solution of the initial-boundary value problem for the one-dimensional wave equation in the domain $(t,s) \in [0,2T] \times [0,2T]$ with known source and zero initial and boundary data. Solving this problem, we determine $w(t,s)$ in the domain $(t,s)$, $t + s \le 2T$. In particular, $w(T,T)$ gives the assertion. □

Next, we formulate the previous result by using the boundary form $\mathcal{B}^{2T}$.

**Lemma 4.16** *Let $f, h \in \mathring{C}^\infty(\Sigma^{2T})$. Then*

$$\int_{\mathcal{M}} u^f(T)\overline{u^h(T)}\, dV_g = \tag{4.65}$$

$$= \frac{1}{2}\int_{-T}^{T} sign(\tau)\mathcal{B}^{2T}[Y_{T+\tau}(f), Y_{T-\tau}(h)]d\tau,$$

*where $Y_\tau$ is the time-delay operator,*

$$(Y_\tau f)(\mathbf{x},t) = f(\mathbf{x}, t-\tau). \tag{4.66}$$

**Proof.** We have

$$\mathcal{B}^{2T}[Y_{T+\tau}(f), Y_{T-\tau}(h)] = \int_{\Sigma^{2T}} (\Lambda^{2T} f(\eta - (T+\tau))\overline{h(\eta - (T-\tau))} -$$

$$- f(\eta - (T+\tau))\overline{\Lambda^{2T} h(\eta - (T-\tau))})\, dS_g d\eta.$$

Changing coordinates from $(\tau, \eta)$ to $(s, t)$,

$$\tau = (s-t)/2, \quad \eta - T = (t+s)/2,$$

and using Blagovestchenskii identity (4.64), we obtain representation (4.65). □

Since we know $\widetilde{\mathcal{B}}^{2T} = S_{\kappa_0}\mathcal{B}^{2T}$, rather then $\mathcal{B}^{2T}$, that is, a form that is gauge-equivalent to $\mathcal{B}^{2T}$, we can compute for any $f, h \in \mathring{C}^\infty(\Sigma^{2T})$ the inner product

$$\int_{\mathcal{M}} u^{\varrho f}(T)\overline{u^{\varrho h}(T)}\, dV_g = \frac{1}{2}\int_{-T}^{T} \operatorname{sign}(\tau)\widetilde{\mathcal{B}}^{2T}[Y_{T+\tau}(f), Y_{T-\tau}(h)]\, d\tau, \tag{4.67}$$

where

$$\varrho = \kappa_0^{-1} \in C^\infty(\partial\mathcal{M}; \mathbf{R}_+) \tag{4.68}$$

is an unknown function. This result is in distinct similarity to the result of section 3.2.4, where we compute the analogous inner product with the same unknown function $\varrho$.

**4.2.6.** As in the case of the spectral problem, we are now able to prove a result which is a direct analogue of Lemma 3.12

**Lemma 4.17** *Let $\Gamma \subset \partial\mathcal{M}$ be an open set and $\tau \in (0, T)$. Given the boundary form $\widetilde{\mathcal{B}}^{2T}$, it is possible to construct boundary sources $f_j \in C_0^\infty(\Gamma \times [0, \tau])$, $j = 1, 2, \ldots,$ such that*

$$v_j(\mathbf{x}) = u^{\varrho f_j}(\mathbf{x}, \tau)$$

*form an orthonormal basis of $L^2(\mathcal{M}(\Gamma, \tau))$.*

We remind the reader that $\mathcal{M}(\Gamma, \tau)$ is the domain of influence given in Definition 3.8.

**Proof.** The proof of this lemma is the same as that of Lemma 3.12, since it is based only on the knowledge of inner products (4.67). □

Following the trend developed in section 3.3.4 we obtain the following result.

**Lemma 4.18** *Let $f, h \in \mathring{C}^\infty(\Gamma \times [0, T])$ and $\Gamma \subset \partial\mathcal{M}$ be an open set. Then, given the boundary form $\widetilde{B}^{2T}$, it is possible to find the inner product*

$$\langle P_{\Gamma,\tau} u^{ef}(t), u^{eh}(s)\rangle = \int_{\mathcal{M}(\Gamma,\tau)} u^{ef}(\mathbf{x}, t)\, \overline{u^{eh}(\mathbf{x}, s)}\, dV_g, \qquad (4.69)$$

*for any $0 < t, s, \tau \leq T$.*

**Proof.** The proof is identical to the proof of Lemma 3.13. □

**4.2.7.** To proceed with the reconstruction of the manifold $\mathcal{M}^T$ and the metric $g$ on it, we use Gaussian beams described in sections 3.5–3.6.

**Lemma 4.19** *Let $\mathbf{z}_0, \mathbf{y} \in \partial\mathcal{M}$ and $0 < s < \min(l(\mathbf{z}_0), T)$. Then boundary form $\widetilde{B}^{2T}$ determines $\min(d(\gamma_{\mathbf{z}_0,\nu}(s), \mathbf{y}), T)$.*

**Proof.** The proof of Lemma 4.19, as well as the proof of Lemma 3.28 is based on the knowledge of the inner product $\langle P_{\Gamma,\tau} u^{ef}(t), u^{eh}(s)\rangle$. However, in the case of Lemma 4.19, we have a restriction $\tau, s, t < T$. Otherwise, the proof of Lemma 4.19 is analogous to that of Lemma 3.28. □

For $\mathbf{x} \in \mathcal{M}^T$, let

$$r_{\mathbf{x}}^T(\mathbf{z}) = \min(d(\mathbf{x}, \mathbf{z}), T), \;\; \mathbf{z} \in \partial\mathcal{M}, \qquad (4.70)$$

be the truncated boundary distance function (see Fig. 4.4).

These functions determine the mapping $R^T : \mathcal{M}^T \to C(\partial\mathcal{M})$, by the formula

$$R^T(\mathbf{x}) = r_{\mathbf{x}}^T.$$

Clearly, $R^T$ is the direct analogue of mapping $R$ introduced in section 3.8.2. The image $R^T(\mathcal{M}^T)$ is called the set of the truncated boundary distance functions. Using previous considerations, we can prove an analogue of Theorem 3.29.

Figure 4.4: Mapping $R^T$

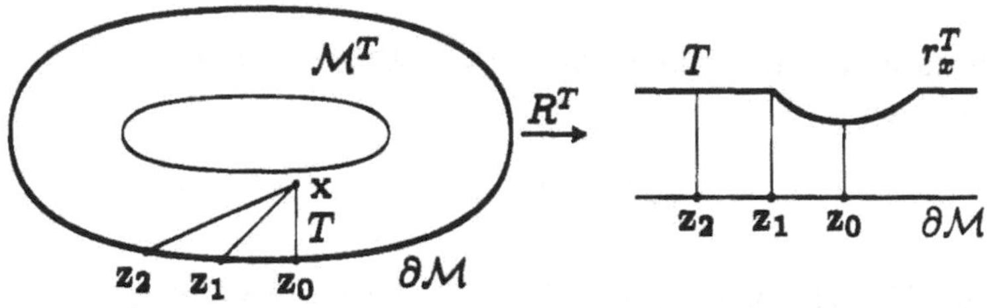


**4.2.8.**

**Lemma 4.20** *Mapping $R^T : \mathcal{M}^T \to R^T(\mathcal{M}^T)$ is a homeomorphism, if $R^T(\mathcal{M}^T) \subset C(\partial\mathcal{M})$ is considered as a topological space with topology inherited from $C(\partial\mathcal{M})$.*

**Proof.** *Step 1.* Due to the triangular inequality, for any $\mathbf{x}, \mathbf{y} \in \mathcal{M}^T$,

$$\|r_{\mathbf{x}}^T - r_{\mathbf{y}}^T\|_{C(\partial\mathcal{M})} \leq d(\mathbf{x}, \mathbf{y}),$$

so that $R^T : \mathcal{M}^T \to C(\partial\mathcal{M})$ is continuous. As

$$s = \min_{\mathbf{z}' \in \partial\mathcal{M}} r_{\mathbf{x}}^T(\mathbf{z}') < T$$

for $\mathbf{x} \in \mathcal{M}^T$, there is a point $\mathbf{z} \in \partial\mathcal{M}$ such that $r_{\mathbf{x}}^T(\mathbf{z}) = s$, i.e. $d(\mathbf{x}, \mathbf{z}) = s \leq d(\mathbf{x}, \mathbf{z}')$, for $\mathbf{z}' \in \partial\mathcal{M}$. Hence, $\mathbf{x} = \gamma_{\mathbf{z},\nu}(s)$. The same arguments, as in the proof of Lemma 3.30, show now that $R^T$ is injective and, henceforth,

$$R^T : \mathcal{M}^T \to R^T(\mathcal{M}^T)$$

is bijective.

*Step 2.* Consider the restriction of $R^T$ onto $\overline{\mathcal{M}}^\tau$, $\tau < T$, where

$$\overline{\mathcal{M}}^\tau = \{\mathbf{x} \in \mathcal{M} : \ d(\mathbf{x}, \partial\mathcal{M}) \leq \tau\}.$$

Then the same considerations, as in the proof of Lemma 3.30, show that

$$R^T : \overline{\mathcal{M}}^\tau \to R^T(\overline{\mathcal{M}}^\tau)$$

is a homeomorphism, so that

$$R^T : \mathcal{M}^\tau \to R^T(\mathcal{M}^\tau)$$

is also a homeomorphism. As for $\mathbf{x} \in \mathcal{M}^\tau$, $\mathbf{y} \in \mathcal{M}^T \setminus \overline{\mathcal{M}^{\tau_1}}$, $\tau < \tau_1 < T$, we have

$$\|r_{\mathbf{x}}^T - r_{\mathbf{y}}^T\| \geq \tau_1 - \tau,$$

the set $R^T(\mathcal{M}^\tau)$ is open in $R^T(\mathcal{M}^T)$. Using the fact that

$$R^T(\mathcal{M}^T) = \bigcup_{\tau<T} R^T(\mathcal{M}^\tau),$$

we see that $R^T : \mathcal{M}^T \to R^T(\mathcal{M}^T)$ is a homeomorphism. □

Summarizing the previous considerations, we prove an analogue of Theorem 3.29.

**Theorem 4.21** *Let a boundary form $\widetilde{\mathcal{B}}^{2T}$ be gauge-equivalent to the boundary form $\mathcal{B}^{2T}$ of a Schrödinger operator. Then $\widetilde{\mathcal{B}}^{2T}$ determines the set $R^T(\overline{\mathcal{M}^T}) \subset C(\partial\mathcal{M})$.*

**Proof.** By means of Lemma 4.19 it is possible to construct the set $R^T(\mathcal{M}^T)$. Lemma 4.20 implies that $R^T(\overline{\mathcal{M}}^T) = \overline{R^T(\mathcal{M}^T)}$. □

Our next goal is to provide $R^T(\mathcal{M}^T)$ with a differentiable structure such that $R^T : \mathcal{M}^T \to R^T(\mathcal{M}^T)$ turns out to be a diffeomorphism. To this end, we use the boundary normal coordinates and the boundary distance coordinates in the same way, as in sections 3.8.3 and 3.8.4. We remind the reader that the boundary normal coordinates on the set $\mathcal{M} \setminus \omega$ are the functions $(s(\mathbf{x}), \mathbf{z}(\mathbf{x}))$, where $s(\mathbf{x}) = d(\mathbf{x}, \partial\mathcal{M})$ and $\mathbf{z}(\mathbf{x}) \in \partial\mathcal{M}$ is the unique boundary point nearest to $\mathbf{x}$. Using the boundary normal coordinates, we immediately obtain the following result:

**Lemma 4.22** *The set of the truncated boundary distance functions $R^T(\mathcal{M}^T)$ determines the sets $R^T(\mathcal{M}^T \cap \omega)$ and $R^T(\mathcal{M}^T \setminus \omega)$. The functions $(Z^T(r^T), S^T(r^T)) \in \partial\mathcal{M} \times \mathbf{R}$,*

$$S^T(r^T) = \min_{\mathbf{z}\in\partial\mathcal{M}} r^T(\mathbf{z}), \tag{4.71}$$

$$r^T(Z^T(r^T)) = S^T(r^T)$$

*determine a coordinate system on the set* $R^T(\mathcal{M}^T \setminus \omega)$*. Moreover,* $R^T : \mathcal{M}^T \setminus \omega \to R^T(\mathcal{M}^T \setminus \omega)$ *is a diffeomorphism.*

Analogously, outside a neighbourhood of the boundary, we can define an analog of the boundary distance coordinates on $R^T(\mathcal{M}^T)$. To this end, we use the evaluation functions

$$E_{\mathbf{z}} : R^T(\mathcal{M}^T) \to \mathbf{R}, \tag{4.72}$$

$$E_{\mathbf{z}}(r^T) = r^T(\mathbf{z}), \tag{4.73}$$

where $\mathbf{z} \in \partial\mathcal{M}$.

**Lemma 4.23** *For any* $r_0^T \in R^T(\mathcal{M}^T)$, $\min_{\mathbf{z}\in\partial\mathcal{M}} r_0^T(\mathbf{z}) > 0$, *there are* $m$ *points* $\mathbf{z}_1, \ldots, \mathbf{z}_m \in \partial\mathcal{M}$, *such that the functions* $E_j = E_{\mathbf{z}_j}$, $j = 1, \ldots, m$, *are local coordinates in a neighborhood of* $r_0^T$*. These coordinates, together with the boundary normal coordinates, define a differentiable structure of* $R^T(\mathcal{M}^T)$*, which makes the mapping* $R^T : \mathcal{M}^T \to R^T(\mathcal{M}^T)$ *a diffeomorphism.*

**Proof.** The proof is analogous to the proof of Lemma 3.32. We need only to take into account that the points $\mathbf{z}_1, \ldots, \mathbf{z}_m \in \partial\mathcal{M}$ can be taken from an arbitrarily small neighborhood of the point $\mathbf{z}_0$, where $\mathbf{z}_0 \in \partial\mathcal{M}$ satisfies the relation $r_0^T(\mathbf{z}_0) = S^T(r_0^T)$. This fact follows from the proof of Lemma 2.14. This implies that the evaluation functions satisfy the inequality,

$$E_{\mathbf{z}_j}(r^T) < T, \tag{4.74}$$

when $r^T$ are close to $r_0^T$. Thus, from the definition of the truncated boundary distance functions, we see that

$$E_{\mathbf{z}_j}(r^T) = d((R^T)^{-1}(r^T), \mathbf{z}_j), \tag{4.75}$$

and the arguments, given in the proof of Lemma 3.32, yield the claim.
□

**4.2.9.** Summarizing the previous considerations, we see that the knowledge of $\widetilde{\mathcal{B}}^{2T}$ makes it possible to construct a differentiable manifold $R^T(\mathcal{M}^T)$, which is diffeomorphic to $\mathcal{M}^T$. Moreover, we have constructed coordinate functions on $R^T(\mathcal{M}^T)$, which correspond to the boundary normal and boundary distance coordinates. As in section 3.8.5, we can use the evaluation functions $E_{\mathbf{z}}$ and the boundary distance coordinates to equip $R^T(\mathcal{M}^T)$ with a metric $\widetilde{g}^T$, so that the mapping $R^T : (\mathcal{M}, g) \to (R^T(\mathcal{M}^T), \widetilde{g}^T)$ becomes an isometry. Indeed, using the same constructions as in section 3.8.5, we obtain the metric tensor $\widetilde{g}_{ij}^T$ in the boundary distance coordinates on the set $R^T(\mathcal{M}^T \setminus \partial\mathcal{M})$. Representing the metric tensor $\widetilde{g}_{ij}^T$ in the boundary normal coordinates and using its continuity, we find $\widetilde{g}_{ij}^T$ on the whole manifold $R^T(\mathcal{M}^T)$.

Summarizing the above considerations, we obtain the following result.

**Lemma 4.24** *Let $\widetilde{\mathcal{B}}^{2T}$ be a given form that is gauge-equivalent to the boundary form $\mathcal{B}^{2T}$ of a Schrödinger operator, $-\Delta_g + q$ on some manifold $\mathcal{M}$. Then it is possible to reconstruct the Riemannian manifold $(\mathcal{M}^T, g)$.*

**4.2.10.** Following the method developed in section 3.9.1, we construct the unknown function $\varrho$ on the boundary. To this end, we use Gaussian beams. By Lemma 4.18, at any time $t < T$, we can find the norms of the Gaussian beams $u_{\epsilon,\varrho}(\mathbf{x}, t) = u^{\varrho f}(\mathbf{x}, t)$, where the boundary source $f = f_{\epsilon,\mathbf{z}_0,t_0}(\mathbf{z}, t)$ is given by formula (3.41) with $V = 1$. This boundary source generates a Gaussian beam that starts at time $t_0$ from the point $\mathbf{z}_0 \in \partial\mathcal{M}$. The Gaussian beam moves in the direction normal to the boundary. Lemma 3.24 and Lemma 2.58 imply that, when $t < T$,

$$\lim_{\epsilon \to 0} \|P_{\mathbf{z}_1,\tau_1} u_{\varrho,\epsilon}(\cdot, t)\|^2 = \frac{\varrho^2(\mathbf{z}_0)(g(\mathbf{z}_0, 0))^{1/2}}{\sqrt{\det(\Im H_0)}}. \tag{4.76}$$

This formula can be used to find $\varrho(\mathbf{z}_0) = \kappa_0(\mathbf{z}_0)^{-1}$. Indeed, all the terms, except $\varrho(\mathbf{z}_0)$, in the right-hand side of this equation are determined by the already reconstructed metric $g$, while the left-hand side is determined by $\widetilde{\mathcal{B}}^{2T}$ due to Lemma 4.18.

**4.2.11.** To compete the reconstruction of the Schrödinger operator $-\Delta_g + q$, it remains to find $q$.

**Lemma 4.25** *The boundary form $\mathcal{B}^{2T}$ of a Schrödinger operator, $-\Delta_g + q$ uniquely determines the potential $q$ on $\mathcal{M}^T$.*

**Proof.** *Step 1.* Let $u_\epsilon(\mathbf{x},t) = u^f(\mathbf{x},t)$ be the Gaussian beam with $f = f_{\epsilon,\mathbf{z}_0,t_0}(\mathbf{z},t)$, given by formula (3.41) with $V = 1$. By Lemma 4.18, it is possible to find the inner products

$$\langle P_{\mathbf{z}_0,t-t_0} u_\epsilon(t), u^h(t')\rangle, \tag{4.77}$$

for any $t, t' < T$ and any $u^h$ being a solution of initial-boundary value problem (4.63). Repeating the proof of Lemma 3.26 with $\varphi_j$ replaced by $u^h(t')$, we obtain the following representation of $u^h(\mathbf{x},t')$ in the boundary normal coordinates $(s,\mathbf{z})$,

$$u^h(s,\mathbf{z},t') = \lim_{\epsilon\to 0} i\epsilon^{-(m+2)/4} 2^{(1-m)/2} \pi^{(2-m)/2} [\det(-iH(t))]^{1/2} \cdot \tag{4.78}$$

$$\cdot u_{00}(t)^{-1} g^{-1/2}(\mathbf{z}_0, s) \langle P_{\mathbf{z}_0,t-t_0} u_\epsilon(t), u^h(t')\rangle,$$

where $s = t - t_0$, $\mathbf{z} = \mathbf{z}_0$. Here the functions $H(t)$ and $u_{00}(t)$ are given by formulae (3.54), (3.57), correspondingly, and depend only on the known metric $g$. Then, by means of formula (4.78), we can determine the values of any wave $u^h(\mathbf{x},t')$ for $\mathbf{x} \in \mathcal{M}^T$ and $t' < T$.

*Step 2.* Let $u^h(\mathbf{x},t')$ be equal to the Gaussian beam $u_\delta(\mathbf{x},t) = u^f(\mathbf{x},t)$, where $f = f_{\delta,\mathbf{z}_1,t_1}(\mathbf{z},t)$ is given by formula (3.41) with $\epsilon$ replaced by $\delta$ and $V = 1$. By step 1, we can determine the values of $u_\delta(\mathbf{x},t)$ when $\mathbf{x} \in \mathcal{M}^T$ and $t < T$. Consider the Gaussian beam $u_\delta$ on the normal geodesic $\gamma_{\mathbf{z}_1,\nu}$. Then,

$$u_\delta(s,\mathbf{z}_1,s+t_1) = (\delta\pi)^{-m/4}(u_{0,0}(s) + \delta u_{1,0}(s) + \mathcal{O}(\delta^2)) \tag{4.79}$$

where $u_{1,0}(s)$ is given by formula (2.216),

$$u_{1,0}(s) = -\frac{1}{2} u_{0,0}(s) \int_0^s q(\mathbf{x}(t'))\, dt' + \widetilde{u}^1_{1,0}(t). \tag{4.80}$$

The term $\widetilde{u}^1_{1,0}(t)$ is given by formulae (2.212)–(2.215) and depends only on the metric $g$. Hence, formula (4.80) makes it possible to find

$$\int_0^s q(\mathbf{x}(t'))\,dt', \quad \mathbf{x}(t') = \gamma_{\mathbf{z}_1,\nu}(t').$$

Clearly, this formula determines $q(\mathbf{x})$ in $\mathcal{M}^T \setminus \omega$. Since $q$ is continuous, we can continue it onto $\mathcal{M}^T$. □

**4.2.12.** Summarizing the previous considerations, we obtain the proof of Theorem 4.14. Returning to Problem 8, we obtain the following

**Corollary 4.26** *Let $\mathcal{A}$ and $\widetilde{\mathcal{A}}$ be operators of form (3.1)–(3.3) on manifolds $\mathcal{M}$ and $\widetilde{\mathcal{M}}$ with $\partial\mathcal{M} = \partial\widetilde{\mathcal{M}}$. Let the corresponding boundary forms $\mathcal{B}^{2T}$ and $\widetilde{\mathcal{B}}^{2T}$ be gauge equivalent. Then,*

$$\mathcal{M}^T = \widetilde{\mathcal{M}}^T$$

*and*

$$\widetilde{a}(\mathbf{x}, D) = a_\kappa(\mathbf{x}, D) \ \text{in} \ \mathcal{M}^T,$$

*for some $\kappa \in C^\infty(\mathcal{M}^T)$.*

Moreover, when $T$ is large enough, we can reconstruct the whole manifold $\mathcal{M}$ and the operator $\mathcal{A}$ on it.

**Corollary 4.27** *Let $\mathcal{A}$ and $\widetilde{\mathcal{A}}$ satisfy conditions of Corollary 4.26. If, in addition,*

$$T > T^* = \max\{d(\mathbf{x}, \partial\mathcal{M}), \ \mathbf{x} \in \mathcal{M}\},$$

*then*

$$\mathcal{M} = \widetilde{\mathcal{M}}$$

*and*

$$\widetilde{\mathcal{A}} \in \sigma(\mathcal{A}).$$

## 4.3. Inverse problems for heat and wave equations and continuation of data

**4.3.1.** In the previous section, we considered the inverse boundary value problem for the wave equation with data given by the boundary form $\mathcal{B}^{2T}$ or the Dirichlet-Robin map $\Lambda^{2T}$. In particular, we have shown that, when $T > T^*$, these data completely determine the manifold and the gauge equivalence class of the operator. Hence, the knowledge of $\Lambda^{2T}$, $T > T^*$, makes it possible to find $\Lambda^{2t}$ for any $t > 0$, that is, to determine $\Lambda = \Lambda^{\infty}$. To find $\Lambda^{2t}$, we can solve the inverse boundary value problem for the wave equation and find the manifold $\mathcal{M}$ and operator $\mathcal{A}$. Then we can solve direct problem (4.7)–(4.9) for any $f$ and, henceforth, determine $\Lambda^{2t}$ for any $t > 0$. However, there is an alternative way to determine $\Lambda^{2t}$, $t > T^*$, and thus $\Lambda$, from $\Lambda^{2T}$, $T > T^*$. This method, which we will describe in this section, does not require solving the inverse problem. Furthermore, by means of the procedure described in section 4.1, we can then obtain from $\Lambda$ the boundary spectral data. This gives another method to solve the inverse boundary value problem for the wave equation with data given on a time interval $[0, 2T]$, $T > T^*$.

At the end of this section, we will consider the continuation problem for the heat equation. Then, it is sufficient to know the corresponding Dirichlet-Robin map on an arbitrary small time interval $[0, t_0]$, $t_0 > 0$.

We will concentrate on the wave and heat equations that correspond to the Schrödinger operator, $-\Delta_g + q$. However, these procedures can be generalized for the general operators if we use gauge transformations. For simplicity, we will assume that the boundary measure $dS_g$ is known. Therefore, the knowledge of the operator $\Lambda^{2T}$ determines the corresponding quadratic form $\mathcal{B}^{2T}$.

**4.3.2.** Our continuation procedure uses the following controllability result.

**Theorem 4.28** *Let $T_0 > 2T^*$. Then the linear subspace,*

$$\{(u^f(T_0), \partial_t u^f(T_0)) \in H_0^1(\mathcal{M}) \times L^2(\mathcal{M}) : \ f \in C_0^\infty(\Sigma^{T_0})\},$$

*is dense in $H_0^1(\mathcal{M}) \times L^2(\mathcal{M})$.*

**Proof.** Assume that, on the contrary, there exists a pair $(\phi, \psi) \in L^2(\mathcal{M}) \times H^{-1}(\mathcal{M})$ such that

$$\langle \partial_t u^f(\cdot, T_0), \phi \rangle - \langle u^f(\cdot, T_0), \psi \rangle = 0, \tag{4.81}$$

for all $f \in C_0^\infty(\Sigma^{T_0})$. To show that $\phi = \psi = 0$, we consider the following initial-boundary value problem for the wave equation,

$$(\partial_t^2 - \Delta_g + q)e = 0, \quad \text{in} \quad Q^{T_0}, \tag{4.82}$$

$$e|_{\Sigma^{T_0}} = 0, \quad e|_{t=T_0} = \phi, \quad \partial_t e|_{t=T_0} = \psi. \tag{4.83}$$

We note that, in initial-boundary value problem (4.82)–(4.83), the initial data is given at the final time, $t = T_0$, rather than at $t = 0$. However, the basic properties of solutions described in section 2.3, remain intact. In particular, $e \in C^1([0, T]; H^{-1}(\mathcal{M}))$.

Let $u^f(\mathbf{x}, t)$ be a solution of the initial-boundary value problem (4.63). Multiplying wave equation (4.63) by $e$ and (4.82) by $u^f$ and integrating by parts, we get the following equation,

$$0 = \int_{Q^{T_0}} [u^f \overline{(\partial_t^2 - \Delta_g + q)e} - (\partial_t^2 - \Delta_g + q)u^f)\overline{e}]\, dV_g\, dt =$$

$$= \int_{\Sigma^{T_0}} f\, \overline{\partial_\nu e}\, dS_g\, dt,$$

where we have taken into account initial and boundary conditions (4.63) and (4.83). Since $f \in C_0^\infty(\Sigma^{T_0})$ is arbitrary,

$$\partial_\nu e|_{\Sigma^{T_0}} = 0.$$

The double cone $K_{\partial\mathcal{M}, T_0}$, corresponding to the boundary layer $\Sigma^{T_0}$,

$$K_{\partial\mathcal{M}, T_0} = \{(x, t) \in Q^{T_0} : \ d(\mathbf{x}, \partial\mathcal{M}) \leq T_0/2 - |t - T_0|\},$$

contains a neighborhood of $\mathcal{M} \times \{T_0/2\}$. Using the same arguments as in section 2.5.32, we can approximate the function $e \in C^1([0, T]; H^{-1}(\mathcal{M}))$ with smooth, mollified functions $e_s = e * \psi_s$. By Theorem 3.11 we see that $e_s = 0$ in neighborhood of $\mathcal{M} \times \{T_0/2\}$, and letting $s \to 0$, we obtain that

$$e|_{t=T_0/2} = 0, \quad \partial_t e|_{t=T_0/2} = 0. \tag{4.84}$$

Hence, $e(\mathbf{x},t)$ is the solution of wave equation (4.82) with homogeneous initial conditions (4.84) and homogeneous Dirichlet boundary conditions, $e|_{\Sigma^{T_0}} = 0$.

This implies that $e(t) = 0$ for $t \in [0, T_0]$. Therefore,

$$\phi = e|_{t=T_0} = 0, \quad \psi = \partial_t e|_{t=T_0} = 0.$$

□

**4.3.3.** By Lemma 4.3, $\Lambda^t$ determines the energy form $E(u^f,t)$ of the wave $u^f(t)$, $f \in C_0^\infty(\Sigma^t)$,

$$E(u^f,t) = \frac{1}{2}\int_{\mathcal{M}} (|\partial_t u^f(\mathbf{x},t)|^2 + |\operatorname{Grad} u^f(\mathbf{x},t)|_g^2 +$$

$$+q(\mathbf{x})|u^f(\mathbf{x},t)|^2)\, dV_g(\mathbf{x}),$$

where we take into account that $u^f(t)|_{\partial\mathcal{M}} = f(t) = 0$. Consider the corresponding quadratic form $\mathcal{E}(\psi,\phi)$,

$$\mathcal{E}(\psi,\phi) = \frac{1}{2}\int_{\mathcal{M}} (|\operatorname{Grad}\psi(\mathbf{x})|_g^2 + q(\mathbf{x})|\psi(\mathbf{x})|^2 + |\phi(\mathbf{x})|^2)\, dV_g(\mathbf{x}) \tag{4.85}$$

on $H_0^1(\mathcal{M}) \times L^2(\mathcal{M})$. If the Schrödinger operator, $-\Delta_g + q$, is positive definite, then this form is equivalent to the canonical quadratic form on $H_0^1(\mathcal{M}) \times L^2(\mathcal{M})$,

$$\|(\psi,\phi)\|^2_{H_0^1(\mathcal{M})\times L^2(\mathcal{M})} = \frac{1}{2}\int_{\mathcal{M}} (|\operatorname{Grad}\psi|_g^2 + |\phi|^2)\, dV_g. \tag{4.86}$$

However, when the Schrödinger operator, $-\Delta_g + q$, is not positive definite, the quadratic form on the right-hand side of (4.85) is also not positive definite. To obtain a form equivalent to the canonical form on $H_0^1(\mathcal{M}) \times L^2(\mathcal{M})$, we introduce the form

$$\mathcal{H}_a(\psi,\phi) = \mathcal{E}(\psi,\phi) + a\|\psi\|^2_{L^2(\mathcal{M})}, \tag{4.87}$$

where $a \geq 0$ is large enough. Previous considerations lead to the following lemma.

**Lemma 4.29** *Forms $\mathcal{E}$ and $\mathcal{H}_a$ have the following properties:*
*i) The Dirichlet-Robin map $\Lambda^t$ determines*

$$E(u^f, t) = \mathcal{E}(u^f(t), \partial_t u^f(t))$$

*for any $f \in C_0^\infty(\Sigma^t)$.*
*ii) The Dirichlet-Robin map $\Lambda^{2t}$ determines $\mathcal{H}_a(u^f(t), \partial_t u^f(t))$ for any $a \geq 0$ and $f \in C_0^\infty(\Sigma^t)$.*
*Moreover, when $-\Delta_g + q$ is positive definite, the quadratic form $\mathcal{E}$ is equivalent to the canonical form (4.86) with $\psi = u^f(t)$ and $\phi = \partial_t u^f(t)$. For sufficiently large $a \geq 0$, the form $\mathcal{H}_a$ is equivalent to the canonical form (4.86) even for a non-positive operator $-\Delta_g + q$.*

**Proof.** The claim follows from Lemma 4.15 and Lemma 4.3. □

Lemma 4.29 elucidates the difference between the positive definite Schrödinger operators and the general ones. Because in our construction of the continuation of Dirichlet-Robin map $\Lambda^t$ we use the norm in $H_0^1(\mathcal{M}) \times L^2(\mathcal{M})$, this lemma explains why the general case is harder than the positive definite case.

### 4.3.4.

**Theorem 4.30** *Let the boundary $\partial\mathcal{M}$ and boundary measure $dS_g$ be known. Then the Dirichlet-Robin map $\Lambda^{2T}$, $T > T^*$ that corresponds to the Schrödinger operator $-\Delta_g + q$ determines the operator $\Lambda$.*

This theorem will be proven in this and the next sections. In this section we will give the proof for the positive definite operators and, in the next one, we will describe changes necessary to generalize it to the non-positive definite case. **Proof.** We will first show that it is possible to extend $\Lambda^{2T}$ onto a larger time interval, $0 \leq t \leq 2T + \epsilon$, where $0 < \epsilon < T - T^*$, i.e. to find $\Lambda^{2T+\epsilon} f$ for any $f \in C_0^\infty(\Sigma^{2T+\epsilon})$ .
Iterating this procedure, we will determine $\Lambda^{2t}$ for any $t > 0$.

*Step 1.* Represent $f$ in the form,

$$f = h + g, \quad h \in C_0^\infty(\Sigma^T), \quad g \in C_0^\infty(\partial\mathcal{M} \times [\epsilon, 2T + \epsilon]),$$

so that

$$\Lambda^{2T+\epsilon} f = \Lambda^{2T+\epsilon} h + \Lambda^{2T+\epsilon} g.$$

As the coefficients of the wave equation are time independent,

$$\Lambda^{2T+\epsilon}g\big|_{[\epsilon,2T+\epsilon]} = Y_\epsilon \Lambda^{2T}(Y_{-\epsilon}g).$$

As $Y_{-\epsilon}g \in C_0^\infty(\Sigma^{2T})$, the function $\Lambda^{2T+\epsilon}g$ is known. Therefore, to find $\Lambda^{2T+\epsilon}f$, it is sufficient to find $\Lambda^{2T+\epsilon}h$.

*Step 2.* Let $T_0 = T - \epsilon > T^*$. By Theorem 4.28, there is a sequence $\{f_j\}_{j=1}^\infty$, $f_j \in C_0^\infty(\Sigma^{2T_0})$ such that

$$\lim_{j\to\infty} \|u^{f_j}(2T_0) - u^h(2T-\epsilon)\|_1 = 0,$$

$$\lim_{j\to\infty} \|\partial_t u^{f_j}(2T_0) - \partial_t u^h(2T-\epsilon)\|_0 = 0. \tag{4.88}$$

For a positive definite Schrödinger operator, equations (4.88) are equivalent to the equation

$$\lim_{j\to\infty} \mathcal{E}\left(u^{f_j}(2T_0) - u^h(2T-\epsilon), \partial_t u^{f_j}(2T_0) - \partial_t u^h(2T-\epsilon)\right) = 0,$$

or, equivalently,

$$\lim_{j\to\infty} E\left(u^{(Y_\epsilon f_j - h)}, 2T-\epsilon\right) = 0. \tag{4.89}$$

By Lemma 4.29 i), equation (4.89) can be verified, if $\Lambda^{2T}$ is given. Therefore, we can find a proper sequence $\{f_j\}$ that satisfies (4.88).

*Step 3.* Consider the waves $u^h(\mathbf{x},t)$ and

$$Y_\epsilon u^{f_j}(\mathbf{x},t) = u^{f_j}(\mathbf{x},t-\epsilon)$$

for $t > 2T - \epsilon$. They satisfy the homogeneous wave equation (4.63) and homogeneous boundary conditions,

$$u^h|_{\partial\mathcal{M}\times(2T-\epsilon,\infty)} = 0, \quad Y_\epsilon u^{f_j}|_{\partial\mathcal{M}\times(2T-\epsilon,\infty)} = 0,$$

In view of (4.88), Theorem 2.30 yields that

$$\lim_{j\to\infty} \|\partial_\nu(u^h - u^{Y_\epsilon f_j})\|_{L^2(\partial\mathcal{M}\times[2T-\epsilon,2T+\epsilon])} = 0. \tag{4.90}$$

Therefore,

$$\Lambda^{2T+\epsilon}h|_{\partial\mathcal{M}\times[2T-\epsilon,2T+\epsilon]} = \lim_{j\to\infty} \Lambda^{2T+\epsilon}(Y_\epsilon f_j)|_{\partial\mathcal{M}\times[2T-\epsilon,2T+\epsilon]}$$

$$= Y_\epsilon\left(\lim_{j\to\infty} \Lambda^{2T} f_j|_{\partial\mathcal{M}\times[2T-2\epsilon,2T]}\right).$$

□

**4.3.5.** We turn now to the general Schrödinger operator. In the proof of Theorem 4.30 for the positive definite case, the only part that fails in the general case is that we cannot verify equation (4.89) by means of the quadratic form $\mathcal{E}(\psi, \phi)$. Therefore, the statement of this theorem will be proven when we show the following result.

**Lemma 4.31** *Let* $\Lambda^{2T}, T > T^*$, *be known and let* $0 < \epsilon < T - T^*$, $T_1 = 2T - \epsilon$. *Then, for any sequence* $\{g_j\}_{j=1}^{\infty}$, $g_j \in C_0^\infty(\Sigma^{T_1})$ *it is possible to find if*

$$\lim_{j\to\infty} \left(\|u^{g_j}(T_1)\|_1^2 + \|\partial_t u^{g_j}(T_1)\|_0^2\right) = 0. \tag{4.91}$$

**Proof.** Clearly, it is enough to show the assertion for the real-valued functions $g_j$.

*Step 1.* We will show that equations (4.91) are equivalent to the following conditions:

$$\lim_{j\to\infty} E\left(u^{g_j}, T_1\right) = 0, \tag{4.92}$$

$$\lim_{j\to\infty} \Lambda^{2T} g_j|_{\partial\mathcal{M}\times[T_1,2T]} = 0, \tag{4.93}$$

and, for any $h \in C_0^\infty(\Sigma^{T_1})$,

$$\lim_{j\to\infty} E\left(u^{g_j}, u^h, T_1\right) = 0, \tag{4.94}$$

where $E(u^f, u^h, t)$ is the bilinear form that corresponds to the quadratic form $E$,

$$E(u^f, u^h, t) = \frac{1}{2}\int_{\mathcal{M}} \Big\{(\operatorname{Grad} u^f(t), \operatorname{Grad} u^h(t))_g + q u^f(t) u^h(t) +$$

$$+ \partial_t u^f(t)\, \partial_t u^h(t)\Big\}\, dV_g + \frac{1}{2}\int_{\partial\mathcal{M}} \sigma u^f(t) u^h(t) dS_g.$$

Because conditions (4.92)–(4.94) can be verified if $\Lambda^{2T}$ is known, this will prove the lemma.

*Step 2.* Let $P_-$, $P_0$, $P_+$ be the orthoprojectors in $L^2(\mathcal{M})$ onto the spaces spanned by the eigenvectors corresponding to the negative, zero, and positive eigenvalues of the operator $-\Delta_g + q$.

Our first goal is to show that $P_- u^{g_j}(T_1) \to 0$. Assume that this is not true. Then, there is a subsequence of $\{g_j\}$ (still denoted by $g_j$) such that for some $c_0 > 0$,

$$\langle AP_- u^{g_j}(T_1), u^{g_j}(T_1)\rangle \le -c_0, \text{ for any } j = 1, 2, \dots \qquad (4.95)$$

Let $p < \infty$ be the dimension of the range of $P_-$. Then, by equation (4.95)

$$\sum_{k=1}^{p} \lambda_k \left|u_k^j\right|^2 \le -c_0 < 0.$$

where $u_k^j$ are the Fourier coefficients of $u^{g_j}(T_1)$. Then there are $\sigma_k \in \{0,1\}$, $k = 1, \dots, p$, such that for some subsequence of $\{g_j\}$

$$\langle Au^{g_j}(T_1), \sum_{k=1}^{p}(-1)^{\sigma_k}\varphi_k\rangle \le -c_1, \qquad (4.96)$$

where $c_1 > 0$ We denote

$$u_- = \sum_{k=1}^{p}(-1)^{\sigma_k}\varphi_k.$$

Take $\delta > 0$. By Theorem 4.28, there is $h = h_\delta \in C_0^\infty(\Sigma^{T_1})$ such that

$$\|\partial_t u^h(T_1)\|_0^2 + \|u^h(T_1) - u_-\|_1^2 \le \delta^2. \qquad (4.97)$$

By condition (4.94), for sufficiently large $j$,

$$|E\left(u^{g_j}, u^h, T_1\right)| \le \delta^2. \qquad (4.98)$$

Clearly,

$$\langle AP_- u^{g_j}(T_1), u_-\rangle = E\left(u^{g_j}, u^h, T_1\right) +$$

$$+\left\langle Au^{g_j}(T_1), (u_- - u^h(T_1))\right\rangle - \left\langle \partial_t u^{g_j}(T_1), \partial_t u^h(T_1)\right\rangle.$$

By using spectral representation of $\mathcal{A}$, we see that $\langle \mathcal{A}u, u\rangle$ is equivalent to $H_0^1(\mathcal{M})$-norm in the range of $P_+$. Thus, it follows from (4.95)–(4.98) that, when $\delta$ is sufficiently small,

$$-\langle \mathcal{A}P_- u^{g_j}(T_1), u_-\rangle \leq \tag{4.99}$$

$$\leq c\delta \left\{ \delta + \|\partial_t u^{g_j}(T_1)\|_0 + [\langle \mathcal{A}P_+ u^{g_j}(T_1), u^{g_j}(T_1)\rangle]^{\frac{1}{2}} \right\}.$$

Now condition (4.92) implies that, for sufficiently large $j$,

$$\langle \mathcal{A}P_+ u^{g_j}(T_1), u^{g_j}(T_1)\rangle + \|\partial_t u^{g_j}(T_1)\|_0^2 \leq \tag{4.100}$$

$$\leq -\langle \mathcal{A}P_- u^{g_j}(T_1), u^{g_j}(T_1)\rangle + \delta^2.$$

Using inequality (4.95) together with (4.99) and (4.100), we obtain that

$$-\langle \mathcal{A}P_- u^{g_j}(T_1), u^{g_j}(T_1)\rangle \leq c\delta,$$

which contradicts to (4.95) when $\delta$ is small enough. Therefore,

$$\lim_{j\to\infty} \|P_- u^{g_j}(T_1)\| = 0. \tag{4.101}$$

*Step 3.* Using (4.101) and taking into account (4.100), we see that

$$\lim_{j\to\infty} \langle \mathcal{A}P_+ u^{g_j}(T_1), u^{g_j}(T_1)\rangle + \|\partial_t u^{g_j}(T_1)\|_0^2 = 0. \tag{4.102}$$

*Step 4.* It remains to show that

$$\lim_{j\to\infty} \|P_0 u^{g_j}(T_1)\| = 0. \tag{4.103}$$

Consider the function $u^{g_j}(\mathbf{x}, t)$ for $t > T_1$. Then,

$$u^{g_j}(t) = P_0 u^{g_j}(t) + (P_- + P_+)u^{g_j}(t).$$

Due to (4.101)–(4.102), it follows from Theorem 2.30 that

$$\lim_{j\to\infty} \partial_\nu[(P_- + P_+)u^{g_j}]\big|_{\partial\mathcal{M}\times[T_1,2T]} = 0.$$

However, condition (4.93) yields that also

$$\lim_{j\to\infty} \partial_\nu u^{g_j}|_{\partial\mathcal{M}\times[T_1,2T]} = 0.$$

Therefore,

$$\lim_{j\to\infty} \partial_\nu P_0 u^{g_j}|_{\partial\mathcal{M}\times[T_1,2T]} = 0.$$

As $P_0 u^{g_j}(t)$ belongs to the finite dimensional eigenspace of $\mathcal{A}$, corresponding to the eigenvalue $\lambda = 0$, and the Neumann boundary values of the eigenfunctions in this space are linearly independent, we see that

$$\lim_{j\to\infty} \|P_0 u^{g_j}(T_1)\| = 0.$$

This proves Theorem 4.30. □

**4.3.6.** In this section, we will consider the continuation problem for the heat equation.

**Theorem 4.32** *Let $t_0 > 0$. The parabolic Dirichlet-Robin map $L^{t_0}$ determines the operator $L$.*

**Proof.** To construct $Lf$ for any $f \in C_0^\infty(\Sigma)$, it is enough to find $Lh$ for any $h \in C_0^\infty(\partial\mathcal{M} \times (0, t_1))$, $t_1 < t_0$. Indeed, we can decompose $f$ as a sum $f = h_1 + \cdots + h_k$ where supp $(h_j) \subset \partial\mathcal{M} \times (a_j, b_j)$, $b_j - a_j < t_1$ and use the linearity of the system. Since the solution $v^h$ of initial-boundary value problem (4.14) has a representation

$$v^h(t) = \sum_{j=1}^{\infty} e^{-\lambda_j(t-t_1)}\langle v^h(t_1), \varphi_j\rangle\varphi_j,$$

we see that $v^h(t)$ is an analytic $H^2(\mathcal{M})$-valued function of $t$, $t \in (t_1, \infty)$. Thus, we can determine $\partial_\nu v^h|_{\partial\mathcal{M}\times(t_1,\infty)}$ from $\partial_\nu v^h|_{\partial\mathcal{M}\times(t_1,t_0)}$ by means of the analytic continuation. Thus $Lh$ is found. □

## 4.4. Inverse problems with data given on a part of the boundary

**4.4.1.** In many applications, the data is given only on a part of the boundary. In this section, we will study the inverse boundary spectral problem for the Schrödinger operator when the normal derivatives of the eigenfunctions are given on an open set $\Gamma \subset \partial\mathcal{M}$. We will prove the following result.

**Theorem 4.33** *Assume that $(\mathcal{M}, \mathcal{A})$ and $(\widetilde{\mathcal{M}}, \widetilde{\mathcal{A}})$ are two pairs of manifolds and Schrödinger operators. Assume that there are open sets $\Gamma \subset \partial\mathcal{M}$ and $\widetilde{\Gamma} \subset \partial\widetilde{\mathcal{M}}$, such that the boundary spectral data on $\Gamma$ and $\widetilde{\Gamma}$ coincide. Namely,*

$$\Gamma = \widetilde{\Gamma}, \ \lambda_j = \widetilde{\lambda}_j, \ \partial_\nu \varphi_j|_\Gamma = \partial_\nu \widetilde{\varphi}_j|_\Gamma, \ j = 1, 2, \ldots,$$

*where $\lambda_j$, $\varphi_j$ and $\widetilde{\lambda}_j$, $\widetilde{\varphi}_j$ are the eigenvalues and the normalized eigenfunctions of $\mathcal{A}$ and $\widetilde{\mathcal{A}}$, correspondingly. Then the manifolds $(\mathcal{M}, g)$ and $(\widetilde{\mathcal{M}}, \widetilde{g})$ are isometric and the operators $\mathcal{A}$ and $\widetilde{\mathcal{A}}$ are equal, i.e., $q = \widetilde{q}$.*

The result can be easily extended to the general operators. If their boundary spectral data on $\Gamma = \widetilde{\Gamma}$ are equal up to a gauge transformation, the manifolds are isometric and the operators are gauge equivalent.

**4.4.2.** As in Chapter 3, we will actually describe a procedure of constructing an isometric copy of $(\mathcal{M}, g)$ and finding the potential $q$. The construction of $\mathcal{M}$ will be given by iterating local constructions. First, we will construct the manifold $\mathcal{M}$ near the given set $\Gamma \subset \partial\mathcal{M}$.

Let $\mathbf{z} \in \Gamma$ and $\gamma_{\mathbf{z},\nu}$ be the normal geodesic starting at $\mathbf{z}$. The analog of the function $\tau_{\partial\mathcal{M}}$, defined in section 2.1.16, is the function $\tau_\Gamma$,

$$\tau_\Gamma(\mathbf{z}) = \{\sup s > 0 : d(\gamma_{\mathbf{z},\nu}(s), \Gamma) = s\}.$$

We note that the function $\tau_\Gamma$ is, in general, not continuous but only upper semicontinuous. Let

$$\Omega_\Gamma = \{(\mathbf{z}, s) \in \Gamma \times \mathbf{R}_+ : \quad s < \tau_\Gamma(\mathbf{z})\}$$

Figure 4.5: Set $\mathcal{M}_\Gamma$

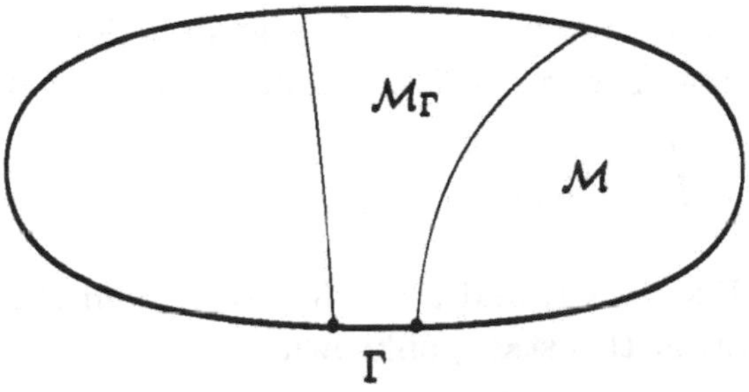


be the largest open set that lies under the graph of $\tau_\Gamma$. Clearly,

$$\Gamma \times (0, \inf_{s\in\Gamma} \tau_{\partial\mathcal{M}}) \subset \Omega_\Gamma.$$

The mapping,

$$\exp_{\partial\mathcal{M}} : \Omega_\Gamma \to \mathcal{M}, \tag{4.104}$$

is a diffeomorphism between $\Omega_\Gamma$ and $\mathcal{M}_\Gamma$,

$$\mathcal{M}_\Gamma = \exp_{\partial\mathcal{M}}(\Omega_\Gamma) \subset \mathcal{M},$$

(see Fig 4.5). Let $\widetilde{g} = (\exp_{\partial\mathcal{M}})^* g$ be a metric on $\Omega_\Gamma$, so that $\exp_{\partial\mathcal{M}}$ is an isometry between $(\Omega_\Gamma, \widetilde{g})$ and $(\mathcal{M}_\Gamma, g)$. In the following, we denote $\mathcal{M}^1 = \mathcal{M}_\Gamma$. Our first aim is to construct $(\Omega_\Gamma, \widetilde{g})$.

**4.4.3.** Consider the initial boundary value problem for the wave equation that corresponds to the Schrödinger operator,

$$\partial_t^2 u - \Delta_g u + qu = 0, \quad \text{in} \quad Q,$$

$$u|_\Sigma = f, \tag{4.105}$$

$$u|_{t=0} = 0, \quad \partial_t u|_{t=0} = 0.$$

As in Chapter 3, we will use an arbitrary positive smooth measure $d\mu$ on $\Gamma$. Then there exists a function $\varrho \in C^\infty(\partial\mathcal{M})$, $\varrho > 0$, such that

$$d\mu = \varrho\, dS_g. \tag{4.106}$$

Since we know the normal derivatives of the eigenfunctions on $\Gamma$, we can use Lemma 3.6 and Theorem 3.7 to compute the inner product

$$\int_{\mathcal{M}} u^{\varrho f}(t)\overline{u^{\varrho h}(s)}\, dV_g, \tag{4.107}$$

for any $f, h \in C_0^\infty(\Gamma \times (0,\infty))$ and $t, s \geq 0$. We remind the reader that the function $\varrho$ is, at this stage, unknown.

**4.4.4.** As in the case when the boundary spectral data are given on the whole boundary $\partial\mathcal{M}$, we can prove a result that is analogous to Lemma 3.12.

**Lemma 4.34** *Let $\Gamma_1 \subset \Gamma$ be an open set and $\tau > 0$. Given the boundary spectral data on $\Gamma$, it is possible to construct boundary sources $f_j \in C_0^\infty(\Gamma_1 \times (0,\infty))$, $j = 1, 2, \ldots$, such that*

$$v_j = u^{\varrho f_j}(\tau)$$

*form an orthonormal basis of $L^2(\mathcal{M}(\Gamma_1, \tau))$.*

**Proof.** The proof of this lemma is completely the same as for Lemma 3.12. □

As in section 3.3.4, we obtain

**Lemma 4.35** *Let $f, h \in C_0^\infty(\Gamma \times (0,\infty))$ and $\Gamma_1 \subset \Gamma$ be an open set. Then, given the boundary spectral data on $\Gamma$, it is possible to find the inner product*

$$\langle P_{\Gamma_1,\tau} u^{\varrho f}(t), u^{\varrho h}(s)\rangle = \int_{\mathcal{M}(\Gamma_1,\tau)} u^{\varrho f}(\mathbf{x}, t)\,\overline{u^{\varrho h}(\mathbf{x}, s)}\, dV_g, \tag{4.108}$$

*for any $t, s, \tau > 0$.*

**Proof.** The proof is identical to the proof of Lemma 3.13. □

**4.4.5.** To proceed further with constructing the manifold $\mathcal{M}_\Gamma$, we need to construct the function $\tau_\Gamma$. To this end, we observe that, for $\Gamma_1 \subset \Gamma$, $s, t > 0$ we have $\mathcal{M}(\Gamma_1, s) \subset \mathcal{M}(\Gamma, t)$ if and only if

$$\|P_{\Gamma,t} u^{ef}(s)\| = \|u^{ef}(s)\| \tag{4.109}$$

for all $f \in C_0^\infty(\Gamma_1 \times (0, s))$. This is an immediate corollary of Theorem 3.10. On the other hand, $s < \tau_\Gamma(\mathbf{z}_0)$ if and only if, for any $t < s$ and any neighborhood $\Gamma_1 \subset \Gamma$ of $\mathbf{z}_0$, we have $\mathcal{M}(\Gamma_1, s) \not\subset \mathcal{M}(\Gamma, t)$. Hence, the boundary spectral data on $\Gamma$ determines $\tau_\Gamma$ and, therefore, $\Omega_\Gamma$. Using some local coordinates $\mathbf{z} = (z^1, \ldots, z^{(m-1)})$ on $\Gamma$, we obtain local coordinates $(z^1, \ldots, z^{(m-1)}, s)$ on $\Omega_\Gamma$.

**4.4.6.** To construct the metric $\widetilde{g}$ on $\Omega_\Gamma$, we use the technique of Gaussian beams.

**Lemma 4.36** *Let $\mathbf{z}, \mathbf{y} \in \Gamma$ and $(\mathbf{y}, s) \in \Omega_\Gamma$. Then the boundary spectral data on $\Gamma$ determines $d(\gamma_{\mathbf{y},\nu}(s), \mathbf{z})$.*

**Proof.** The proof of the lemma is identical to that of Lemma 3.30. □

Thus, we are able to construct the analogs of the evaluation functions $E(\mathbf{z})$, i.e., the functions

$$E_{\mathbf{z}}^\Gamma : \Omega_\Gamma \to \mathbf{R}, \tag{4.110}$$

$$E_{\mathbf{z}}^\Gamma(\mathbf{y}, s) = d(\mathbf{z}, \gamma_{(\mathbf{y},\nu)}(s)), \tag{4.111}$$

where $\mathbf{z} \in \Gamma$.

Evaluating differentials $d_{(\mathbf{y},s)} E_{\mathbf{z}}^\Gamma$ at a point $(\mathbf{y}_0, s_0) \in \Omega_\Gamma$ and using the same considerations as in section 3.8.5, with $r$ replaced by $(\mathbf{y}, s)$, we can find the metric tensor $\widetilde{g}_{ij}(\mathbf{y}_0, s_0)$. As $(\Omega_\Gamma, \widetilde{g})$ is isometric to $(\mathcal{M}_\Gamma, g)$, we obtain the following result.

**Lemma 4.37** *Let the boundary spectral data be given on $\Gamma$. Then it possible to construct the Riemannian manifold $(\mathcal{M}_\Gamma, g)$.*

**4.4.7.** As the metric tensor $g$ on $\mathcal{M}_\Gamma$ and, therefore $dS_g$ are already found, we can now find the function $\varrho$. Thus, for any $f, h \in C_0^\infty(\Gamma \times (0,\infty))$ and any $t, s \geq 0$, we can evaluate the inner product,

$$\int_{\mathcal{M}} u^f(t)\overline{u^h(s)}\, dV_g. \tag{4.112}$$

Next we use Gaussian beams to find the potential $q|_{\mathcal{M}_\Gamma}$.

**Lemma 4.38** *The boundary spectral data of a Schrödinger operator given on a subset on $\Gamma \subset \partial\mathcal{M}$, determine uniquely the restrictions on $\mathcal{M}_\Gamma$ of the potential $q$ and eigenfunctions $\varphi_j$, $j = 1, 2, \ldots$.*

**Proof.** This assertion can be proven by the same arguments as presented in section 3.9.2. □

**4.4.8.** Summarizing the previous considerations, we have shown that the boundary spectral data on $\Gamma$ determine the Riemannian manifold $(\mathcal{M}_\Gamma, g)$ and the restrictions on $\mathcal{M}_\Gamma$ of the potential $q$ and eigenfunctions $\varphi_j$.

To continue the construction, let $\mathcal{D} \subset \mathcal{M}_\Gamma$ be an open domain with smooth boundary $\partial\mathcal{D}$. Consider the manifold $\mathcal{M} \setminus \mathcal{D}$ with boundary $\partial(\mathcal{M}\setminus\mathcal{D}) = \partial\mathcal{M} \cup \partial\mathcal{D}$. Let $\mathcal{A}_\mathcal{D}$ be the Dirichlet Schrödinger operator $-\Delta_g + q$, on $\mathcal{M} \setminus \mathcal{D}$. We are going to find the boundary spectral data of $\mathcal{A}_\mathcal{D}$ on $\partial\mathcal{D}$.

**Lemma 4.39** *Assume that we are given a part $\Gamma$ of $\partial\mathcal{M}$ and the boundary spectral data on $\Gamma$ of the Schrödinger operator on $\mathcal{M}$. Assume, in addition, that we know the Riemannian manifold $(\mathcal{M}_\Gamma, g)$ and the restrictions of the eigenfunctions $\varphi_j|_{\mathcal{M}_\Gamma}$, $j = 1, 2, \ldots$. Then these data determine the boundary spectral data on $\partial\mathcal{D}$ of the Schrödinger operator $\mathcal{A}_\mathcal{D}$.*

**Proof.** We denote by $\varphi_j^\mathcal{D}$ the eigenfunctions of $\mathcal{A}_\mathcal{D}$ in $\mathcal{M}\setminus\mathcal{D}$. By continuing these functions by zero into $\mathcal{D}$, we can consider the functions $\varphi_j^\mathcal{D}$ as functions on $\mathcal{M}$. Thus, when

$$\varphi_{j,k}^\mathcal{D} = \langle \varphi_j^\mathcal{D}, \varphi_k \rangle$$

are the Fourier coefficients of $\varphi_j^{\mathcal{D}}$, then

$$\sum_{k=1}^{\infty} \varphi_{j,k}^{\mathcal{D}} \varphi_k(\mathbf{x}) = \begin{cases} \varphi_j^{\mathcal{D}}(\mathbf{x}), & \mathbf{x} \in \mathcal{M} \setminus \mathcal{D}, \\ 0, & \mathbf{x} \in \mathcal{D}. \end{cases} \tag{4.113}$$

Let $u^f(T)$, $f \in C_0^\infty(\Gamma \times (0,T))$ be the wave generated at $t = T$ by the boundary source $f$. Let

$$u^{(f,\chi)}(\mathbf{x}, T) = \chi(\mathbf{x}) u^f(\mathbf{x}, T),$$

where

$$\chi \in C^\infty(\mathcal{M}), \quad \chi|_{\mathcal{D}} = 0, \quad \chi|_{\mathcal{M} \setminus \mathcal{M}_\Gamma} = 1. \tag{4.114}$$

Because we know the Fourier coefficients of $u^f(T)$ and also the eigenfunctions $\varphi_j$ on $\mathcal{M}_\Gamma$, we can find the Fourier coefficients $u_k^{(f,\chi)}(T)$ of $u^{(f,\chi)}(\mathbf{x}, T)$,

$$u_k^{(f,\chi)}(T) = \langle u^{(f,\chi)}(T), \varphi_k \rangle.$$

Therefore, we can evaluate Rayleigh quotients for these functions

$$Q(f, \chi) = \frac{\langle (-\Delta_g + q)(\chi(\mathbf{x}) u^f(\mathbf{x}, T)), \chi(\mathbf{x}) u^f(\mathbf{x}, T) \rangle}{\langle \chi(\mathbf{x}) u^f(\mathbf{x}, T), \chi(\mathbf{x}) u^f(\mathbf{x}, T) \rangle}. \tag{4.115}$$

Let now

$$T > 2 \max\{d(\mathbf{x}, \Gamma) : \ \mathbf{x} \in \mathcal{M}\}.$$

Considerations analogous to those in the proof of Theorem 4.28 show that the waves $u^f(\mathbf{x}, T)$, $f \in C_0^\infty(\Gamma \times (0,T))$, are dense in $H_0^1(\mathcal{M})$. Therefore, the functions $\chi(\mathbf{x}) u^f(\mathbf{x}, T)$, where $\chi$ runs through the set of functions that satisfy (4.114), are dense in $H_0^1(\mathcal{M} \setminus \mathcal{D})$.

Using these observations, we can use the Courant-Hilbert principle to find the eigenvalues $\lambda_j^{\mathcal{D}}$ of the Schrödinger operator and the Fourier coefficients $\langle \varphi_j^{\mathcal{D}}, \varphi_k \rangle$. Indeed, by applying this principle, we get

$$\lambda_1^{\mathcal{D}} = \inf_{(f,\chi)} Q(f, \chi),$$

and there is a sequence $(f_p, \chi_p)$ such that

$$\lim_{p \to \infty} Q(f_p, \chi_p) = \lambda_1^{\mathcal{D}}.$$

Moreover, for this sequence $u_k^{(f_p,\chi_p)} \to \varphi_{1,k}^{\mathcal{D}}$ in $L^2(\mathcal{M}\setminus\mathcal{D})$. Therefore, we can find

$$\varphi_{1,k}^{\mathcal{D}} = \lim_{p\to\infty} u_k^{(f_p,\chi_p)}(T).$$

Clearly, the set of projections

$$\{\chi u^f(T) - \langle \chi u^f(T), \varphi_1^{\mathcal{D}}\rangle \varphi_1^{\mathcal{D}} \ : \ f \in C_0^\infty(\Gamma \times [0,T]), \\ \chi \text{ satisfies } (4.114)\},$$

is dense in the space

$$\{u \in H_0^1(\mathcal{M}\setminus\mathcal{D}) : \ \langle u, \varphi_1^{\mathcal{D}}\rangle = 0\}.$$

Thus, we can apply the Courant-Hilbert principle to this set and find $\lambda_2^{\mathcal{D}}$ and the Fourier coefficients $\varphi_{2,k}^{\mathcal{D}}$.

Continuing this procedure, we determine all eigenvalues $\lambda_j^{\mathcal{D}}$ and Fourier coefficients $\varphi_{j,k}^{\mathcal{D}}, j,k = 1,2,\ldots$ of the eigenfunctions $\varphi_j^{\mathcal{D}}(\mathbf{x})$. As $\varphi_k|_{\mathcal{M}_\Gamma}$ are known, the Fourier coefficients determine the eigenfunctions $\varphi_j^{\mathcal{D}}(\mathbf{x})$ for $\mathbf{x} \in \mathcal{M}_\Gamma \setminus \mathcal{D}$. Therefore, we find the boundary spectral data $\{\lambda_j^{\mathcal{D}}, \partial_\nu \varphi_j^{\mathcal{D}}|_{\partial\mathcal{D}}\}$. □

**4.4.9.** In this section, we will complete the proof of Theorem 4.33. Consider the manifold $\mathcal{M}\setminus\mathcal{D}$ and the Schrödinger operator $\mathcal{A}_{\mathcal{D}}$ on it. Then, the boundary spectral data of $\mathcal{A}_{\mathcal{D}}$ is given on a part, $\partial\mathcal{D}$, of the boundary $\partial(\mathcal{M}\setminus\mathcal{D})$ of $\mathcal{M}\setminus\mathcal{D}$. Using the same constructions, as in sections 4.4.3, 4.4.4, with $\Gamma$ replaced by $\partial\mathcal{D}$ and $\mathcal{M}$ replaced by $\mathcal{M}\setminus\mathcal{D}$, we find a manifold $\mathcal{M}_{\partial\mathcal{D}} \subset \mathcal{M}\setminus\mathcal{D}$ and the restrictions of the metric $g$, potential $q$, and all eigenfunctions $\varphi_j^{\mathcal{D}}$ on $\mathcal{M}_{\partial\mathcal{D}}$.

Now $\varphi_{j,k}^{\mathcal{D}}$ can be considered in two ways. On one hand, they are the Fourier coefficients of the zero-continuations of the eigenfunctions $\varphi_j^{\mathcal{D}}$ with respect to the basis $\{\varphi_j\}$ of $L^2(\mathcal{M})$. On the other hand, they are the Fourier coefficients of $\varphi_k|_{\mathcal{M}\setminus\mathcal{D}}$, i.e.

$$\varphi_k(\mathbf{x}) = \sum_{j=1}^{\infty} \varphi_{j,k}^{\mathcal{D}} \varphi_j^{\mathcal{D}}(\mathbf{x}), \quad \mathbf{x} \in \mathcal{M}\setminus\mathcal{D}.$$

As we know $\varphi_j^{\mathcal{D}}(\mathbf{x})$, $\mathbf{x} \in \mathcal{M}_{\partial\mathcal{D}}$, we can find $\varphi_k(\mathbf{x})$ in $\mathbf{x} \in \mathcal{M}_{\partial\mathcal{D}}$.

So far, for any $\mathcal{D} \subset \mathcal{M}_\Gamma$, we have constructed a manifold $\mathcal{M}_{\partial\mathcal{D}} \subset \mathcal{M} \setminus \mathcal{D}$, and the eigenfunctions $\varphi_k$ and metric tensor $g$ on it. Let $\mathcal{D}$ and $\mathcal{D}'$ be subsets of $\mathcal{M}$. In the manifolds $\mathcal{M}_{\partial\mathcal{D}}$ and $\mathcal{M}_{\partial\mathcal{D}'}$ we identify the points $\mathbf{x} \in \mathcal{M}_{\partial\mathcal{D}}$ and $\mathbf{x}' \in \mathcal{M}_{\partial\mathcal{D}'}$, such that $\varphi_j(\mathbf{x}) = \varphi_j(\mathbf{x}')$ for all $j = 1, 2, \ldots$. In this case, the points $\mathbf{x}$ and $\mathbf{x}'$ correspond to the same point on $\mathcal{M}$. Analogously, we identify points on $\mathcal{M}_{\partial\mathcal{D}}$ and $\mathcal{M}_\Gamma = \mathcal{M}^1$ that correspond to the same point on $\mathcal{M}$. Using these identifications, we can construct the manifold $\mathcal{M}^2 \subset \mathcal{M}$,

$$\mathcal{M}^2 = \bigcup_{\mathcal{D} \subset \mathcal{M}_\Gamma} \mathcal{M}_{\partial\mathcal{D}} \cup \mathcal{M}^1.$$

It also follows from the previous considerations that we have constructed the restrictions of the metric $g$, potential $q$ and eigenfunctions $\varphi_k$ on $\mathcal{M}^2$.

Continuing this procedure, we construct an increasing sequence of manifolds $\mathcal{M}^n$ and the restrictions of the metric, potential, and eigenfunctions on these manifolds. Next we show that, for sufficiently large $n$,

$$\mathcal{M}^n = \mathcal{M}^{int}.$$

To this end, consider a compact manifold $\mathcal{N}$ without boundary, such that $\mathcal{M} \subset \mathcal{N}$. By compactness of $\mathcal{N}$, there is $\delta > 0$ such that

$$\delta < \min\{\min_{\mathbf{z} \in \partial\mathcal{M}} \tau_{\partial\mathcal{M}}(\mathbf{z}), \min_{(\mathbf{y},\mathbf{w}) \in S\mathcal{N}} \tau(\mathbf{y}, \mathbf{w})\}. \tag{4.116}$$

Here $\tau_{\partial\mathcal{M}}(\mathbf{z})$ and $\tau(\mathbf{y}, \mathbf{w})$ are the critical values of the functions that correspond to the boundary exponential mapping on $\mathcal{M}$ and the exponential mapping on $\mathcal{N}$ (see sections 2.1.16 and 2.1.11).

Next, we consider the set

$$\mathcal{M} \setminus \mathcal{M}^\delta = \{x \in \mathcal{M} : \ d(x, \partial\mathcal{M}) \geq \delta\}.$$

Due to definition (4.116) of $\delta$, $\mathcal{M} \setminus \mathcal{M}^\delta$ a manifold with smooth boundary that is homotopic to $\mathcal{M}$ and, therefore, connected. Thus, there is a constant $T_\Gamma > 0$, such that any $\mathbf{x} \in \mathcal{M}$ can be connected with $\Gamma$ by a smooth path $\mu \subset \mathcal{M}$ of the length $L$, $L \leq T_\Gamma$. Moreover, if $\mu$ is parametrized with its arclength, then the following conditions are satisfied.

i) $\mu(0) = \mathbf{z} \in \Gamma$, $\mu(L) = \mathbf{x}$,

Figure 4.6: Path $\mu$

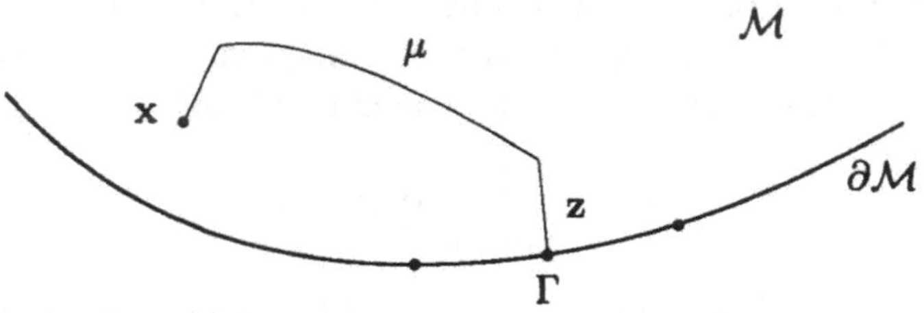


ii) $\mu[0,\delta]$ coincides with the normal geodesic to $\mathbf{z}$,

iii) If $d(\mathbf{x},\partial\mathcal{M}) \le \delta$, then $\mu[L-(\delta-d(\mathbf{x},\partial\mathcal{M})),L]$ coincides with the continuation of the normal geodesic from the boundary to the point $\mathbf{x}$,

iv) When $\delta < s < L-(\delta-d(\mathbf{x},\partial\mathcal{M}))$, then $\mu(s) \in \mathcal{M}\setminus\mathcal{M}^\delta$.

(See Fig. 4.6). Let

$$\mathbf{x}_k = \mu(k\delta),\ k=1,\dots,K,\ K=\left[\frac{L}{\delta}\right],\quad \mathbf{x}_{K+1}=\mathbf{x}.$$

By previous constructions, $\mathbf{x}_1 \in \mathcal{M}^1$. Assume that $\mathbf{x}_k \in \mathcal{M}^k$, $k<K$. Then, for sufficiently small $\rho>0$ and $\mathcal{D}=B_\rho(\mathbf{x}_k)$, we have $\mathcal{D}\subset\mathcal{M}^k$. It follows from definition (4.116) of $\delta$, that

$$\inf_{\mathbf{z}\in\partial\mathcal{D}} \tau_{\partial\mathcal{D}}(\mathbf{z}) \ge \delta-\rho.$$

Thus, $B_\delta(\mathbf{x}_k)\subset\mathcal{M}^{k+1}$ and, in particular, $\mathbf{x}_{k+1}\in\mathcal{M}^{k+1}$. By induction, we see that $\mathbf{x}=\mathbf{x}_{K+1}\subset\mathcal{M}^{K+1}$, which proves the assertion. □

## 4.5. Inverse problems for operators in $\mathbf{R}^m$.

**4.5.1.** Our previous considerations, both in Chapter 3 and the first four sections of this chapter, deal with inverse boundary problems for second order elliptic operators on manifolds with boundary. As we explained in section 3.1.2, considering inverse problems on manifolds makes it possible to avoid additional difficulties due to possible coordinate transformations. These difficulties occur even if we study inverse problems for operators in domains of $\mathbf{R}^m$. As we know, even for the inverse problem for operators on manifolds, boundary data determine an operator only up to an arbitrary gauge transformation. Clearly this also holds true for operators in domains of $\mathbf{R}^m$. Moreover, in the case of $\mathbf{R}^m$, there is a particular, Cartesian, coordinate system, which is considered to be fixed. Because we use the fixed Cartesian system of coordinates, there appears an additional group of transformations which conserves the boundary spectral data.

Indeed, let $X$ be a diffeomorphism of $\Omega$, which is identical on the boundary. We denote the set of these diffeomorphisms by $\mathcal{X}$,

$$\mathcal{X} = \{X : \Omega \to \Omega : \; X \text{ is a diffeomorphism}, \;\; X|_{\partial\Omega} = \mathrm{id}_{\partial\Omega}\}.$$

**Definition 4.40** *Let $X \in \mathcal{X}$. Consider the transformation $S_X$,*

$$S_X : L^2(\Omega, dV) \to L^2(\Omega, dV_X), \tag{4.117}$$

*where $dV_X = X^* dV = det(dX)dV$. It is defined by formula*

$$S_X u(\mathbf{x}) = u(X(\mathbf{x})). \tag{4.118}$$

*These transformations form a group $G_{\mathcal{X}}$,*

$$G_{\mathcal{X}} = \{S_X : \;\; X \in \mathcal{X}\} \tag{4.119}$$

*with respect to composition.*

*Moreover, let $G_\Omega$ be the group of transformations generated by the group $G_{\mathcal{X}}$ together with the group $G$ of gauge transformations,*

$$G_\Omega = \{S = S_1 \circ S_2 \circ \cdots \circ S_l : \;\; S_i \in G_{\mathcal{X}} \text{ or } \;\; S_i \in G\}. \tag{4.120}$$

**Lemma 4.41** *Each element $S$ of the group $G_\Omega$ can be represented in the form*

$$S = S_X \circ S_\kappa. \tag{4.121}$$

*Similarly, it can be represented in the form*

$$S = S_{\widetilde{\kappa}} \circ S_X. \tag{4.122}$$

**Proof.** The claim follows from the fact that $G_X$ and $G$ are groups and the commutation relation

$$S_\kappa \circ S_X = S_X \circ S_{\widetilde{\kappa}}, \tag{4.123}$$

where

$$\widetilde{\kappa}(\mathbf{x}) = \kappa(X^{-1}(\mathbf{x})). \tag{4.124}$$

□

The group $G_\Omega$ also generates a group of transformations $\mathcal{A}_{\kappa,X}$ of an elliptic operator $\mathcal{A}$ of form (2.30), (2.32) defined as

$$\mathcal{A}_{\kappa,X} = S_X \circ S_\kappa \circ \mathcal{A} \circ S_{\kappa^{-1}} \circ S_{X^{-1}}. \tag{4.125}$$

For the operator $\mathcal{A}_{\kappa,X}$ we also use the notation

$$\mathcal{A}_{\kappa,X} = \mathcal{A}_S$$

for $S = S_X \circ S_\kappa$. Because we have fixed the Cartesian system of coordinates, we do not need to consider the coefficients of the operator $\mathcal{A}$ as coordinate invariant tensors. Thus, for convenience, we introduce new notations for the differential expression of the operator,

$$a(\mathbf{x}, D)u = (-\rho^{-1}\partial_i a^{ij}\partial_j + q)u, \tag{4.126}$$

where

$$\rho = mg^{1/2}, \quad a^{ij} = mg^{1/2}g^{ij} \tag{4.127}$$

are smooth real functions. Note that $m$ and $g^{ij}$ can be considered tensors whereas $\rho$ and $a^{ij}$ do not have invariant meaning. The corresponding operator $\mathcal{A}$ is self-adjoint in $L^2(\Omega, d\mu_\rho)$,

$$d\mu_\rho = \rho dx^1 \wedge \cdots \wedge dx^m = mdV_g.$$

Dealing with operators in Euclidean space $\mathbf{R}^m$ it is usual to consider $\partial\Omega$ to be equipped with the canonical metric determined by the embedding $i$, $i : \partial\Omega \to \mathbf{R}^m$. Instead of the boundary derivatives $\partial_\nu \varphi_k|_{\partial\Omega}$, $k = 1, 2, \dots$, it is natural to use oblique normal derivatives

$$\partial_a \varphi_k|_{\partial\Omega} \equiv a^{ij} n_i \partial_j \varphi_k|_{\partial\Omega} \tag{4.128}$$

of the normalized eigenfunctions. Here $\mathrm{n} = (n_1, \dots, n_m)$ is the inward unit normal to $\partial\Omega$ with respect to the Euclidean metric. We call the collection

$$\{\partial\Omega,\ \lambda_j,\ \partial_a \varphi_k|_{\partial\Omega} :\ j = 1, 2, \dots\} \tag{4.129}$$

the oblique boundary spectral data. Now

$$\langle \mathbf{n}, \mathbf{n} \rangle_g = g^{ij} n_i n_j, \quad n_i = \langle \mathbf{n}, \mathbf{n} \rangle_g^{1/2} \nu_i,$$

where $\nu_i$ are the components of the unit normal co-vector of the boundary with respect to the metric $g$. We obtain for the boundary derivative $\partial_a$ the formula

$$\partial_a \varphi_j|_{\partial\Omega} = \eta \partial_\nu \varphi_j|_{\partial\Omega} \tag{4.130}$$

where

$$\eta = \rho \langle \mathbf{n}, \mathbf{n} \rangle_g^{1/2}.$$

**4.5.2.** Rewriting Theorem 3.3 for the case of operators in $\Omega$, we obtain the following result.

**Theorem 4.42** *Let $\mathcal{A}$ and $\widetilde{\mathcal{A}}$ be two operators of form (4.126) in a bounded domain $\Omega \subset \mathbf{R}^m$ with $C^\infty$-smooth boundary that are self-adjoint in $L^2(\Omega, dV)$ and $L^2(\Omega, d\widetilde{V})$, correspondingly. Then, the oblique boundary spectral data of $\mathcal{A}$ and $\widetilde{\mathcal{A}}$ are gauge equivalent, i.e.,*

$$\lambda_j = \widetilde{\lambda}_j \text{ and } \partial_a \varphi_j|_{\partial\Omega} = \kappa_0\, \partial_{\widetilde{a}} \widetilde{\varphi}_j|_{\partial\Omega} \tag{4.131}$$

*for some $\kappa_0 \in C^\infty(\partial\Omega)$, $\kappa_0 > 0$, if and only if*

$$\widetilde{\mathcal{A}} = \mathcal{A}_S \tag{4.132}$$

*for some $S = S_X \circ S_\kappa$ with $\kappa|_{\partial\Omega} = \kappa_0$. Moreover, the corresponding measures satisfy $d\widetilde{V} = \kappa^{-2}(X^{-1})^* dV$.*

**Proof.** *Step 1.* First, we show that operators $\mathcal{A}$ and $\mathcal{A}_S$ have gauge equivalent oblique boundary data. We use the notation $\mathcal{A}_S = \widetilde{\mathcal{A}}$. Moreover, we denote by $g$, $\nu$ and $\widetilde{g}$, $\widetilde{\nu}$ the metric tensor and the unit normal vector for operators $\mathcal{A}$, $\widetilde{\mathcal{A}}$, correspondingly. Obviously, the eigenvalues of $\mathcal{A}$ and $\widetilde{\mathcal{A}}$ coincide. Thus it is sufficient to consider the $\partial_a$-derivatives of eigenfunctions and prove the identity

$$\partial_{\widetilde{a}}\widetilde{\varphi}_j|_{\partial\Omega} = \kappa_0^{-1}\, \partial_a\varphi_j|_{\partial\Omega}, \tag{4.133}$$

where $\varphi_j$ and $\widetilde{\varphi}_j$ are the normalized eigenfunctions of the operators $\mathcal{A}$, $\mathcal{A}_S$, correspondingly. Since transformation $S$ is an isometry from $L^2(\Omega, dV)$ to $L^2(\Omega, d\widetilde{V})$, we have

$$\widetilde{\varphi}_j(\mathbf{x}) = \kappa(X(\mathbf{x}))\varphi_j(X(\mathbf{x})).$$

Moreover, $\widetilde{\nu} = dX(\nu)$ where $dX$ is the differential of the map $X : \Omega \to \Omega$. Because $\varphi_j|_{\partial\Omega} = 0$, this yields

$$\partial_{\widetilde{\nu}}\widetilde{\varphi}_j|_{\partial\Omega} = \kappa\partial_\nu\varphi_j|_{\partial\Omega}.$$

By definition (4.130) we see that

$$\partial_{\widetilde{a}}\widetilde{\varphi}_j = \widetilde{\rho}\,\langle\mathbf{n},\mathbf{n}\rangle_{\widetilde{g}}^{1/2}\,\partial_{\widetilde{\nu}}\widetilde{\varphi}_j = \kappa\widetilde{m}\widetilde{g}^{1/2}\,\langle\mathbf{n},\mathbf{n}\rangle_{\widetilde{g}}^{1/2}\,\partial_\nu\varphi_j.$$

At last, using transformation formula (2.52), we obtain that

$$\partial_{\widetilde{a}}\widetilde{\varphi}_j = \kappa^{-1}m\left(\widetilde{g}^{1/2}\,\langle\mathbf{n},\mathbf{n}\rangle_{\widetilde{g}}^{1/2}\right)\partial_\nu\varphi_j.$$

Hence, to prove equation (4.133), it is enough to show that

$$\widetilde{g}\,\langle\mathbf{n},\mathbf{n}\rangle_{\widetilde{g}} = g\,\langle\mathbf{n},\mathbf{n}\rangle_g \quad \text{on } \partial\Omega, \tag{4.134}$$

where $g$ and $\widetilde{g}$ are determinants of metric tensor. To see this, assume, without loss of generality, that $\mathbf{x} \in \partial\Omega$ has coordinates $(0,\dots,0)$ and $T_{\mathbf{x}}(\partial\Omega) = \text{span}\{\mathbf{e}_1,\dots,\mathbf{e}_{m-1}\}$ where $\mathbf{e}_j = \partial_j$ are Euclidean unit coordinate vectors. Let $\widetilde{\mathbf{x}} = X(\mathbf{x})$. Then, due to $X|_{\partial\Omega} = \text{id}|_{\partial\Omega}$,

$$dX(0) = \begin{bmatrix} I & * \\ 0 & \frac{\partial\widetilde{x}^m}{\partial x^m} \end{bmatrix} \equiv A,$$

where $I$ is $(m-1)\times(m-1)$ identity matrix and $*$ denotes some $(m-1)\times 1$-matrix. Hence, the determinants of metric tensors satisfy

$$\widetilde{g}(0) = g(0)\left[\frac{\partial\widetilde{x}^m}{\partial x^m}\right]^{-2}.$$

However,

$$\langle \mathbf{n}, \mathbf{n} \rangle_{\widetilde{g}} = \langle Ae_m, Ae_m \rangle_g = \left( \frac{\partial \widetilde{x}^m}{\partial x^m} \right)^2 \langle \mathbf{n}, \mathbf{n} \rangle_g .$$

These two identities show the invariance of $g \langle \mathbf{n}, \mathbf{n} \rangle_g$ in the change of coordinates, that is, formula (4.134).

*Step 2.* Next we show that if the oblique boundary spectral data (4.129) of operators $\mathcal{A}$ and $\widetilde{\mathcal{A}}$ are gauge equivalent, then $\widetilde{\mathcal{A}} = \mathcal{A}_S$ with some $S \in G_\Omega$.

To apply earlier results of Chapter 3, we consider domain $\Omega$ as an "abstract" differential manifold $\mathcal{M}_\Omega$. to this end, we identify $\Omega$ and $\mathcal{M}_\Omega$ with a natural diffeomorphism $\psi : \Omega \to \mathcal{M}_\Omega$, which corresponds to the identity operator. On the manifold $\mathcal{M}_\Omega$, the operator $\mathcal{A}$ corresponds to an "abstract" operator $\mathcal{A}_a$, which determines the Riemannian metric $g$. Similarly, the volume element $dV$ corresponds to the volume element $dV_a = (\psi^{-1})^* dV$. By this definition, the map $(\psi^{-1})^* : L^2(\Omega, dV) \to L^2(\mathcal{M}_\Omega, dV_a)$ is an isometry. Similar definitions are introduced for the operator $\widetilde{\mathcal{A}}$.

Due to formula (4.130), the gauge equivalence of $\partial_a \varphi_j$ and $\partial_{\widetilde{a}} \widetilde{\varphi}_j$ implies the gauge equivalence of $\partial_\nu \varphi_j$ and $\partial_{\widetilde{\nu}} \widetilde{\varphi}_j$. Hence, the operators $\mathcal{A}_a$ and $\widetilde{\mathcal{A}}_a$ have gauge equivalent boundary spectral data. Thus, by Theorem 3.3, there is an isometry $\Psi : (\mathcal{M}_\Omega, g) \to (\widetilde{\mathcal{M}}_\Omega, \widetilde{g})$. Moreover, if we identify the manifold $(\widetilde{\mathcal{M}}_\Omega, \widetilde{g})$ with the isometric manifold $(\mathcal{M}_\Omega, g)$, then the abstract operator $\widetilde{\mathcal{A}}_a$ is gauge equivalent to $\mathcal{A}_a$. This means that

$$\widetilde{\mathcal{A}}_a u = (\Psi^{-1})^* (S_k \mathcal{A}_a S_k^{-1})(\Psi^* u), \tag{4.135}$$

where $k$ is a positive function on $\mathcal{M}_\Omega$. Moreover, when $dV_k$ is the measure corresponding to the operator $(\mathcal{A}_a)_k = S_k \mathcal{A}_a S_k^{-1}$, by Lemma 2.26 the gauge transformation

$$S_k : L^2(\mathcal{M}_\Omega, dV) \to L^2(\mathcal{M}_\Omega, dV_k)$$

is an isometry. Thus

$$(\Psi^{-1})^* S_k : L^2(\mathcal{M}_\Omega, dV_a) \to L^2(\widetilde{\mathcal{M}}_\Omega, d\widetilde{V}_a)$$

is an isometry. Next, we give an interpretation of the these observations for $\mathcal{A}$ and $\widetilde{\mathcal{A}}$ on $\Omega$.

First, the commuting diagram

$$\begin{array}{ccc} \Omega & \longrightarrow & \Omega \\ \downarrow{\scriptstyle \psi} & & \downarrow{\scriptstyle \widetilde{\psi}} \\ \mathcal{M}_\Omega & \xrightarrow[\Psi]{} & \widetilde{\mathcal{M}}_\Omega \end{array}$$

induces a diffeomorphism $X = \widetilde{\psi}^{-1} \circ \Psi \circ \psi : \Omega \to \Omega$. Second, since

$$\mathcal{A}_a u = (\psi^{-1})^* \mathcal{A}(\psi^* u), \quad \widetilde{\mathcal{A}}_a u = (\widetilde{\psi}^{-1})^* \widetilde{\mathcal{A}}(\widetilde{\psi}^* u),$$

the formula (4.135) in terms of the operators $\mathcal{A}$ and $\widetilde{\mathcal{A}}$ takes the form

$$\widetilde{\mathcal{A}} = \mathcal{A}_{\kappa,X}$$

where $\kappa(x) = k(\psi(x))$. Moreover, all mappings in the diagram

$$\begin{array}{ccc} L^2(\Omega, dV) & \longrightarrow & L^2(\Omega, d\widetilde{V}) \\ {\scriptstyle (\psi^{-1})^*}\downarrow & & {\scriptstyle (\widetilde{\psi}^{-1})^*}\downarrow \\ L^2(\mathcal{M}_\Omega, dV_a) & \xrightarrow[(\Psi^{-1})^* S_k]{} & L^2(\widetilde{\mathcal{M}}_\Omega, d\widetilde{V}_a) \end{array}$$

are isometries, and thus

$$S_{\kappa,X} = S_X \circ S_\kappa : L^2(\Omega, dV) \to L^2(\Omega, d\widetilde{V}) \tag{4.136}$$

is an isometry. Thus, the orthonormalized eigenfunctions $\varphi_j$ and $\widetilde{\varphi}_j$ of operators $\mathcal{A}$ and $\widetilde{\mathcal{A}}$ can be chosen such that

$$\widetilde{\varphi}_j = S_{\kappa,X} \varphi_j. \tag{4.137}$$

Relation (4.137), in turn, implies relation (4.133). Due to (4.131) we see that $\kappa|_{\partial\Omega} = \kappa_0$. Finally, since the transformation (4.136) is an isomorphism, we have the identity $d\widetilde{V} = \kappa^{-2}(X^{-1})^* dV$. □

**4.5.3.** In many practical applications of inverse problems, we do *a priori* have some additional information that specifies the structure of the operator under consideration. As we have already mentioned in section 3.1.7, it is possible to use this additional information to decrease the admissible group of transformations. In the forthcoming sections, we consider several examples of this type of problem when, in some cases, it is even possible to prove the uniqueness of the corresponding inverse problem. This should not be a big surprise because, for example, we have shown that the boundary spectral data determine a Schrödinger operator uniquely. Before going to considerations of particular examples, let us make two remarks:

i) In this section, we deal only with the inverse boundary spectral problem. Using the results of sections 4.1–4.3, it is easy to translate the results obtained for the inverse boundary spectral problem onto the hyperbolic and parabolic inverse boundary problems, i.e., onto the cases i)-v) described in section 4.1.

ii) Our further considerations are limited to few examples that are suitable for practical applications and often considered in the literature on inverse boundary problems. In our selection, we also consider some cases when the admissible group becomes only the group of diffeomorphisms $G_{\mathcal{X}}$ (or its subgroup) or, vice versa, only the group of gauge transformations $G$ (or its subgroup).

**4.5.4.** In this section, we consider the inverse boundary spectral problem for the anisotropic conductivity operator $\mathcal{A}$ in a smooth domain $\Omega \subset \mathbf{R}^m$. We call an operator $\mathcal{A}$ of the form (4.126) an anisotropic conductivity operator if $q \equiv 0$ and $\rho \equiv 1$, i.e.,

$$\mathcal{A}u = -\partial_i a^{ij} \partial_j u. \tag{4.138}$$

This means, in particular, that $\mathcal{A}$ is self-adjoint in $L^2(\Omega)$ with canonical, Euclidean, volume element $dV_{\text{can}}$.

**Theorem 4.43** *Let $\mathcal{A}$ and $\widetilde{\mathcal{A}}$ be two anisotropic conductivity operators in $\Omega \subset \mathbf{R}^m$. Then their oblique boundary spectral data $\{\lambda_j, \partial_a \varphi_j|_{\partial\Omega};\ j = 1, 2, \dots\}$ and $\{\widetilde{\lambda}_j, \partial_{\widetilde{a}} \widetilde{\varphi}_j|_{\partial\Omega};\ j = 1, 2, \dots\}$ coincide if and only if*

$$\widetilde{\mathcal{A}} = S_X \mathcal{A}, \tag{4.139}$$

*for some $X \in \mathcal{X}_0$. Here $\mathcal{X}_0$ is the subgroup of $\mathcal{X}$ that consists of the unimodular, i.e., volume preserving, diffeomorphisms $X$, such that*

$$X^*(dV_{can}) = dV_{can}, \quad dV_{can} = dx^1 \wedge \cdots \wedge dx^m. \tag{4.140}$$

**Proof.** The case "if" of the assertion can easily be seen to be true. Next, we prove the "only if" part of the assertion. In view of Theorem 4.42,

$$\widetilde{\mathcal{A}} = \mathcal{A}_{\kappa,X}$$

for some $\kappa$ with $\kappa|_{\partial\Omega} = 1$ and $X \in \mathcal{X}$. As $q = \widetilde{q} = 0$, equation (2.52) implies that $\kappa$ is the solution of the Dirichlet problem

$$\partial_i a^{ij} \partial_j (\kappa^{-1}) = 0, \quad (\kappa^{-1})|_{\partial\Omega} = 1.$$

Hence, $\kappa = 1$ and $\widetilde{\mathcal{A}} = S_X(\mathcal{A})$.

Next, we show that $X \in \mathcal{X}_0$. To this end, we use representation (2.29) for differential expressions $a(\mathbf{x}, D)$ and $\widetilde{a}(\mathbf{x}, D)$,

$$a(\mathbf{x}, D) = -m^{-1} g^{-1/2} \partial_i (m g^{1/2} g^{ij} \partial_j),$$

$$\widetilde{a}(\mathbf{x}, D) = -\widetilde{m}^{-1} \widetilde{g}^{-1/2} \partial_i (\widetilde{m} \widetilde{g}^{1/2} \widetilde{g}^{ij} \partial_j).$$

Here $\partial_j$ are differentiations in Cartesian coordinates and $m$, $g$ as well as $\widetilde{m}$, $\widetilde{g}$ are uniquely determined. As $\widetilde{a}(\mathbf{x}, D)$ is the $S_X$-transform of $a(\mathbf{x}, D)$, we have

$$m = X^* \widetilde{m}, \quad dV_g = X^* dV_{\widetilde{g}},$$

i.e.,

$$m(\mathbf{x}) = \widetilde{m}(X(\mathbf{x})), \quad g^{1/2}(\mathbf{x}) dV_{can} = \widetilde{g}^{1/2}(X(\mathbf{x})) X^*(dV_{can}). \tag{4.141}$$

An operator $\mathcal{A}$ of form (4.126), with $q = 0$, is an anisotropic conductivity operator if and only if $m g^{1/2} \equiv 1$. Thus, by multiplying relations (4.141) and using the fact that both $\mathcal{A}$ and $\widetilde{\mathcal{A}}$ are anisotropic conductivity operators, we see

$$X^*(dV_{can}) = dV_{can}. \tag{4.142}$$

Thus $X \in \mathcal{X}_0$. □

**4.5.5.** The rest of section 4.5 is devoted to the consideration of the inverse boundary spectral problem for a general isotropic operator

$$\mathcal{A}_{\epsilon,\mu,q}u = -\epsilon \operatorname{Div} \mu \operatorname{Grad} u + qu, \tag{4.143}$$

with $\epsilon, \mu, q \in C^\infty(\overline{\Omega})$, $\epsilon, \mu > 0$. Here, Div and Grad are divergence and gradient in Euclidean space. Besides a general operator (4.143), we study the special cases of an Euclidean Schrödinger operator, $-\Delta + q$, i.e., $\epsilon = \mu = 1$, and an operator $\mathcal{A}_{\epsilon,\mu} = -\epsilon \operatorname{Div} \mu \operatorname{Grad}$, i.e. $q = 0$. The last example plays an important role in the heat conduction problems, e.g., in parabolic equation,

$$w_t - \epsilon \operatorname{Div} \operatorname{Grad} w = 0.$$

Our next aim is to study the inverse problem for the isotropic operator $\mathcal{A}_{\epsilon,\mu,q}$ when the oblique boundary spectral data

$$\{\partial\Omega,\ \lambda_j,\ \partial_a\varphi_j|_{\partial\Omega};\ j = 1, 2, \dots\} \tag{4.144}$$

is given. Observe that for the isotropic operator $\mathcal{A}_{\epsilon,\mu,q}$,

$$\partial_a\varphi_j|_{\partial\Omega} = \mu\partial_{\mathbf{n}}\varphi_j|_{\partial\Omega}\,.$$

We will use *a priori* knowledge of the form of the operator to reduce the group of admissible coordinate transformations preserving oblique boundary spectral data to a trivial one. Our plan is to first reconstruct the Riemannian manifold $(\mathcal{M}, g)$, find the values of the function $\epsilon\mu$ on the manifold $\mathcal{M}$ and then construct the unique embedding from $\mathcal{M}$ to $\Omega$. In other words, we first construct the abstract Riemannian manifold and use *a priori* information to give a unique interpretation for this manifold in the Euclidean space.

We first note that operator (4.143) has a representation of form (4.126) with

$$\rho = \epsilon^{-1}, \quad a^{ij} = \mu\delta^{ij}, \quad m = \epsilon^{(m-2)/2}\mu^{m/2}, \quad g^{ij} = (\epsilon\mu)\delta^{ij}. \tag{4.145}$$

The gauge transformation $S_\kappa$ of operator $\mathcal{A}_{\epsilon,\mu,q}$ transforms $\epsilon \to \epsilon_\kappa$ and $\mu \to \mu_\kappa$ where, by formulae (2.52) and (4.145),

$$\epsilon_\kappa = \epsilon\kappa^2, \quad \mu_\kappa = \mu\kappa^{-2}. \tag{4.146}$$

Then the gauge transformation $S_\kappa$ with $\kappa$ given by formula

$$\kappa = \epsilon^{(m-2)/4}\mu^{m/4} \tag{4.147}$$

transforms operator $\mathcal{A}_{\epsilon,\mu,q}$ to the Schrödinger operator $-\Delta_g+\widetilde{q}$ with

$$\widetilde{q} = q - \kappa\epsilon\,\mathrm{Div}\,\mu\,\mathrm{Grad}\,(\kappa^{-1}) \tag{4.148}$$

(see formula (2.52)). By means of Theorem 3.37, we can reconstruct from the oblique boundary spectral data the abstract Riemannian manifold $(\mathcal{M}, g)$ and the Schrödinger operator on it. By choosing proper coordinates on $\partial\mathcal{M}$, we can identify it with $\partial\Omega$. Thus, we have on $\partial\Omega$ two metrics, the Euclidean one, and the one induced by $g$. Using formula (4.145) for the metric tensor $g^{ij}$ we see that, by comparing the metric induced by $g$ on the boundary and the Euclidean metric on $\partial\Omega$, we can find $\epsilon\mu|_{\partial\Omega}$.

**Lemma 4.44** *The oblique boundary spectral data of $\mathcal{A}_{\epsilon,\mu,q}$ of form (4.144) determines uniquely the boundary values $\epsilon|_{\partial\Omega}$ and $\mu|_{\partial\Omega}$.*

**Proof.** Next, we consider $(\Omega, (\epsilon\mu)^{-1}\delta_{ij})$ as a Riemannian manifold that is isometric to $(\mathcal{M}, g)$ and the operator $\mathcal{A}_{\epsilon,\mu,q}$ as an operator on this manifold. Then $-\Delta_g + \widetilde{q}$ is the unique Schrödinger operator corresponding to $\mathcal{A}_{\epsilon,\mu,q}$. Let $\varphi_j^S$ be the normalized eigenfunctions of $-\Delta_g+\widetilde{q}$. As $S_\kappa$ is the gauge transform transforming $\mathcal{A}_{\epsilon,\mu,q}$ to the unique Schrödinger operator corresponding to it,

$$\varphi_j = \kappa^{-1}\varphi_j^S. \tag{4.149}$$

Thus by formulae (4.130) and (4.145) the oblique boundary spectral data determines

$$\partial_a\varphi_j|_{\partial\Omega} = \epsilon^{-1}(\epsilon\mu)^{\frac{1}{2}}\partial_\nu\varphi_j\Big|_{\partial\Omega} = \tag{4.150}$$

$$= \epsilon^{-\frac{1}{2}}\mu^{\frac{1}{2}}\kappa_0^{-1}\partial_\nu\varphi_j^S\Big|_{\partial\Omega} = \widetilde{\kappa}_0\partial_\nu\varphi_j^S|_{\partial\Omega},$$

where $\widetilde{\kappa}_0 = \epsilon^{-1/2}\mu^{1/2}\kappa_0^{-1}$ and $\kappa_0 = \kappa|_{\partial\Omega}$. Because these boundary values (4.149) are gauge equivalent to the boundary spectral data of the Schrödinger operator $-\Delta_g+\widetilde{q}$ via gauge transformation with $\widetilde{\kappa}_0$, Lemma 3.36 yields that we can determine $\widetilde{\kappa}_0$. By formula (4.147) this function is

$$\widetilde{\kappa}_0 = (\epsilon|_{\partial\Omega})^{-m/4}(\mu|_{\partial\Omega})^{(2-m)/4}.$$

This function, together with $\epsilon\mu|_{\partial\Omega}$, determines uniquely $\epsilon|_{\partial\Omega}$ and $\mu|_{\partial\Omega}$. □

**4.5.6.** Our next considerations are based upon the following.

**Lemma 4.45** *Assume that we are given the oblique boundary spectral data of an operator $A_{\epsilon,\mu,q}$ of form (4.144). Then it is possible to determine uniquely a Riemannian manifold $(\mathcal{M}, g)$ such that there is an isometry $\Psi : (\Omega, (\epsilon\mu)^{-1}\delta_{ij}) \to (\mathcal{M}, g)$ and a function $\alpha$ on $\mathcal{M}$ such that $\alpha(\Psi(\mathbf{x})) = \epsilon(\mathbf{x})\mu(\mathbf{x})$.*

**Proof.** Again, by choosing proper coordinates on $\partial\mathcal{M}$, we identify $\partial\mathcal{M}$ with $\partial\Omega$. As we already know, given $\{\lambda_k, \partial_a\varphi_k|_{\partial\Omega}\}$ it is possible to reconstruct $(\mathcal{M}, g)$ and $\epsilon\mu|_{\partial\Omega}$. Thus, we have found $\alpha|_{\partial\mathcal{M}}$. Then, by representation (4.145), the metric $g$ on $\mathcal{M}$ is conformally Euclidean, that is, the metric tensor, in the some coordinates, has the form $g_{jk}(x) = \sigma(x)\delta_{jk}$ where $\sigma(x) > 0$. Hence (see Notes to Chapter 4) the function $\beta = \frac{1}{2}\ln(\alpha)$, when $m = 2$, and $\beta = \alpha^{(m-2)/4}$, when $m \geq 3$, should satisfy the so-called scalar curvature equation

$$\Delta_g\beta - k_g = 0 \quad (m = 2), \tag{4.151}$$

$$\frac{4(m-1)}{m-2}\Delta_g\beta - k_g\beta = 0 \quad (m \geq 3), \tag{4.152}$$

where $k_g$ is the scalar curvature of $(\mathcal{M}, g)$. The idea of these equations is that if $\beta$ satisfies, e.g., equation (4.152) in the case $m \geq 3$, then the metric $\beta^{4/(m-2)}g$ has zero scalar curvature. Together with boundary data $\beta|_{\partial\mathcal{M}}$ being given, we obtain Dirichlet boundary value problem for $\beta$ and, henceforth $\alpha$ in $\mathcal{M}$.

Clearly, Dirichlet problem for equation (4.151) has a unique solution that gives $\alpha$ when $m = 2$. In the case $m \geq 3$, it is necessary to check that 0 is not an eigenvalue of the operator $\frac{4(m-1)}{m-2}\Delta_g - k_g$ with Dirichlet boundary condition. As we know, $\beta = \alpha^{(m-2)/4}$ is a positive solution of the Dirichlet problem for equation (4.152) with boundary condition $\beta|_{\partial\mathcal{M}} = \alpha^{(m-2)/4}|_{\partial\mathcal{M}}$. Assume that there is another possible solution of this problem,

$$\widetilde{\beta} = v\beta, \quad v > 0, \quad v|_{\partial\mathcal{M}} = 1. \tag{4.153}$$

Then both $(\mathcal{M}, \beta^{4/(m-2)}g)$ and $(\mathcal{M}, \widetilde{\beta}^{4/(m-2)}g)$ have zero scalar curvatures. Denoting $g_1 = \beta^{4/(m-2)}g$, $g_2 = \widetilde{\beta}^{4/(m-2)}g$, we obtain that $v$ should satisfy the scalar curvature equation

$$\frac{4(m-1)}{m-2}\Delta_{g_1}v - k_{g_1}v = 0.$$

Here, however, $k_{g_1} = 0$. Together with boundary condition (4.153), this equation implies that $v \equiv 1$, i.e. $\beta = \widetilde{\beta}$. This immediately yields that 0 is not the eigenvalue of the Dirichlet operator (4.152) because, otherwise, we could obtain a positive solution $\widetilde{\beta} = \beta + c_0\psi_0$, where $\psi_0$ is the Dirichlet eigenfunction, corresponding to zero eigenvalue, and $|c_0|$ is sufficiently small. □

**4.5.7.** Our next goal is to embed the abstract manifold $(\mathcal{M}, g)$ with conformally Euclidean metric into $\Omega$ with metric $(\epsilon\mu)^{-1}\delta_{ij}$. To achieve this goal, we use the *a priori* knowledge that such embedding exists and the fact that we have already constructed $\alpha$ corresponding to $\epsilon\mu$ on $\mathcal{M}$.

**Lemma 4.46** *Let $(\mathcal{M}, g)$ be a Riemannian manifold, $\alpha(\mathbf{x})$ a positive smooth function on $\mathcal{M}$, and $\psi : \partial\Omega \to \partial\mathcal{M}$ a diffeomorphism. Assume also that there is a diffeomorphism $\Psi : \Omega \to \mathcal{M}$ such that*

$$\Psi|_{\partial\Omega} = \psi, \quad \Psi^* g = (\alpha(\Psi(\mathbf{x})))^{-1}\delta_{ij}.$$

*Then, if $\Omega$, $(\mathcal{M}, g)$, $\alpha$, and $\psi$ are known, it is possible to construct the diffeomorphism $\Psi$ by solving ordinary differential equations.*

**Proof.** *Step 1.* Let $\zeta = (\mathbf{z}, \tau)$ be the boundary normal coordinates on $\mathcal{M} \setminus \omega$. Our goal is to construct the coordinate representation for $\Psi^{-1} = \mathbf{x}$,

$$\begin{aligned} \mathbf{x} \ &: \ \mathcal{M} \setminus \omega \to \Omega, \\ & \mathbf{x}(\mathbf{z}, \tau) = (x^1(\mathbf{z}, \tau), \dots, x^m(\mathbf{z}, \tau)). \end{aligned}$$

Denote by $h_{ij}(\mathbf{x}) = \alpha(\Psi(\mathbf{x}))^{-1}\delta_{ij}$ the metric tensor in $\Omega$. Let $\Gamma_{i,jk} = g_{ip}\Gamma^p_{jk}$ be the Christoffel symbols of $(\Omega, h_{ij})$ in the Euclidean coordinates and let $\widetilde{\Gamma}_{\sigma,\mu\nu}$ be Christoffel symbols of $(\mathcal{M}, g)$, in $\zeta$-coordinates. Next, we consider functions $h_{ij}$, $\Gamma_{k,ij}$, etc. as functions on $\mathcal{M} \setminus \omega$ in $(\mathbf{z}, \tau)$-coordinates evaluated at the point $\mathbf{x} = \mathbf{x}(\mathbf{z}, \tau)$, e.g., $\Gamma_{k,ij}(\mathbf{z}, \tau) = \Gamma_{k,ij}(\mathbf{x}(\mathbf{z}, \tau))$. Then, since $\Psi$ is an isometry, the transformation rule of Christoffel symbols with respect to the change of coordinates implies

$$\widetilde{\Gamma}_{\sigma,\mu\nu} = \Gamma_{k,ij}\frac{\partial x^i}{\partial \zeta^\mu}\frac{\partial x^j}{\partial \zeta^\nu}\frac{\partial x^k}{\partial \zeta^\sigma} + h_{ij}\frac{\partial x^i}{\partial \zeta^\sigma}\frac{\partial^2 x^j}{\partial \zeta^\mu \partial \zeta^\nu}, \tag{4.154}$$

where

$$h_{ij} = \alpha^{-1}\delta_{ij}. \tag{4.155}$$

Using equations (4.154) and (4.155) we can write $\frac{\partial^2 x^j}{\partial\zeta^\mu\partial\zeta^\nu}$ in the form

$$\frac{\partial^2 x^j}{\partial\zeta^\mu\partial\zeta^\nu}(\zeta) = \alpha(\zeta)\delta^{jp}\left(\widetilde{\Gamma}_{\sigma,\mu\nu}\frac{\partial\zeta^\sigma}{\partial x^p} -\right.$$

$$\left. - \frac{1}{2}\frac{\partial\alpha^{-1}}{\partial\zeta^\sigma}\left[\frac{\partial\zeta^\sigma}{\partial x^n}\delta_{pi} + \frac{\partial\zeta^\sigma}{\partial x^i}\delta_{pn} - \frac{\partial\zeta^\sigma}{\partial x^p}\delta_{ni}\right]\frac{\partial x^i}{\partial\zeta^\mu}\frac{\partial x^n}{\partial\zeta^\nu}\right). \tag{4.156}$$

As $\alpha$ and $\widetilde{\Gamma}_{\sigma,\mu\nu}$ are known as a function of $\zeta$, the right-hand side of (4.156) can be written in the form

$$\frac{\partial^2 x^j}{\partial\zeta^\mu\partial\zeta^\nu} = \mathcal{F}^j_{\mu,\nu}\left(\zeta, \frac{\partial \mathbf{x}}{\partial\zeta}\right), \tag{4.157}$$

where $\mathcal{F}^j_{\mu,\nu}$ are known functions. Choose $\nu = m$, so that

$$\frac{\partial^2 x^j}{\partial\zeta^\mu\partial\zeta^m} = \frac{d}{d\tau}\left(\frac{\partial x^j}{\partial\zeta^\mu}\right).$$

Then, equation (4.157) becomes a system of ordinary differential equations along normal geodesics for the matrix $[\frac{\partial x^j}{\partial\zeta^\mu}(\tau)]^m_{j,\mu=1}$. Moreover, since diffeomorphism $\Psi : \partial\Omega \to \partial\mathcal{M}$ is given, the boundary derivatives $\frac{\partial x^j}{\partial\zeta^\mu}$, $\mu = 1, \ldots, m-1$, are known for $\zeta^m = \tau = 0$. By relation (4.155),

$$\frac{\partial x^j}{\partial\zeta^m} = \frac{\partial x^j}{\partial\tau} = \alpha^{-1}\frac{\partial x^j}{\partial \mathrm{n}} = \alpha^{-1}n^j$$

for $\zeta^m = \tau = 0$ where $\mathrm{n} = (n^1, \ldots, n^m)$ is the Euclidean unit inward normal vector. Thus, $\frac{\partial x^j}{\partial\tau}(\mathbf{z}, 0)$ are also known. Solving a system of ordinary differential equations (4.157) with these initial conditions at $\tau = 0$, we can construct $\frac{\partial x^j}{\partial\zeta^\mu}(\mathbf{z},\tau)$ everywhere on $\mathcal{M}\setminus\omega$. In particular, taking $\mu = m$, we find $\frac{dx^j}{d\tau}(\mathbf{z},\tau)$. Using again the fact that $(x^1(\mathbf{z},0), \ldots, x^m(\mathbf{z},0)) = \psi(\mathbf{z})$ are known, we obtain the functions $x^j(\mathbf{z},\tau)$, $\mathbf{z}$ fixed, $0 \le \tau \le \tau_{\partial\mathcal{M}}(\mathbf{z})$, i.e., reconstruct all normal geodesics on $\Omega$ with respect to metric $h_{ij}$. Clearly, this gives us the embedding of $(\mathcal{M}, g)$ onto $(\Omega, h_{ij})$. □

**4.5.8.** Summarizing the previous considerations, we come to the following result:

**Theorem 4.47** *Let the oblique boundary spectral data of the operator $\mathcal{A}_{\epsilon,\mu,q}$ of the form (4.129) be known. Then, it is possible to reconstruct the functions $\epsilon\mu$ and $\widetilde{q}$ in $\Omega$, where $\widetilde{q}$ is of form (4.148).*

Next, we return to the question of the admissible group of transformations.

**Corollary 4.48** *The oblique boundary spectral data of the operator $\mathcal{A}_{\epsilon,\mu,q}$ of the form (4.129) determine the orbit of the group $\sigma_0(\mathcal{A}_{\epsilon,\mu,q})$ of an operator $\mathcal{A}_{\epsilon,\mu,q}$ of form (4.138) with respect to the normalized gauge transformations. It is also clear that gauge transformations $S_\kappa \in G$ transform isotropic operators into isotropic operators.*

We remind the reader that the orbit of $\mathcal{A}$ in normalized gauge transformations defined in Example 1 of section 3.1.7 is

$$\sigma_0(\mathcal{A}) = \{S_\kappa \mathcal{A} : \ \kappa > 0, \ \kappa|_{\partial\Omega} = 1\}.$$

**Proof.** Obviously, all operators in the orbit $\sigma_0(\mathcal{A})$ have the same oblique boundary spectral data. As $\epsilon\mu \in C^\infty(\overline{\Omega})$ is determined by the boundary spectral data uniquely, gauge transformations change $\epsilon$ and $\mu$ according to formula (4.146) such that $\epsilon\mu$ is invariant. So, by applying the gauge transformations, we can make one parameter, for example $\epsilon$, to be any function with given boundary value $\epsilon|_{\partial\Omega} = \epsilon_0$. After fixing $\epsilon$, the function $\mu$ is determined uniquely. Because $\widetilde{q}$ on $\mathcal{M}$ and, henceforth, on $\Omega$ is known, equations (4.147) and (4.148) make it possible to find $q$ from $\widetilde{q}$, $\epsilon$ and $\mu$. Hence, we can determine the orbit $\sigma(\mathcal{A}_{\epsilon,\mu,q})$ of all gauge transformations. However, by Lemma 4.44, $\{\lambda_k, \mu|_{\partial\Omega}\, \partial_n\, \varphi_k|_{\partial\Omega}\}$ uniquely determine $\epsilon|_{\partial\Omega}$ and $\mu|_{\partial\Omega}$. Thus, we can find the orbit $\sigma_0(\mathcal{A}_{\epsilon,\mu,q})$ of the normalized gauge transformations. □

**Corollary 4.49** *Let $\mathcal{A}_{\epsilon,\mu,q}$ be an operator of form (4.143).*

*i) Assume that we are given $\epsilon$ or $\mu$ and a collection that is gauge equivalent to the oblique boundary spectral data. Then we can determine $\epsilon$, $\mu$, and $q$ uniquely.*

*ii) Assume that $q$ and the oblique boundary spectral data of $\mathcal{A}_{\epsilon,\mu,q}$ are given. Then we can determine $\epsilon$ and $\mu$ uniquely.*

**Proof.** In the case i), we use Corollary 4.48 to find $\epsilon\mu$, and $\widetilde{q}$. As $\epsilon$ or $\mu$ is known, this determines both $\epsilon$ and $\mu$. Thus, equations (4.147) and (4.148) determine $\kappa$ and $q$.

In the case ii), we use Theorem 4.47 to find $\epsilon\mu \in C^\infty(\overline{\Omega})$ and $\widetilde{q} \in C^\infty(\overline{\Omega})$. Thus, we have found the Schrödinger operator $-\Delta_g + \widetilde{q}$ on $\Omega$. As $\epsilon|_{\partial\Omega}$, $\mu|_{\partial\Omega}$ are found from the boundary spectral data, we can determine $\kappa_0 = \kappa|_{\partial\Omega}$. As the gauge transform $S_{\kappa^{-1}}$ transforms $-\Delta_g + \widetilde{q}$ to $\mathcal{A}_{\epsilon,\mu,q}$, formula (2.52) implies that $w = \kappa$ satisfies

$$\begin{aligned} -\Delta_g w + (\widetilde{q} - q)w &= 0, \\ w|_{\partial\Omega} &= \kappa_0. \end{aligned} \tag{4.158}$$

Next we show that this equation has a unique solution. As (4.158) is know to have a positive solution $\kappa$, the potential $\widetilde{q} - q$ can be written as

$$\widetilde{q} - q = \rho^{-\frac{1}{2}} \Delta_g(\rho^{\frac{1}{2}}), \quad \rho = \kappa^2.$$

An easy computation shows that equation (4.158) is then equivalent to

$$\begin{aligned} \mathrm{Div}_g \rho\, \mathrm{Grad}_g(\rho^{-\frac{1}{2}} w) &= 0 \\ \rho^{-\frac{1}{2}} w|_{\partial\Omega} &= 1 \end{aligned}$$

As this Dirichlet problem is uniquely solvable, so is (4.158). Thus, by solving this equation, we find $\kappa$. Using representation (4.147) for $\kappa$, we can determine $\epsilon$ and $\mu$ on $\Omega$. □

**Notes.** Chapter 4 contains a wide range of different subjects. The aim of these notes is to give historical background and references and also to point out the references to other methods and results.

Relations between different types of inverse problems and the inverse boundary spectral problem described in section 4.1 cannot be found from the existing literature in a complete form. We point out that the transformations of data described in section 4.1 do not involve analytic continuation. The reader should note that not all types of inverse boundary value problems can be reduced to the inverse boundary spectral problem. A particularly important example is given by inverse problems with fixed-frequency data. More precisely, in these problems one knows the Dirichlet-Robin mapping $\Lambda_k$ at one value of the spectral parameter $k$, which is often taken to be 0. These problems are more difficult than the inverse boundary spectral problems, because the fixed-frequency data contains less information than the boundary spectral data. For the corresponding results applicable to Riemannian manifolds, we refer to [LeU], [Sy1], [Na2], [LaU], [Lo] and [U].

Since the boundary control method is of hyperbolic character, its applicability to the dynamic inverse problems of the type considered in section 4.2 was clear from the very beginning. In fact, starting from the original paper of Belishev [Be1], the method was developed to study hyperbolic inverse problems, see also [BeBl2], [BeKa1], [BeKu1] and [Be3]. Later, new ideas based on the WKB-method were imported into the boundary control method to obtain the so-called amplitude formulae, see [BeKu2], [BeKa2], [Be2] and [BeIsPSh]. The different approach developed in section 4.2 was also applied to study the inverse boundary problem for the wave equation with attenuation in [KuLa4].

As the hyperbolic inverse boundary value problem often occurs in applications, they have been studied for a long time and there are several methods to solve them. Quite different from the dynamical inverse problem, which is studied in section 4.2, is the kinematic inverse problem. In this problem, we are given the travel times, i.e., the geodesic distances, between the boundary points. The kinematic inverse problem is, in fact, a geometric problem that makes it close to the approach discussed in this book. However, in this problem, one is given only a part of the data contained in the response operator. This makes the kinematic inverse problem very difficult. The kinematic inverse problem is discussed in detail in [Sh1] and [Sh2]. See also [StU2] and references in [Sh1], [Sh2].

Returning to the hyperbolic inverse boundary-value problems, most methods are based on geometrical ideas and finite speed of wave propagation, sometimes implicitly, see, e.g., textbooks [LRoSi], [Ro1], [Is1] and references therein. Other results devoted to the uniqueness in these problems can be found in, e.g., [Ro2], [We1]–[We2], [RkSy], [La3]. Although stability in the hyperbolic inverse boundary problem is a less studied subject than uniqueness, there are already many interesting results in this direction. These results can be found in [AlSy], [Rk], [Is3], [Ya], [PuYa], and [StU1].

The direct continuation of the hyperbolic boundary data, discussed in section 4.3, can be found in [KuLa4]. The continuation of the parabolic boundary data is a simple observation that is scattered through numerous papers. For its applications see, e.g., [AvSe].

The inverse boundary value problem with data given on a part of the boundary, discussed in section 4.4, was studied in [KaKu1] in the spectral case and in [KuLa2] in the dynamical case. For results in the fixed frequency case see, e.g., [GrU].

The Courant-Hilbert minimax principle, called also the variational principle, used in section 4.4 is a classical result of functional analysis and can be found in various textbooks, see, e.g., [CuHi], section 3.4, [BiSo], section 9.2, [EgSb], section 4.1, or [EdEv].

The result, obtained in section 4.5 go back to [Ku1] and [Ku2]. The scalar curvature equation can be found in [KzWa]. The embedding procedure of the manifold into $\mathbf{R}^m$ used in this section is based on the properties of the Christoffel symbols described in, e.g., [Ei], section 1.7, [Kl], section 4.1 and [NoFo], section 2.6.

As inverse boundary value problems for elliptic equations in $\mathbf{R}^m$ are of significant importance for practical applications, they have attracted much interest in the mathematical community. In this book, we have already discussed the inverse boundary spectral problems. Another class of inverse problems for elliptic equations is the fixed-frequency problems. At the moment, the most powerful method to study these problems is based on the use of exponentially growing solutions. The origin of this method goes back to Faddeev, who introduced the famous Faddeev Green's functions and exponentially growing solutions to study inverse problems of quantum scattering, see [Fa1]–[Fa3], [Nw]. The inverse boundary value problem at fixed frequency was first formulated by Calderon. He used the exponentially growing solutions to study the linearized version of this problem [Cl]. Further developments in this problem were obtained in [KoVo1], [KoVo2]. The crucial step was made by Sylvester and Uhlmann, who solved the inverse conductivity problem in dimensions three or more [SyU]. Similar results for the scattering problems were simultaneously obtained by Novikov and Khenkin [NvKh]. A further important development belonged to Nachman, who solved this problem in dimension two [Na2], see also [Nv2]. Since that time, the method was employed to prove a number of uniqueness results, e.g., [NaSyU], [Na1], [Nv2], [Rm2], [BrU], [NkSuU], [NkTs]. Reconstruction algorithms and numerical implementation of this method were also studies in, e.g., [Na1]–[Na2], [SiMuIa]. For related methods, see also [SCnII], [BuGST]. The corresponding stability results have been developed in, e.g., [Al], [AlSy], [Is2], [BaBaRu].

An important development obtained by this method is the solution of the inverse problems for systems of equations. An inverse problem for the Maxwell system was solved by Ola, Päivärinta and Somersalo [OPaS], see also [CoPa], [OS], and for the elasticity system by Nakamura and Uhlmann [NkU1]–[NkU2], see also [AIkTYa], [IkNkYa].

As for the other theoretical methods for elliptic inverse boundary value problems, we refer to, e.g., the layer stripping [So], [SCnII], [Sy2] and Carleman estimate methods, see, e.g., [Kb], [ImYa], [PuYa], [Ya], [Is1]–[Is4] and references therein.

An extended overview of various contemporary methods for solving multidimensional inverse problems is given in the lecture series and presentations at the EMS Summer School on inverse problems (ICMS, Edinburgh, July 25–Aug. 6, 2000) [KuSo].

# Bibliography

[Ag1] Agranovich, M.S. Elliptic boundary problems. *Encyclopedia Math. Sci.*, **79**, Partial differential equations, IX, 1–144, Springer, Berlin, 1997.

[Ag2] Agranovich, M.S. Elliptic boundary problems. *Encyclopedia Math. Sci.*, **79**, Partial differential equations, IX, 1–144, Springer, Berlin, 1997.

[Al] Alessandrini, G. Stable determination of conductivity by boundary measurements. *Appl. Anal.* **27** (1988), no. 1-3, 153–172.

[AlSy] Alessandrini, G., Sylvester, J. Stability for a multidimensional inverse spectral theorem. *Comm. Part. Diff. Equations* **15** (1990), no. 5, 711–736.

[Am] Ambarzumjan W.A. Uber eine Frage dere Eigenwerttheorie. *Zeitschrift für Fhysik* **53** (1929), 690-695.

[AIkTYa] Ang, D. D., Ikehata, M., Trong, D. D., Yamamoto, M. Unique continuation for a stationary isotropic Lam system with variable coefficients. *Comm. Part. Diff. Equations* **23** (1998), no. 1-2, 371–385.

[AvSe] Avdonin, S. Seidman, T. Identification of $q(x)$ in $u_t = \Delta u - qu$ from boundary observations. *SIAM J. Control Optim.* **33** (1995), no. 4, 1247–1255.

[BaBuMo] Babich, V.M., Buldyrev, V.S., Molotkov, I.A. *The Space-time Ray Method, Linear and Nonlinear waves.* (Russian) Leningrad Univ., Leningrad, 1985. 272 pp.

[BaUl] Babich, V.M., Ulin, V.V. The complex space-time ray method and "quasiphotons." (Russian) *Zap. Nauchn. Sem. LOMI* **117** (1981), 5–12.

[BaPa] Babich V.M, Pankratova T.F. The discontinuities of the Green's function for a mixed problem for the wave equation with a variable coefficient. (Russian) *Problems of Mathematical Physics*, **6**(1973), 9–27.

[BkGo] Bakushinsky, A.; Goncharsky, A. *Ill-posed problems: theory and applications.* Mathematics and its Applications, **301**. Kluwer, Dordrecht, 1994. 256 pp.

[BaBaRu] Barselo B., Barselo J., and Ruiz, A. Stability of the inverse conductivity problem in the plane for less regular conductivities. *J. Diff. Equat.* to appear.

[Be1] Belishev, M. An approach to multidimensional inverse problems for the wave equation. (Russian) *Dokl. Akad. Nauk SSSR* **297** (1987), no. 3, 524–527; translated in *Soviet Math. Dokl.* **36** (1988), no. 3, 481–484.

[Be2] Belishev, M. Boundary control in reconstruction of manifolds and metrics (the BC method). *Inverse Problems* **13** (1997), **5**, R1–R45.

[Be3] Belishev, M. Wave bases in multidimensional inverse problems. (Russian) *Mat. Sb.* **180** (1989), 584–602.

[BeBl1] Belishev, M., Blagoveščenskii, A. S. Direct method to solve a nonstationary inverse problem for the wave equation. *Ill-Posed Problems of Math. Phys and Anal.* (Russian) Krasnoyarsk, 1988, 43-49.

[BeBl2] Belishev, M., Blagoveščenskii, A.S. *Dynamical Inverse Problems of the Wave Theory.* (Russian) St. Petersburg Univ. Press, 1999. 226p.

[BeRFi] Belishev, M., Ryzhov, V., Filippov, V. A spectral variant of the BC-method: theory and numerical experiment. (Russian)*Dokl. Akad. Nauk* **337** (1994), no. 2, 172–176; translated in *Phys. Dokl.* **39** (1994), no. 7, 466–470.

[BeIsPSh] Belishev, M., Isakov, V., Pestov, L., Sharafutdinov, V. On the reconstruction of a metric from external electromagnetic measurements. (Russian) *Dokl. Akad. Nauk* **372** (2000), no. 3, 298–300.

[BeKa1] Belishev, M., Kachalov, A. Boundary control and quasiphotons in a problem of the reconstruction of a Riemannian manifold from dynamic data. (Russian) *Zap. Nauchn. Sem. POMI* **203** (1992), 21–50; translated in *J. Math. Sci.* **79**(1996), no.4, 1172–1190.

[BeKa2] Belishev, M., Kachalov, A. Boundary control in the inverse spectral problem for an inhomogeneous string (Russian), *Zap. Nauchn. Sem. LOMI* **179** (1989), 14–22 translation in *J. Soviet Math.*, **50**(1990), no.6, 1944–1951.

[BeKa3] Belishev, M., Kachalov, A. Operator integral in the multidimensional inverse spectral problem. Reconstruction of a Riemannian manifold from dynamic data. (Russian) *Zap. Nauchn. Sem. POMI* **215** (1994), 9–37; translated in *J. Math. Sci. (New York)*, **85**(1997), no.1, 1559–1577.

[BeKu1] Belishev, M., Kurylev, Y. A nonstationary inverse problem for the multidimensional wave equation "in the large." (Russian) *Zap. Nauchn. Sem. LOMI* **165** (1987), 21–30; translated in *J. Soviet Math.* **50** (1990), no. 6, 1944–1951.

[BeKu2] Belishev, M., Kurylev, Y. Boundary control, wave field continuation and inverse problems for the wave equation. Multidimensional inverse problems. *Comput. Math. Appl.* **22** (1991), no. 4-5, 27–52.

[BeKu3] Belishev, M., Kurylev, Y. To the reconstruction of a Riemannian manifold via its spectral data (BC-method). *Comm. Part. Diff. Equations* **17** (1992), no. 5-6, 767–804.

[Br1] Berezanskii, Y. M. On the uniqueness of determination of the Schrödinger equation from its spectral function. (Russian) *Doklady Akad. Nauk SSSR* **93** (1953), no.4, 591–594.

[Br2] Berezanskii, Y. M. On the inverse problem of spectral analysis for the Schrödinger operator. (Russian) *Doklady Akad. Nauk SSSR* **105** (1955), no. 2, 197–200.

[Br3] Berezanskii, Y. M. The uniqueness theorem in the inverse problem of spectral analysis for the Schrödinger equation. (Russian) *Trudy Moskov. Mat. Obsch.*, **7** (1958), 1–62.

[BiSo] Birman M.S., Solomyak M.Z. *Spectral Theory of Self-adjoint Operators in Hilbert Space.* D. Reidel, Dordrecht, 1987. 343 pp.

[Bl1] Blagoveščenskii, A. A one-dimensional inverse boundary value problem for a second order hyperbolic equation. (Russian) *Zap. Nauchn. Sem. LOMI*, **15** (1969), 85–90.

[Bl2] Blagoveščenskii, A. The local method of solution of the nonstationary inverse problem for an inhomogeneous string. (Russian) *Trudy Mat. Inst. Steklova*, **115** (1971), 28–38.

[Bl3] Blagoveščenskii, A. Inverse boundary problem for the wave propagation in an anisotropic medium. (Russian) *Trudy Mat. Inst. Steklova*, **65** (1971), 39–56.

[Bo] Börg, G. Eine Umkehrung der Sturm-Liouvillischen Eigenwertaufgabe. Bestimmugm der Differentialalgeleichung durch die Eigenwerte. *Acta Math.* **78** (1946), 1–96.

[BrU] Brown, R., Uhlmann, Gunther U. Uniqueness in the inverse conductivity problem for nonsmooth conductivities in two dimensions. *Comm. Part. Diff. Equations* **22** (1997), no. 5-6, 1009–1027.

[BuZa] Burago, Y., Zalgaller, V. *An Introduction to Riemannian Geometry.* (Russian) Nauka, Moscow, 1994. 319 pp.

[BuGST] Burov, V., Goryunov, A., Saskovets, A., Tikhonova, T. Inverse scattering problems in acoustics (review). (Russian) *Soviet Phys. Acoust.* **32** (1986), no. 4, 273–282; translated in *Akust. Zh.* **32** (1986), no. 4, 433–449.

[Cl] Calderon, A.-P. On an inverse boundary value problem. *Seminar on Numerical Analysis and its Applications to Continuum Physics (Rio de Janeiro, 1980)*, pp. 65–73, Soc. Brasil. Mat., Rio de Janeiro, 1980.

[Cr] Carleman T, *Les Fonctions Quasi-analytiques.* Gauthier-Villars, Paris, 1926.

[ChSa] Chadan, K., Sabatier, P. *Inverse Problems in Quantum Scattering Theory.* Second edition. Springer-Verlag, New York-Berlin, 1989. 499 pp.

[ChCoPaRn] Chadan, K., Colton, D., Päivärinta, L., Rundell, W. *An Introduction to Inverse Scattering and Inverse Spectral Problems.* With a foreword by M. Cheney. SIAM Monographs on Math. Model. and Comput., SIAM, Philadelphia, PA, 1997. 198 pp.

[Cv] Chavel, I. *Riemannian Geometry–a Modern Introduction.* Cambridge Tracts in Mathematics, **108**. Cambridge University Press, Cambridge, 1993. 386 pp.

[CoKs] Colton, D., Kress, R. *Inverse Acoustic and Electromagnetic Scattering Theory.* Applied Mathematical Sciences, **93**. Springer-Verlag, Berlin, 1998. 334 pp.

[CoPa] Colton, D., Päivärinta, L. The uniqueness of a solution to an inverse scattering problem for electromagnetic waves. *Arch. Rational Mech. Anal.* **119** (1992), no. 1, 59–70.

[CuHi] Courant R., Hilbert D. *Methods of Mathematical Physics.* Vol. I. Wiley, New York, 1989. 560 pp.

[CnIN] Cheney, M. Isaacson, D. Newell, J. Electrical impedance tomography. *SIAM Rev.* **41** (1999), no. 1, 85–101 (electronic).

[ElsNkTa] Eller, M., Isakov, V. Nakamura, G., Tataru, D. Uniqueness and stability in the Cauchy problem for Maxwell's and elasticity systems, preprint.

[EgSb] Egorov, Y., Shubin, M. Foundations of the Classical Theory of Partial Differential Equations. *Encyclopaedia Math. Sci.*, **30**, Springer, Berlin, 1992.

[Ei] Eisenhart, L. *Riemannian Geometry.* Princeton Landmarks in Mathematics. Princeton University Press, Princeton, NJ, 1997. 306 pp.

[EnHaNu] Engl, Heinz W., Hanke, M., Neubauer, A. *Regularization of Inverse Problems.* Mathematics and its Applications, **375**. Kluwer, Dordrecht, 1996., 321 pp.

[EdEv] Edmunds, D., Evans, W. *Spectral Theory and Differential Operators.* Oxford Mathematical Monographs, Oxford University Press, New York, 1987. 574 pp.

[Ev] Evans, L. *Partial Differential Equations.* Graduate Studies in Mathematics, **19**. American Mathematical Society, Providence, RI, 1998. 662 pp.

[Fa1] Faddeev, L.D. Growing solutions of the Schrödinger equation. (Russian) *Dokl. Akad. Nauk SSSR* **165** (1965), no. 3, 514-517.

[Fa2] Faddeev, L.D. Factorization of the $S$-matrix for the multidimensional Schödinger operator. (Russian) *Soviet Physics Dokl.* **11** (1966), 209–211.

[Fa3] Faddeev, L.D. The inverse problem in the quantum theory of scattering. II. (Russian) *Current problems in mathematics*, **3** 93–180, Akad. Nauk SSSR Vsesojuz. Inst. Naučn. i Tehn. Informacii, Moscow, 1974.

[GaHuLa] Gallot, S., Hulin, D., Lafontaine, J. *Riemannian Geometry.* Universitext. Springer-Verlag, Berlin, 1990. 284 pp.

[Ge] Gel' fand, I.M. Some aspects of functional analysis and algebra. *Proc. Intern. Cong. Math.*, **1** (1954) (Amsterdam 1957), 253–277.

[GeLe] Gel' fand, I.M., Levitan, B.M. On the determination of a differential equation from its spectral function. (Russian) *Izvestiya Akad. Nauk SSSR. Ser. Mat.* **15** (1951), 309–360.

[GrU] Greenleaf, A., Uhlmann, G. Local uniqueness for Dirichlet-to-Neumann map via the two-plane transform, *Duke Math. J.*, to appear.

[G] Groetsch, C. W. *Inverse Problems in the Mathematical Sciences.* Vieweg Mathematics for Scientists and Engineers. Friedr. Vieweg & Sohn, Braunschweig, 1993, 152 pp.

[GrKlMe] Gromoll, D., Klingenberg, W., Meyer, W. *Riemannsche Geometrie im Groen.* (German) Zweite Auflage. Lecture Notes in Mathematics, **55**. Springer-Verlag, Berlin-New York, 1975. 287 pp.

[Ha] Haehner P., A periodic Faddeev-type solution operator. *J. Differ. Equations* **128** (1996), 300-308 .

[Hi] Hille, E. *Analytic Function Theory.* Vol. II. Ginn and Co., Boston, Mass.-New York-Toronto, Ont. 1962, 496 pp.

[Ho1] Hörmander, L. *Linear Partial Differential Operators.* Die Grundlehren der mathematischen Wissenschaften, **116**, Springer-Verlag, Berlin-Göttingen-Heidelberg 1963, 287 pp.

[Ho2] Hörmander, L. *The Analysis of Linear Partial Differential Operators.* Vol. I–IV. Springer-Verlag, Berlin, 1985.

[Ho3] Hörmander, L. A uniqueness theorem for second order hyperbolic differential equations. *Comm. Part. Diff. Equations* **17** (1992), no. 5-6, 699–714.

[Ho4] L. Hörmander, On the uniqueness of the Cauchy problem under partial analyticity assumptions. In *Geometrical Optics and Related Topics* (Eds. F. Colombini and N. Lerner), pp. 179–219, Birkhäuser, Boston 1997.

[IkNkYa] Ikehata, M., Nakamura, G., Yamamoto, M. Uniqueness in inverse problems for the isotropic Lamé system. *J. Math. Sci. Univ. Tokyo* **5** (1998), no. 4, 627–692.

[ImYa] Imanuvilov, O., Yamamoto, M. Lipschitz stability in inverse parabolic problems by the Carleman estimate. *Inverse Problems* **14** (1998), no. 5, 1229–1245.

[Is1] Isakov, V. *Inverse Problems for Partial Differential Equations.* Applied Mathematical Sciences, **127**. Springer-Verlag, New York, 1998. 284 pp.

[Is2] Isakov, V. Carleman type estimates in an anisotropic case and applications. *J. Differential Equations* **105** (1993), no. 2, 217–238.

[Is3] Isakov, V. Uniqueness and stability in multi-dimensional inverse problems. *Inverse Problems* **9** (1993), 579–621.

[Is4] Isakov, V. Carleman type estimates and their applications. In *New Analytic and Geometrical Methods in Inverse Problems*

(Eds. Y. Kurylev and E. Somersalo), Springer lect. Notes, to appear.

[IsSu] Isakov, V., Sun, Z. Stability estimates for hyperbolic inverse problems with local boundary data. *Inverse Problems* 8 (1992), no. 2, 193–206.

[Iz] Izosaki, H. Some remarks on multidimensional Börg–Levinson theorem. *J. Math. Kyoto Univ.*, **31** (1991), 743–753.

[Ki] Kirsch, A. *An Introduction to the Mathematical Theory of Inverse Problems.* Applied Mathematical Sciences, **120**. Springer-Verlag, New York, 1996. 282 pp.

[KcKr] Kac, I.S., Krein, M.G. Spectral function of a string. In Atkinson, F. *Discrete and Continuous Boundary Problems.* (Russian) Mir, Moscow 1968 749 pp.

[Ka] Katchalov A. A system of coordinate for describing the "quasiphoton". (Russian) *Zap. Nauch. Semin. LOMI,* **140** (1984), 73–76.

[KaKu1] Kachalov, A., Kurylev, Y. The multidimensional inverse Gel'fand problem with incomplete boundary spectral data. (Russian) *Dokl. Akad Nauk SSSR,* **346** (1996), no. 5, 587–589.

[KaLa] Katchalov A., Lassas, M. Gaussian beams and inverse boundary spectral problems, in *New Analytic and Geometrical Methods in Inverse Problems* (Eds. Y. Kurylev and E. Somersalo), Springer lect. Notes, to appear.

[KaKu2] Katchalov, A., Kurylev, Y. Multidimensional inverse problem with incomplete boundary spectral data. *Comm. Part. Diff. Equations* **23** (1998), no. 1-2, 55–95.

[KaPo1] Katchalov A., Popov M., Application of the Gaussian beam summation method to computation of wave fields. (Russian) *Dokl. Akad Nauk SSSR,* **258** (5)(1981), 1097–1101.

[KaPo2] Katchalov A., Popov M. Application of Gaussian beams to elasticity theory. *Geophys. J. R. Astr. Soc.*, **81**(1985), 205–214.

[KaPo3] Katchalov A., Popov M. Gaussian beam methods and theoretical seismograms. Geophys. *J. R. Astr. Soc.*, **93** (1988), 465–475.

[Kl] Klingenberg, W. *Riemannian geometry.* de Gruyter Studies in Mathematics, 1. Walter deGruyter & Co., Berlin-New York, 1982, 396 pp.

[Kr1] Krein, M.G. Solution of the inverse Sturm-Liouville problem. (Russian) *Doklady Akad. Nauk SSSR*, **76** (1951), 21–24.

[Kr2] Krein, M.G. Determination of the density of an inhomogeneousstring from its spectrum. (Russian) *Doklady Akad. Nauk. SSSR*, **76**(1951), no.3, 345–348.

[Kr3] Krein, M.G. On inverse problems for an inhomogeneousstring. (Russian) *Doklady Akad. Nauk. SSSR*, **82** (1951), no.5, 669-672.

[Kr4] Krein, M.G. On a method of effective solution of the inverse boundary problem. (Russian) *Dokl. Akad. Nauk SSSR*, **94** (1954), no. 6, 987–990.

[KKuLa] Katsuda A., Kurylev Y., Lassas M. Stability on inverse boundary spectral problem, (Eds. Y. Kurylev and E. Somersalo), Springer lect. Notes, to appear.

[KzWa] Kazdan, J., Warner, F. Existence and conformal deformation of metrics with prescribed Gaussian and scalar curvatures. *Ann. of Math.* (2), **101** (1975), 317–331.

[Kb] Klibanov, M. Carleman estimates and inverse problems in the last two decades. *Surveys on Solution Methods for Inverse Problems,* 119–146, Springer, Vienna, 2000.

[KoVo1] Kohn, R., Vogelius, M. Determining conductivity by boundary measurements. *Comm. Pure Appl. Math.*, **37** (1984), no. 3, 289–298.

[KoVo2] Kohn, R., Vogelius, M. Determining conductivity by boundary measurements. II. Interior results. *Comm. Pure Appl. Math.* **38** (1985), no. 5, 643–667

[Ku1] Kurylev, Y. Admissible groups of transformations that preserve the boundary spectral data in multidimensional inverse problems. (Russian) *Dokl. Akad. Nauk*, **327** (1992), no. 3, 322–325; translated in *Soviet Phys. Dokl.* **37** (1993), no. 11, 544–545.

[Ku2] Kurylev, Y. Multi-dimensional inverse boundary problems by BC-method: groups of transformations and uniqueness results. *Math. Comput. Modelling* **18** (1993), no. 1, 33–45.

[Ku3] Kurylev, Y. An inverse boundary problem for the Schrödinger operator with magnetic field. *J. Math. Phys.* **36** (1995), no. 6, 2761–2776.

[Ku4] Kurylev, Ya. A multidimensional Gelfand-Levitan inverse boundary problem. *Differential Equations and Mathematical Physics* (Birmingham, AL, 1994), 117–131, Int. Press, Boston, MA, 1995.

[Ku5] Kurylev Y. Multidimensional Gel'fand inverse problem and boundary distance map, *Inverse Problems Related with Geometry* (ed. H.Soga) (1997), 1-15.

[KuLa1] Kurylev, Y., Lassas, M. Gelf'and inverse problem for a quadratic operator pencil. *J. Funct. Anal.* **176** (2000), no. 2, 247–263.

[KuLa2] Kurylev, Y., Lassas, M. Hyperbolic inverse problem with data on a part of the boundary. *Differential Equations and Mathematical Physics (Birmingham, AL, 1999)*, 259–272, AMS/IP Stud. Adv. Math., **16**, Amer. Math. Soc., Providence, 2000.

[KuLa3] Kurylev, Y., Lassas, M. The multidimensional Gel'fand inverse problem for non-selfadjoint operators. *Inverse Problems* **13** (1997), no. 6, 1495–1501.

[KuLa4] Kurylev, Y., Lassas, M. Hyperbolic inverse problem and unique continuation of Cauchy data of solutions along the boundary, *Proc. Roy. Soc. Edinburgh, Ser. A*, to appear.

[KuPe] Kurylev, Y., Peat, K. Hausdorff moments in two-dimensional inverse acoustic problems. *Inverse Problems* **13** (1997), no. 5, 1363–1377.

[KuSo] Kurylev Y., Somersalo S. (Eds.) *New Analytic and Geometric Methods in Inverse Problems* Lect. Notes in Math., Springer, to appear.

[KuSr] Kurylev, Y., Starkov, A. Directional moments in the acoustic inverse problem. *Inverse Problems in Wave Propagation (Minneapolis, MN, 1995)*, 295–323, IMA Vol. Math. Appl., 90, Springer, NewYork, 1997.

[KwKuSg] Kawashita M., Kurylev Y., Soga H. Harmonic moments and an inverse problem for the heat equation. *SIAM J. Math. Anal.* **32** (2000), no. 3, 522–537.

[Ld] Ladyzhenskaya, O.A. *The Boundary Value Problems of Mathematical Physics.* Applied Mathematical Sciences, **49**. Springer-Verlag, New York-Berlin, 1985. 322 pp.

[LsLiTr] Lasiecka, I., Lions, J.-L., Triggiani, R. Nonhomogeneous boundary value problems for second order hyperbolic operators. *J. Math. Pures Appl.* (9) **65** (1986), no. 2, 149–192.

[LsTr1] Lasiecka, I., Triggiani, R. *Control Theory for Partial Differential Equations: Continuous and Approximation Theories. I. Abstract Parabolic Systems.* Encyclopedia of Mathematics and its Applications, **74**. Cambridge University Press, Cambridge, 2000. 648 pp.

[LsTr2] Lasiecka, I., Triggiani, R. *Control Theory for Partial Differential Equations: Continuous and Approximation Theories. II. Abstract Hyperbolic-like Systems over a Finite Time Horizon.* Encyclopedia of Mathematics and its Applications, **75**. Cambridge University Press, Cambridge, 2000, 645–1067.

[La1] Lassas, M. Non-selfadjoint inverse spectral problems and their applications to random bodies. *Ann. Acad. Sci. Fenn. Math. Diss.* **103** (1995), 108 pp.

[La2] Lassas, M. Inverse boundary spectral problem for non-selfadjoint Maxwell's equations with incomplete data. *Comm. Part. Diff. Equations* **23** (1998), no. 3-4, 629–648.

[La3] Lassas M. Inverse boundary spectral problem for a hyperbolic equation with first order perturbation, *Applicable Analysis* **70** (1999), 219-231.

[LaU] Lassas M., Uhlmann G. Determining Riemannian manifold from boundary measurements, *Ann. Sci. École Norm. Sup.*, to appear.

[LxPh] Lax, P., Phillips, R. *Scattering Theory.* Pure and Applied Mathematics, **26**. Academic Press, Inc., Boston, MA, 1989. 309 pp.

[LRoSi] Lavrent'ev, M.M., Romanov, V.G., Shishatskii, S.P. *Ill-posed Problems of Mathematical Physics and Analysis.* Translations of Mathematical Monographs, **64**. American Mathematical Society, Providence, 1986. 290 pp.

[LeU] Lee, J., Uhlmann, G. Determining anisotropic real-analytic conductivities by boundary measurements. *Comm. Pure Appl. Math.* **42** (1989), no. 8, 1097–1112.

[Lv] Levinson N. The inverse Sturm-Liouville problem, *Mat. Tidsskr. B* (1949), 25-30.

[LeSa] Levitan, B., Sargsjan, I. *Sturm-Liouville and Dirac Operators.* Mathematics and its Applications, **59**. Kluwer, Dordrecht, 1991. 350 pp.

[Le] Levitan, B. *Inverse Sturm-Liouville Problems.* VSP, Zeist, 1987. 240 pp.

[Lo] Lionheart, W. Conformal uniqueness results in anisotropic electrical impedance imaging. *Inverse Problems* **13** (1997), no. 1, 125–134.

[LiMg] Lions, J.-L., Magenes, E. *Non-homogeneous Boundary Value Problems and Applications.* I. Springer-Verlag, New York-Heidelberg, 1972. 357 pp.

[Ma1] Marchenko, V.M. On differential operators of second order (Russian), *Dokl. Akad. Nauk SSSR* **72** (1950), no. 3, 457–460.

[Ma2] Marchenko, V.M. On the one-dimensional linear differential equations of second order (Russian), *Trudy Moskov Mat. Obsch.*, 1(1952), 327-420 2 (1953), 3-82.

[Ma3] Marchenko, V.A. *Spectral Theory of Sturm-Liouville Operators* (Russian) Naukova Dumka, Kiev, 1972. 219 pp.

[Ma4] Marchenko, V.A. *Sturm-Liouville Operators and Applications. Operator Theory* Advances and Applications, 22. Birkhäuser Verlag, Basel-Boston, Mass., 1986. 367pp.

[Me] Mellin, A. Intertwining operators in inverse scattering. *In New Analytic and Geometrical Methods in Inverse Problems* (Eds. Y. Kurylev and E. Somersalo), Springer lect. Notes, to appear.

[Na1] Nachman, A. Reconstructions from boundary measurements. *Ann. of Math.* (2) **128** (1988), no. 3, 531–576.

[Na2] Nachman, A. Global uniqueness for a two-dimensional inverse boundary value problem. *Ann. of Math.* (2) **143** (1996), no. 1, 71–96.

[NaSyU] Nachman, A., Sylvester, J. Uhlmann, G. An $n$-dimensional Borg-Levinson theorem. *Comm. Math. Phys.* **115** (1988), no. 4, 595–605.

[NkU1] Nakamura, G., Uhlmann, G., Global uniqueness for an inverse boundary problem arising in elasticity. *Invent. Math.* **118** (1994), no. 3, 457–474.

[NkU2] Nakamura, G.; Uhlmann, G. Inverse elastic scattering at a fixed energy. *J. Inverse Ill-Posed Probl.* **7** (1999), no. 3, 283–288.

[NkSuU] Nakamura, G., Sun, Z., Uhlmann, G. Global identifiability for an inverse problem for the Schrödinger equation in a magnetic field. *Math. Ann.* **303** (1995), no. 3, 377–388.

[NkTs] Nakamura, G., Tsuchida, T. Uniqueness for an inverse boundary value problem for Dirac operators. *Comm. Part. Diff. Equations* **25** (2000), no. 7-8, 1327–1369.

[Nw] Newton, R. *Inverse Schödinger Scattering in Three Dimensions.* Texts and Monographs in Physics. Springer-Verlag, Berlin-New York, 1989. 170 pp.

[Nz] Nizhnik, L. *The Inverse Nonstationary Scattering Problem* (Russian) Naukova Dumka, Kiev, 1973. 182 pp.

[Nm] Nomofilov, V. Asymptotic solutions of a system of second-order equations concentrated in the neighborhood of a ray. (Russian) *Zap. Nauchn. Sem. LOMI* 104 (1981), 170–179, 239.

[NoFo] Novikov, S.P., Fomenko, A.T. *Basic elements of differential geometry and topology.* Mathematics and its Applications, 60. Kluwer, Dordrecht, 1990. 490 pp.

[Nv1] Novikov, R. A multidimensional inverse spectral problem for the equation $-\Delta\psi+(v(x)-Eu(x))\psi=0$. (Russian) *Funktsional. Anal. i Prilozhen.* 22 (1988), no. 4, 11–22; translated in *Funct. Anal. Appl.* 22 (1989), no. 4, 263–272.

[Nv2] Novikov, R. $\bar{\partial}$-method with nonzero background potential. Application to inverse scattering for the two-dimensional acoustic equation. *Comm. Part. Diff. Equations* 21 (1996), no. 3-4, 597–618.

[NvKh] Novikov, R., Khenkin, G. The $\bar{\partial}$-equation in the multidimensional inverse scattering problem. (Russian) *Uspekhi Mat. Nauk* **42** (1987), no. 3(255), 93–152.

[OPaS] Ola, P., Päivärinta, L., Somersalo, E. An inverse boundary value problem in electrodynamics. *Duke Math. J.* **70** (1993), no. 3, 617–653.

[OS] Ola, P., Somersalo, E. Electromagnetic inverse problems and generalized Sommerfeld potentials. *SIAM J. Appl. Math.* **56** (1996), no. 4, 1129–1145.

[Pe] Petersen, P. *Riemannian Geometry.* Graduate Texts in Mathematics, 171. Springer-Verlag, New York, 1998. 432 pp.

[Po] Popov M. The method of summation of Gaussian beams in the isotropic theory of elasticity. (Russian) *Izv. Akad. Nauk SSSR. Fiz. Zemli,* 9 (1983), 39–50.

[PoTr] Pöschel, J., Trubowitz, E. *Inverse Spectral Theory.* Pure and Applied Mathematics, 130. Academic Press, Boston, MA, 1987. 192 pp.

[PuYa] Puel, J.-P., Yamamoto, M. Generic well-posedness in a multidimensional hyperbolic inverse problem. *J. Inverse Ill-Posed Probl.* 5 (1997), no. 1, 55–83.

[ReRg] Renardy, M., Rogers, R. *An Introduction to Partial Differential Equations.* Texts in Applied Mathematics, **13**. Springer-Verlag, New York, 1993. 428 pp.

[Rk] Rakesh. Reconstruction for an inverse problem for the wave equation with constant velocity. *Inverse Problems* **6** (1990), no. 1, 91–98.

[RkSy] Rakesh, Symes, W. Uniqueness for an inverse problem for the wave equation. *Comm. Part. Diff. Equations* **13** (1988), no. 1, 87–96.

[Rl] Ralston, J. Gaussian beams and propagation of singularities. Studies in Partial Differential Equations, 206–248, *MAA Stud. Math.*, 23, Washington, 1982.

[Rm1] Ramm, A. *Scattering by Obstacles.* Mathematics and its Applications, **21**. D. Reidel Publishing, Dordrecht-Boston, Mass., 1986. 423 pp.

[Rm2] Ramm, A. *Multidimensional Inverse Scattering Problems.* Pitman Monographs and Surveys in Pure and Applied Mathematics, 51. Longman Scientific & Technical, Harlow, 1992. 379 pp.

[Rm3] Ramm, A. Property C with constraints and inverse spectral problems with incomplete data. *J. Math. Anal. Appl.* **180** (1993), no. 1, 239-244.

[Ru] Rauch, J. *Partial Differential Equations.* Graduate Texts in Mathematics, **128**. Springer-Verlag, New York, 1991. 263 pp.

[RuTy] Rauch, J., Taylor, M. Penetrations into shadow regions and unique continuation properties in hyperbolic mixed problems. *Indiana Univ. Math. J.* **22** (1972/73), 277–285.

[Rb] Robbiano, L. A uniqueness theorem adapted to the control of solutions of hyperbolic problems(French) *Comm. Part. Diff. Equations* **16** (1991), no. 4-5, 789–800.

[RbZu] Robbiano, L., Zuily, C. Uniqueness in the Cauchy problem for operators with partially holomorphic coefficients. *Invent. Math.* **131** (1998), no. 3, 493–539.

[Ro1] Romanov, V.G. *Inverse Problems of Mathematical Physics.* VNU Science Press, Utrecht, 1987, 239 pp.

[Ro2] Romanov, V.G. Uniqueness theorems in inverse problems for some second-order equations. (Russian)*Dokl. Akad. Nauk SSSR* **321** (1991), no. 2, 254–257; translated in *Soviet Math. Dokl.* **44** (1992), no. 3, 678–682.

[Ro3] Romanov, V.G. On the problem of determining the coefficients in the lowest order terms of a hyperbolic equation. (Russian) *Sibirsk. Mat. Zh.* **33** (1992), no. 3,156–160; translated in *Siberian Math. J.* **33** (1992), no. 3, 497-500.

[Rd] Rudin, W. *Functional Analysis.* McGraw-Hill, Inc., New York, 1991. 424 pp.

[Sh1] Sharafutdinov, V. *Integral Geometry of Tensor Fields.* Inverse and Ill-posed Problems Series. VSP, Utrecht, 1994. 271 pp.

[Sh2] Sharafutdinov, V. Ray Transform on Riemannian manifolds. In *New Analytic and Geometrical Methods in Inverse Problems* (Eds. Y. Kurylev and E. Somersalo), Springer lect. Notes, to appear.

[SiMuIa] Siltanen, S., Mueller, J. Isaacson, D. An implementation of the reconstruction algorithm of A. Nachman for the 2D inverse conductivity problem. *Inverse Problems* **16** (2000), no. 3, 681–699.

[So] Somersalo, E. Layer stripping for time-harmonic Maxwell's equations with high frequency. *Inverse Problems* **10** (1994), no. 2, 449–466.

[SCnII] Somersalo, E., Cheney, M. Isaacson, D., Isaacson, E. Layer stripping: a direct numerical method for impedance imaging. *Inverse Problems* **7** (1991), no. 6, 899–926.

[Sp] Spivak, M. *A Comprehensive Introduction to Differential Geometry.* Vol. I–V. Second edition. Publish or Perish, 1979.

[StU1] Stefanov, P., Uhlmann, G. Stability estimates for the hyperbolic Dirichlet to Neumann map in anisotropic media. *J. Funct. Anal.* **154** (1998), no. 2, 330–358.

[StU2] Stefanov, P., Uhlmann, G. Rigidity for metrics with the same lengths of geodesics. *Math. Res. Lett.* **5** (1998), no. 1-2, 83–96.

[Sy1] Sylvester, J. An anisotropic inverse boundary value problem. *Comm. Pure Appl. Math.* **43** (1990), no. 2, 201–232.

[Sy2] Sylvester, J. The layer stripping approach to impedance tomography, In *Inverse Problems: Principles and Applications in Geophysics, Technology, and Medicine*, Potsdam 1993, 307–321.

[SyU] Sylvester, J., Uhlmann, G. A global uniqueness theorem for an inverse boundary value problem. *Ann. of Math.* (2) **125** (1987), no. 1, 153–169.

[Ta1] Tataru, D. Unique continuation for solutions to PDEs; between Hörmander's theorem and Holmgren's theorem. *Comm. Part. Diff. Equations* **20** (1995), no. 5-6, 855–884.

[Ta2] Tataru, D. Boundary observability and controllability for evolutions governed by higher order PDE. *J. Math. Anal. Appl.* **193** (1995), no. 2, 632–658.

[Ta3] Tataru, D. Unique continuation for operators with partially analytic coefficients. *J. Math. Pures Appl.* (9) **78** (1999), no. 5, 505–521.

[Tr] Triebel, H. *Interpolation Theory, Function Spaces, Differential Operators.* 2nd edit. Johann Ambrosius Barth, Heidelberg, 1995. 532 pp.

[U] Uhlmann G. Inverse boundary value problems for partial differential equations. *Proceedings of the International Congress of Mathematicians*, Vol. III (Berlin, 1998). Doc. Math. 1998, 77–86.

[We1] Weston, V. Invariant imbedding for the wave equation in three dimensions and the applications to the direct and inverse problems. *Inverse Problems* **6** (1990), no. 6, 1075–1105.

[We2] Weston, V. Invariant imbedding and wave splitting in $R^3$. II. The Green function approach to inverse scattering. *Inverse Problems* 8 (1992), no. 6, 919–947.

[Ya] Yamamoto, M. Uniqueness and stability in multidimensional hyperbolic inverse problems. *J. Math. Pures Appl.* (9) 78 (1999), no. 1, 65–98.

# Table of notation

| Notation | Meaning | See page |
|---|---|---|
| $\mathbf{R}, \mathbf{C}$ | Real and complex numbers | |
| $\Re, \Im$ | Real and imaginary parts | |
| $\mathcal{N}, \mathcal{M}$ | Manifolds, $\partial\mathcal{M} \neq \emptyset$ | 34 |
| $H^s(\mathcal{M}), H^s_0(\mathcal{M})$, | Sobolev spaces | 58, 65 |
| $\lVert\cdot\rVert_s, \lVert\cdot\rVert_{(s,\partial M)}$ | Sobolev $s$-norms | 58 |
| $\langle\cdot,\cdot\rangle,\ \lVert\cdot\rVert$ | Inner product, norm in $L^2$ | 3,60 |
| $C^p, L^p([0,T]; H^s)$ | $H^s$-valued functions | 15, 70 |
| $\vert\vert\vert\cdot\vert\vert\vert$ | Norm of $L^1([0,T]; L^2(\mathcal{M}))$ | 71 |
| $C_0^\infty, \mathring{C}^\infty$ | Smooth function spaces | 196 |
| $T_\mathbf{x}\mathcal{N}, T^*_\mathbf{x}\mathcal{N}, S\mathcal{N}, S^*\mathcal{N}$ | Tangent space, etc. | 35,37,44 |
| $(\mathbf{v},\mathbf{v})_g, (\mathbf{p},\mathbf{q})_g, (\mathbf{p},\mathbf{v})$ | Inner products, duality | 39, 37 |
| $(\cdot,\cdot)$ | Also, inner product of $\mathbf{R}^m$ | 167 |
| $\{\cdot,\cdot\}$ | Poisson brackets | 122 |
| $B_\rho(\mathbf{y})$ | Ball in $(\mathcal{M}, g)$ | 45 |
| $g_{ik}, \Gamma_{i,kl}, R^i_{jkl}$ | Metric, Christoffel, curvature | 39,41,42 |
| $\gamma_{\mathbf{z},\mathbf{w}}, \gamma_{\mathbf{z},\nu}$ | Geodesics | 43 |
| $\exp, \exp_{\partial\mathcal{M}}$ | Exponential mappings | 44,50 |
| $l(\mathbf{z}_0)$ | Length of $\gamma_{\mathbf{z},\nu}$ | 168 |
| $\tau(\mathbf{x},\xi), \tau_{\partial M}(\mathbf{z})$ | Critical values on geodesics | 46,51 |
| $\omega(\mathbf{z}_0), \omega_{\partial\mathcal{M}}$ | Cut loci | 46,51 |
| $\nabla_\mathbf{v}\mathbf{w}, \frac{D\mathbf{w}}{ds}$ | Covariant derivative | 41,42 |
| $\mathcal{M}(\tau_1,\tau_2)$ | Slices | 21 |
| $\mathcal{M}^T, \mathcal{M}(\Gamma,\tau), \mathcal{M}(\mathbf{z},\tau)$ | Domains of influence | 50,156,161 |
| $D^T(\Omega), K_{\Gamma,T}$ | Cone, double cone | 87, 163 |
| $Q^T, \Sigma^T$ | Time-domains | 70 |
| $\mathcal{A}, \mathcal{A}_0, \mathcal{A}_\kappa$ | Operators | 6,7,60 |
| $a(\mathbf{x},D), p(\mathbf{x},D)$ | Differential expressions | 58,116 |
| $dV_g, dV_{can}$ | Volume elements | 40,60 |
| $P_{\Gamma,\tau}, P_{\mathbf{z},\tau}, P_{\tau_1,\tau_2}$ | Projections onto $\mathcal{M}(\Gamma,\tau)$ | 21,160 |
| $u_\epsilon, U^N_\epsilon$ | Gaussian beams | 90, 115 |
| $\theta(\mathbf{x},t), u_n(\mathbf{x},t)$ | Phase and amplitudes | 90 |
| $f^\epsilon, M_\epsilon$ | Boundary source of $u_\epsilon$, etc. | 90,167 |
| $R, r_\mathbf{x}, R^T, r^T_\mathbf{x}$ | Boundary distance functions | 179 |
| $\Pi, \Lambda, \mathcal{B}, L, \Lambda_\lambda$ | Boundary maps/forms | 200 |
| $\sigma(\mathcal{A}), \sigma_0(\mathcal{A})$ | Orbits of gauge group | 67,152 |

# *Index*